Probability and Its Applications

Published in association with the Applied Probability Trust

Editors: J. Gani, C.C. Heyde, P. Jagers, T.G. Kurtz

Probability and Its Applications

Anderson: Continuous-Time Markov Chains.
Azencott/Dacunha-Castelle: Series of Irregular Observations.
Bass: Diffusions and Elliptic Operators.
Bass: Probabilistic Techniques in Analysis.
Berglund/Gentz: Noise-Induced Phenomena in Slow-Fast Dynamical Systems
Chen: Eigenvalues, Inequalities and Ergodic Theory
Choi: ARMA Model Identification.
Costa, Fragoso and Marques: Discrete-Time Markov Jump Linear Systems
Daley/Vere-Jones: An Introduction to the Theory of Point Processes.
Volume I: Elementary Theory and Methods, Second Edition.
de la Peña/Giné: Decoupling: From Dependence to Independence.
Durrett: Probability Models for DNA Sequence Evolution.
Galambos/Simonelli: Bonferroni-type Inequalities with Applications.
Gani (Editor): The Craft of Probabilistic Modelling.
Grandell: Aspects of Risk Theory.
Gut: Stopped Random Walks.
Guyon: Random Fields on a Network.
Kallenberg: Foundations of Modern Probability, Second Edition.
Last/Brandt: Marked Point Processes on the Real Line.
Leadbetter/Lindgren/Rootzén: Extremes and Related Properties of Random Sequences and Processes.
Molchanov: Theory of Random Sets
Nualart: The Malliavin Calculus and Related Topics.
Rachev/Rüschendorf: Mass Transportation Problems. Volume I: Theory.
Rachev/Rüschendorf: Mass Transportation Problems. Volume II: Applications.
Resnick: Extreme Values, Regular Variation and Point Processes.
Schmidli: Stochastic Control in Insurance
Shedler: Regeneration and Networks of Queues.
Silvestrov: Limit Theorems for Randomly Stopped Stochastic Processes
Thorisson: Coupling, Stationarity, and Regeneration.
Todorovic: An Introduction to Stochastic Processes and Their Applications.

Hanspeter Schmidli

Stochastic Control in Insurance

Hanspeter Schmidli, PhD
Institute of Mathematics
University of Cologne
Weyertal 86-90
D-50931 Köln
Germany

Mathematics Subject Classification (2000): 60G51, 90C39, 91B30, 60K15, 49L20, 60F10

British Library Cataloguing in Publication Data
A catalogue record for this book is available from the British Library

Library of Congress Control Number: 2007939040

ISBN: 978-1-84800-002-5 ISBN: 978-1-84800-003-2 (eBook)

Printed on acid-free paper

9 8 7 6 5 4 3 2 1

springer.com

Für Monika, Eliane und Stefan.

Für all die Zeit, die meine Familie auf mich verzichten musste.

Preface

In *operations research* and *physics*, *optimisation* was introduced a long time ago. A physical system always tries to reach a state with maximal entropy. Thus, a physicist needs to solve an optimisation problem in order to find these states. In operations research one looks for business strategies leading to minimal costs or maximal profit. For example, one has to decide how many items one has to produce. On the one hand, there are storage costs if items cannot be sold immediately; on the other hand, a penalty has to be paid if an order cannot be fulfilled because the non-served customer is lost. Another problem is the construction of a computer or telecommunication network. Such a network preferably is small, but at the same time the probability of a buffer overflow should be small, too.

The problems considered in this book start with some stochastic process $\{X_t^u\}$ whose dynamics can be changed via a *control process* $\{u_t\}$. To each initial value x and each admissible control process $\{u_t\}$ we associate a value $V^u(x)$. We are interested in determining the maximal value $V(x) = \sup_u V^u(x)$ called the *value function.* Two questions then arise: What is the value function $V(x)$ and — if it exists at all — what is the optimal control process $\{u_t^*\}$, i.e., the control process leading to the value function $V(x) = V^{u^*}(x)$? In many problems the optimal control process is of feedback form $u_t^* = u^*(X_t^*)$, where $\{X_t^*\}$ is the process following the dynamics if the optimal control is applied.

A well-known approach to these problems — particularly popular in theoretical physics — is the variational approach. One writes the function to optimise as a functional of the strategy applied. By disturbing the "optimal" control function $u^*(x)$ by another function $g(x)$, that is using the control $u(x) = u^*(x) + \varepsilon g(x)$, the functional has a maximum at $\varepsilon = 0$. Holding for all $g(x)$, this usually leads to an integro-differential equation for the value function or for an equation for the optimal control $u^*(x)$.

In this book we will use a complementary approach. Our approach is stochastic and based on *martingale arguments.* The martingale formulation of dynamic programming goes back to Striebel [177]. The method had been available in 1974 but was published 10 years later.

As with the variational approach, we will obtain a *Hamilton–Jacobi–Bellman equation.* However, the derivation of the equation is heuristic only. In order to find the equation, one has to make several assumptions on the unknown value function. Hence, one has to prove that a possible solution to the equation really is the value function. This *verification theorem* often is not difficult to prove using martingale arguments.

A second problem — and usually the hard problem — is to show that the value function really is a solution to the Hamilton–Jacobi–Bellman equation. The most convenient situation is the case where an explicit solution can be found. Then, of course, one undertakes the verification argument for the explicit solution and does not have to bother with possible further solutions to the equation. If an explicit solution cannot be found, one has to solve the equation numerically. But a numerical solution only makes sense if one first verifies that a solution really exists. In nice cases a (local) *contraction argument* yields this property. A contraction argument automatically gives uniqueness, and one will not have to worry about conditions guaranteeing the correct solution. If no contraction argument is at hand, one may have to show directly that the value function solves the Hamilton–Jacobi–Bellman equation. The problem then is usually to verify that the value function is (twice continuously) differentiable. In case the solution to the Hamilton–Jacobi–Bellman equation is not unique one will have to find further properties of the value function in order to obtain the correct solution.

Sometimes further problems may occur. So the solution may not be differentiable. In this case how to solve the equation numerically is not straightforward. One then needs to determine the points where the derivative jumps. Or if the process $\{X_t^u\}$ contains a diffusion term and the solution is not twice differentiable, the martingale arguments do not apply directly to the value function. A possibility to circumvent the problem is to consider *viscosity solutions.* We will only shortly discuss this more advanced tool in the present book.

The theory of martingales is standard in probability theory. The strength of the tool comes from the *martingale convergence theorem* A.1 and the *martingale stopping theorem* A.2. Martingales also occur naturally in financial mathematics where discounted price processes should be martingales under some measure. Thus, today's financial mathematicians are familiar with martingales. In actuarial mathematics Gerber [74] introduced martingale methods for the estimation of the ruin probability. Therefore, many actuaries are also familiar with the concept of martingales.

In this book we do not want to prove general results under quite restrictive conditions. The reader can find such general results, for example, in the monographs [20], [40], [60], [61], [144], [182], or [187]. We will here consider some optimisation problems and then discuss possible methods to approach these problems. In this way we will use different techniques that have been successful for the corresponding situation. The basic approach always will be to find a Hamilton–Jacobi–Bellman equation. However, for different problems we will use different methods to show that the value function really solves the equation. From this point of view this is a practical approach to the problems. In some situations the solution of the equation is quite difficult. We therefore just prove a verification theorem that shows that a solution with the correct boundary conditions is the value function. Often the general theory of differential equations shows that a solution really exists. However, the reader has to be aware that the main problem is to solve the equation.

A natural field of the application of control techniques is *insurance mathematics*. On the one hand, economic problems usually are optimisation problems. On the other hand, an actuary is educated to make decisions. These decisions should be taken in an optimal way. In the field of *mathematical finance* it was observed quite early that there is a need for the application of optimisation techniques; see, for instance, [114], [118], [121], [131], [132], and [182]. The optimisation problems considered in insurance were mainly utility maximisation problems or the determination of Pareto-optimal risk exchanges (see, for instance, [178]) or linear programming (see [27]) where no stochastic control problem is present. A stochastic control problem was formulated by de Finetti [59] and solved by Gerber [72]. We will consider this problem in Sections 1.2 and 2.4. The corresponding problem for a diffusion approximation has been solved by Shreve et al. [168]. A similar problem was treated by Frisque [63]. Another early work on the topic is a series of lectures given by Martin-Löf [130] for the Swedish society of actuaries. The author is not aware of other early work in the application of stochastic control techniques in insurance.

A conference on the *Interplay between Insurance, Finance and Control* (see [8]) initiated a lot of work on stochastic control applied to insurance. The present book is a summary of some of the problems that have been considered since or shortly before this conference. An alternative summary article on stochastic control problems in insurance is Hipp [96], where some of the problems considered in this book also have been treated.

The above-mentioned conference was also my starting point for research in the area. Here I saw the stochastic control approach for the first time and realised that this was the tool to solve the optimal reinsurance problem that we consider in Section 2.3.1. I had this problem in my mind for quite a long time but no clue how to attack it.

The prerequisite knowledge for this book is basic probability theory with a basic knowledge of Brownian motion, Markov processes, martingales, and stochastic calculus. These topics are covered in Appendices A and B. It is recommended that a reader not familiar with these tools also has a look at some of the references given in the appendix in order to obtain some experience. In order to understand the technical details, measure theory is also needed. However, a reader only interested in the application of the optimisation techniques may skip these theoretical aspects. But it is possible to understand intuitively many of the concepts only with the knowledge given in the appendices.

Part of this book was used as material for an optional course at the Laboratory of Actuarial Mathematics in Copenhagen. At the end of the course the students were able to find the Hamilton–Jacobi–Bellman equations and to perform the verification arguments without clearly knowing what the generator of a Markov process is. A reader interested in a broader understanding will find material to deepen the required knowledge in the references given in the bibliographical remarks at the end of the sections and the appendices.

The book is organised as follows. Chapter 1 gives an introduction to stochastic control in discrete time. In this case the results can be stated quite generally. Discrete-time dynamic programming was the starting point for stochastic control and was initiated in operations research a long time ago. The continuous time case is treated in Chapter 2. After the presentation of the Hamilton–Jacobi–Bellman approach, several optimisation problems are solved. Chapter 3 also deals with optimisation in continuous time, but the problems originate from life insurance. Finally, Chapter 4 considers the problem of how asymptotic properties of the value function can be obtained from the Hamilton–Jacobi–Bellman equation. The problem is that the solutions and the optimal controls are not known explicitly but via the Hamilton–Jacobi–Bellman equation only, that is, via a highly nonlinear equation. Several appendices give a short introduction to the theory the book is based on, such as stochastic processes, Markov processes, risk theory, or life insurance mathematics.

Finally, we make some conventions. Throughout the book we work on a probability space $(\Omega, \mathcal{F}, \mathbb{P})$ that is large enough to carry all the stochastic objects defined. We assume that $\mathcal{F}$ is *complete*, i.e., that it contains all $\mathbb{P}$-null sets. The filtrations $\{\mathcal{F}_t\}$ are assumed to be right-continuous, that is, $\mathcal{F}_t = \cap_{s>t}\mathcal{F}_s$. But we do *not*, as usual in books, assume that $\{\mathcal{F}_t\}$ is complete in the continuous-time case. That is, $\mathcal{F}_t$ does not necessarily contain all the $\mathbb{P}$-null sets. If we completed the filtration, in Chapter 4 we would not be able to change the measure on $\mathcal{F}$, because the measure $\mathbb{P}^*$ could not be extended to $\mathcal{F}$.

Unless stated otherwise, the stochastic processes in continuous time are assumed to be *cadlag* (right-continuous with left limits existing). This simplifies some technical problems. For example, the martingale convergence theorem and the optional stopping theorem hold quite generally. In particular, we will choose controls that are cadlag. Instead of the left-continuity that usually is used in books, we will need the left limit of the control for the development of the controlled process. The disadvantage is that the controlled process is observed after the effect of the control and the information has to be taken from the filtration instead of from the controlled process. For example, in the optimal dividend problem the post-dividend process and not the pre-dividend process is observed. The controlled process then has a jump of the size of the claim plus the dividend. Hence, we cannot get the size of the dividend from the surplus. We need additionally to know the claim size to determine the dividend payment. However, because we lose the cadlag property when considering the pre-dividend process, I prefer the presentation with cadlag stochastic processes.

If one of the basic processes is Brownian motion, we often will deal with stochastic differential equations. Then we have the problem of the existence of a unique solution. Sometimes it may even happen that no solution exists on the given probability space, but that there is a probability space on which a solution exists. This is called a weak solution. In order to avoid the problem, we assume that we have chosen a probability space on which at least the stochastic differential equation for the optimal process has a (strong) solution. The reader, however, should be aware that some technical difficulties may arise. Anyway, an insurer has a surplus process and not a probability space. Since the law of the process and not the underlying probability space is important it is no problem to choose the "right" probability space.

To simplify the notation we will omit the expressions *almost surely* or *with probability one.* Unless otherwise stated, we consider all statements to hold almost surely. For example, we say that a stochastic process is "cadlag" rather than saying it is "cadlag a.s." Of course, we could consider a probability space on which all the paths are cadlag. But sometimes it is more convenient also to allow for paths in the probability space that are not cadlag. The reader should always be aware that there might be elements in the sample space Ω for which an assertion does not hold. Because this is more a technical problem. No confusion should arise with this simplification.

A book could never be written without the help and encouragement of many other people. I therefore conclude this preface by thanking Natalie Kulenko and Julia Eisenberg for finding many misprints. I further thank Hansjörg Albrecher, Christian Hipp, and Stefan Thonhauser for pointing out some misprints and some useful references. Many very helpful remarks from an unknown reviewer are acknowledged. The reviewer, which spent a lot of time giving me detailed comments, led to a considerable improvement of the presentation and removed several mistakes present in an earlier version of the book. Last but not least, the biggest thanks go to my family, Monika, Eliane, and Stefan, for accepting that their husband/father was busy writing a book instead of enjoying more time with them.

Hanspeter Schmidli
Cologne, June 2007

Contents

1

Stochastic Control in Discrete Time

We start by considering stochastic processes in discrete time. Optimisation is simpler in discrete time than in continuous time because we can give quite general results like the dynamic programming principle (Lemma 1.1) or the optimal strategy (Corollaries 1.2 and 1.3). We will show in some simple examples how the theory can be applied.

In this chapter we consider processes in discrete time, i.e., the set of possible time points is $I = \mathbb{N}$. We will work on some Polish measurable space $(E, \mathcal{E})$, with $\mathcal{E}$ denoting the Borel-σ-algebra on E. The Borel-σ-algebra is the smallest σ-algebra containing all the open sets. A reader not familiar with metric spaces can just replace E by $\mathbb{N}^d$, $\mathbb{Z}^d$, or $\mathbb{R}^d$ endowed with the Euclidean distance. By $\mathbb{N}^*$ we denote the strictly positive integers.

1.1 Dynamic Programming

1.1.1 Introduction

Let $\{Y_n : n \in \mathbb{N}^*\}$ be an iid sequence of random variables on some Polish space $(E_Y, \mathcal{E}_Y)$. These random variables model the stochastic changes over time. We work with the natural filtration $\{\mathcal{F}_n\} = \{\mathcal{F}_n^Y\}$. At each time point $n \in \mathbb{N}$ a decision is made. We model this decision as a variable U_n from some space $\mathcal{U}$. $\mathcal{U}$ is endowed with some topology we do not mention explicitly here. The stochastic process $U = \{U_n : n \in \mathbb{N}\}$ must be adapted, because the decision can only be based on the present and not on future information. We therefore only allow controls U that are adapted. We may make some restriction to the possible strategies U. Let $\mathfrak{U}$ denote the set of *admissible* strategies, i.e., the adapted strategies $U = \{U_n\}$ that are allowed.

The controlled stochastic process is now constructed in the following way. Let $(E, \mathcal{E})$ be a Polish space, the state space of the stochastic process, and

$x \in E$ be the initial state. We let $X_0 = x$ be the starting value of the process. Note that the initial value is not stochastic. The process at time $n+1$ is

$$X_{n+1} = f(X_n, U_n, Y_{n+1}) ,$$

where $f : E \times \mathcal{U} \times E_Y \to E$ is a measurable function. The interpretation is the following. The next state of the process X only depends on the present state and the present decision. The decisions made at earlier times and the path up to the present state do not matter. The process X therefore looks similar to a Markov process. Note that we do not have a Markov process unless the decision U_n depends on X_n only.

At each time point there is a reward, $r(X_n, U_n)$. A negative value of $r(X_n, U_n)$ can be regarded as a cost. The value connected to some strategy U is then

$$V_T^U(x) = \mathbb{E}\Bigl[\sum_{n=0}^{T} r(X_n, U_n)e^{-\delta n}\Bigr] .$$

The time horizon T can be finite or infinite. The parameter $\delta \geq 0$ is a discounting parameter. If $T = \infty$, we often will have to assume that $\delta > 0$ in order for $V_\infty^U(x)$ to be finite for all $U \in \mathfrak{U}$.

Our goal will be to maximise $V_T^U(x)$. We therefore define the *value function*

$$V_T(x) = \sup_{U \in \mathfrak{U}} V_T^U(x) .$$

In the case $T = \infty$, we just write $V(x)$ and $V^U(x)$ instead of $V_\infty(x)$ and $V_\infty^U(x)$, respectively. We now assume that $V_T(x) \in \mathbb{R}$ for all x. It is clear that if there is a strategy U such that $V_T^U(x) \in \mathbb{R}$, then $V_T(x) > -\infty$. The property $V_T(x) < \infty$ has to be proved for every problem separately. Another (technical) problem is to show that $V_T(x)$ is a measurable function. In many problems it can be shown that $V_T(x)$ is increasing or continuous, and hence measurable.

In the following considerations we will need measurability in several steps. We will just assume that we can always make our choices in a measurable way. General conditions for measurability can be found in more advanced textbooks on control in discrete time. We will not worry about this point because the examples we consider later have continuous value functions. Then measurability is granted. In any case, this is a technical issue, and readers not familiar with measure theory should just accept that measurability is not a problem.

1.1.2 Dynamic Programming

It is not feasible to find $V(x)$ by calculating the value function $V_T^U(x)$ for each possible strategy U, particularly not if E and T are infinite. One therefore has to find a different way to characterise the value function $V_T(x)$. In our setup it turns out that the problem can be simplified. We next prove the *dynamic programming principle*, also called *Bellman's equation*. We allow all controls $\{U_n\}$ that are adapted. With $V_t(x)$ and $V_t^U(x)$ we denote the remaining value if t time units are left. For instance, $V_{T-1}(x)$ is the value if we stand at time 1 and $X_1 = x$. We let $V_{-1}(x) = 0$.

Lemma 1.1. *Suppose that $V_T(x)$ is finite. The function $V_T(x)$ fulfils the* dynamic programming principle

$$V_T(x) = \sup_{u\in\mathcal{U}}\{r(x,u) + e^{-\delta}\,\mathbb{E}[V_{T-1}(f(x,u,Y))]\}\;, \tag{1.1}$$

where Y is a generic random variable with the same distribution as Y_n. If $T = \infty$, the dynamic programming principle *becomes*

$$V(x) = \sup_{u\in\mathcal{U}}\{r(x,u) + e^{-\delta}\,\mathbb{E}[V(f(x,u,Y))]\}\;. \tag{1.2}$$

Proof. Let U be an arbitrary strategy. Then $X_1 = f(x, U_0, Y_1)$ and

$$V_T^U(x) = \mathbb{E}[r(x,U_0)] + e^{-\delta}\,\mathbb{E}\Bigl[\sum_{n=0}^{T-1} r(X_{n+1},U_{n+1})e^{-\delta n}\Bigr]\;.$$

Condition on X_1, U_0 (we allow random decisions) and let $\tilde{X}_n = X_{n+1}$, $\tilde{U}_n = U_{n+1}$, and $\tilde{Y}_n = Y_{n+1}$. Then

$$\tilde{X}_{n+1} = f(\tilde{X}_n, \tilde{U}_n, \tilde{Y}_{n+1})$$

and

$$\begin{aligned}\mathbb{E}\Bigl[\sum_{n=0}^{T-1} r(X_{n+1},U_{n+1})e^{-\delta n}\Bigm| X_1,U_0\Bigr] &= \mathbb{E}\Bigl[\sum_{n=0}^{T-1} r(\tilde{X}_n,\tilde{U}_n)e^{-\delta n}\Bigm| X_1,U_0\Bigr]\\ &= V_{T-1}^{\tilde{U}}(X_1) \le V_{T-1}(X_1)\;.\end{aligned}$$

Thus,

$$\begin{aligned}V_T^U(x) &\le \mathbb{E}\Bigl[r(x,U_0) + e^{-\delta}V_{T-1}(X_1)\Bigr]\\ &= \mathbb{E}\Bigl[r(x,U_0) + e^{-\delta}V_{T-1}(f(x,U_0,Y_1))\Bigr]\\ &\le \sup_{u\in\mathcal{U}}\{r(x,u) + e^{-\delta}\,\mathbb{E}[V_{T-1}(f(x,u,Y))]\}\;.\end{aligned}$$

Because U is arbitrary, this shows that

$$V_T(x) \le \sup_{u\in\mathcal{U}}\{r(x,u) + e^{-\delta}\,\mathbb{E}[V_{T-1}(f(x,u,Y))]\}\;.$$

Fix $\varepsilon > 0$ and $u \in \mathcal{U}$. Let us now consider a strategy $\tilde{U}$ such that, conditioned on $X_1 = f(x, u, Y_1)$, $V_{T-1}(X_1) < V_{T-1}^{\tilde{U}}(X_1) + \varepsilon$. Here, we do not address here the problem of whether we can do that in a measurable way because this point usually is clear in the examples, particularly the examples treated in this book. For conditions on measurability, see, for example, [20]. Let $U_0 = u$ and $U_n = \tilde{U}_{n-1}$. Then

$$\begin{aligned} r(x,u) + \mathrm{e}^{-\delta}\,\mathbb{E}[V_{T-1}(f(x,u,Y_1))] &< r(x,u) + \mathrm{e}^{-\delta}\,\mathbb{E}[V_{T-1}^{\tilde{U}}(X_1)] + \varepsilon \\ &= V_T^U(x) + \varepsilon \le V_T(x) + \varepsilon\,. \end{aligned}$$

Thus,

$$\sup_{u \in \mathcal{U}}\{r(x,u) + \mathrm{e}^{-\delta}\,\mathbb{E}[V_{T-1}(f(x,u,Y))]\} \le V_T(x) + \varepsilon\,.$$

Because ε is arbitrary, the result follows.

The proof does not explicitly use the finiteness of T. Thus, we can replace T and $T-1$ by ∞, and (1.2) is proved in the same way. □

The result says that we have to maximise the present reward plus the value of the future rewards. If we do that at each time point, we end up with the optimal value. Equation (1.1) can be solved recursively. We will discuss later how to solve Equation (1.2) numerically.

1.1.3 The Optimal Strategy

We next characterise the optimal strategy.

Corollary 1.2. *Suppose that $T < \infty$, $V_T(x)$ is finite, and that for any $t \le T$ there exists $u_t(x)$ such that $u = u_t(x)$ is maximising the right-hand side of (1.1) for $T = t$. We assume that $u_t : E \to \mathcal{U}$ is measurable for each t. Let $U_n = u_{T-n}(X_n)$. Then*

$$V_T(x) = V_T^U(x)\,.$$

Proof. Clearly, $V_T^U(x) \le V_T(x)$. If $T = 0$, then for any strategy $U' = U_0'$

$$V_0^{U'}(x) = \mathbb{E}[r(x, U_0')] \le r(x, u_0(x)) = V_0^U(x)\,,$$

and $V_0(x) \le V_0^U(x)$ follows. We prove the assertion for $T < \infty$ by induction. Suppose that the assertion is proved for $T = n$. Let U' be an arbitrary strategy for $T = n+1$, and use the tilde sign as in the proof of Lemma 1.1. Then

$$\begin{aligned} V_{n+1}^{U'}(x) &= \mathbb{E}[r(x,U_0') + \mathrm{e}^{-\delta}\,\mathbb{E}[V_n^{\tilde{U}'}(f(x,U_0',Y_1)) \mid U_0']] \\ &\le \mathbb{E}[r(x,U_0') + \mathrm{e}^{-\delta}\,\mathbb{E}[V_n(f(x,U_0',Y_1)) \mid U_0']] \\ &\le r(x,u_{n+1}(x)) + \mathrm{e}^{-\delta}\,\mathbb{E}[V_n(f(x,u_{n+1}(x),Y_1))] \\ &= r(x,u_{n+1}(x)) + \mathrm{e}^{-\delta}\,\mathbb{E}[V_n^{\tilde{U}}(f(x,u_{n+1}(x),Y_1))] = V_{n+1}^U(x)\,. \end{aligned}$$

This proves that $V_{n+1}(x) \le V_{n+1}^U(x)$. □

We can easily see from the proof that if U_n does not maximise the Bellman equation, then it cannot be optimal. In particular, if $u_n(x)$ does not exist for all $n \le T$, then an optimal strategy cannot exist.

If the time horizon is infinite, the proof of the existence of an optimal strategy is slightly more complicated. But the optimal strategy does not explicitly depend on time and is therefore simpler.

Corollary 1.3. *Suppose that $T = \infty$, $V(x) < \infty$, and that for every x there is a $u(x)$ maximising the right-hand side of* (1.2). *Suppose further that $u(x)$ is measurable and that*

$$\lim_{n\to\infty} \sup_{U' \in \mathfrak{U}} \mathbb{E}\Bigl[\sum_{k=n}^{\infty} |r(X'_k, U'_k)| \mathrm{e}^{-\delta k}\Bigr] = 0 , \tag{1.3}$$

where $X'_{n+1} = f(X'_n, U'_n, Y_{n+1})$. Let $U_n = u(X_n)$. Then $V^U(x) = V(x)$.

Proof. We first show that for any strategy U' with a value U'_0 that does not maximise the right-hand side of (1.2) there exists a strategy U'' with $U''_0 = u(x)$ that yields a larger value. Choose $\varepsilon > 0$. For each initial value x there exists a strategy $\tilde{U}''$ such that $V(\tilde{x}) < V^{\tilde{U}''}(\tilde{x}) + \varepsilon$. Also here we refrain from the technical problem of showing that $\tilde{U}''$ can be chosen in a measurable way, because it is simpler to address this problem for the specific examples. Let U'' be the strategy with $U''_0 = u(x)$ and $U''_{n+1} = \tilde{U}''_n$, where the initial capital is $\tilde{x} = f(x, u(x), Y_1)$. Thus,

$$\begin{aligned}
V^{U'}(x) &= \mathbb{E}[r(x, U'_0) + \mathrm{e}^{-\delta}\, \mathbb{E}[V^{\tilde{U}'}(f(x, U'_0, Y_1)) \mid U'_0]] \\
&\le \mathbb{E}[r(x, U'_0) + \mathrm{e}^{-\delta}\, \mathbb{E}[V(f(x, U'_0, Y_1)) \mid U'_0]] \\
&< r(x, u(x)) + \mathrm{e}^{-\delta}\, \mathbb{E}[V(f(x, u(x), Y_1))] = V(x) \\
&< r(x, u(x)) + \mathrm{e}^{-\delta}\, \mathbb{E}[V^{\tilde{U}''}(f(x, u(x), Y_1))] + \varepsilon = V^{U''}(x) + \varepsilon\, .
\end{aligned}$$

If $\varepsilon < V(x) - V^{U'}(x)$, we have that $V^{U'}(x) < V^{U''}(x)$.

Let $\mathfrak{U}_n$ be the set of all strategies U' with $U'_k = u(X_k)$ for $0 \le k \le n$. We just have shown that $V(x) = \sup_{U' \in \mathfrak{U}_0} V^{U'}(x)$. Suppose that $V(x) = \sup_{U' \in \mathfrak{U}_n} V^{U'}(x)$. Let U' be a strategy such that $U'_k = u(X_k)$ for $k \le n$ and U'_{n+1} does not maximise the right-hand side of (1.2) for $x = X_{n+1}$. Let $\tilde{U}'_k = U'_{n+1+k}$. Then by the argument used for $n = 0$, there is a strategy $\tilde{U}''$ with $\tilde{U}''_0 = u(X_{n+1})$ such that $V^{\tilde{U}''}(X_{n+1}) > V^{\tilde{U}'}(X_{n+1})$. Let U'' be the strategy with $U''_k = U'_k$ and $U''_{n+1+k} = \tilde{U}''_k$. Because

$$\begin{aligned}
V^{U'}(x) &= \mathbb{E}\Bigl[\sum_{k=0}^{n} r(X_k, u(X_k))\mathrm{e}^{-\delta k} + \mathrm{e}^{-\delta(n+1)} V^{\tilde{U}'}(X_{n+1})\Bigr] \\
&< \mathbb{E}\Bigl[\sum_{k=0}^{n} r(X_k, u(X_k))\mathrm{e}^{-\delta k} + \mathrm{e}^{-\delta(n+1)} V^{\tilde{U}''}(X_{n+1})\Bigr] ,
\end{aligned}$$

we get $V(x) = \sup_{U' \in \mathfrak{U}_{n+1}} V^{U'}(x)$.

Because for all n we have that $V(x) = \sup_{U' \in \mathfrak{U}_n} V^{U'}(x)$, we are now able to prove that $U_n = u(X_n)$ is optimal. Let $\varepsilon > 0$. There exists $n \in \mathbb{N}$ such that $\mathbb{E}[\sum_{k=n+1}^{\infty} |r(X'_k, U'_k)|e^{-\delta k}] < \varepsilon$ for any strategy U'. Let U' be a strategy in $\mathfrak{U}_n$ such that $V(x) - V^{U'}(x) < \varepsilon$. Then

$$
\begin{aligned}
V(x) < V^{U'}(x) + \varepsilon &= \mathbb{E}\Bigl[\sum_{k=0}^{\infty} r(X'_k, U'_k)e^{-\delta k}\Bigr] + \varepsilon \\
&< \mathbb{E}\Bigl[\sum_{k=0}^{n} r(X'_k, U'_k)e^{-\delta k}\Bigr] + 2\varepsilon \\
&= \mathbb{E}\Bigl[\sum_{k=0}^{n} r(X_k, U_k)e^{-\delta k}\Bigr] + 2\varepsilon \le V^U(x) + 3\varepsilon\;.
\end{aligned}
$$

Because ε is arbitrary, it follows that $V(x) \le V^U(x)$. □

The reader should note that the technical condition (1.3) is always fulfilled if the reward $r(x,u)$ is bounded and $\delta > 0$. Alternatively, if one knows the value function $V(x)$, one could just prove that $V^U(x) = V(x)$. Then condition (1.3) is not needed.

1.1.4 Numerical Solutions for $T = \infty$

In general, if $T < \infty$, then the optimal strategy U_n will depend on the time point n. Therefore, the only way to calculate the value function (and the optimal strategy) is to calculate $V_n(x)$ recursively. The situation is different for $T = \infty$. A first idea is to consider the corresponding finite horizon problem and calculate $V_n(x)$. Letting $n \to \infty$ will yield the value function, at least under condition (1.3). The problem is only that we need some criteria when n is large enough. The result below uses this idea [let $v_0(x) = 0$] and shows that $V(x)$ is the fixed point of a contraction.

Lemma 1.4. *Suppose that $\delta > 0$ and $\sup_{x\in E} |V(x)| < \infty$. Then the operator*

$$\mathcal{V}(v)(x) = \sup_{u\in\mathcal{U}} \{r(x,u) + e^{-\delta}\,\mathbb{E}[v(f(x,u,Y))]\}$$

is a contraction. In particular, $V(x)$ is the only bounded solution to (1.2). If $v_0(x)$ is an arbitrary function and $v_{n+1}(x) = \mathcal{V}(v_n)(x)$, then $\lim_{n\to\infty} v_n(x) = V(x)$. The convergence rate is geometric, i.e.,

$$\sup_{x\in E} |V(x) - v_n(x)| \le e^{-\delta n} \sup_{x\in E} |V(x) - v_0(x)|\;.$$

Remark 1.5. The assumption that $\sup_{x\in E} |V(x)| < \infty$ is quite strong. In some cases it may happen that it is enough to find the value function on a bounded interval and that outside this interval the value function can be calculated as a function of the values inside the interval. Such an example is given in Section 1.2. Another situation may arise when $\mathcal{V}$ is locally a contraction, and some value $v(x_0)$ is known. In this case $v_n(x)$ will first converge close to x_0. If the value is close enough on some interval around x_0, then it will also converge close to this interval. In this way it also is possible to calculate $V(x)$ even though $\mathcal{V}$ is not globally a contraction. An example in continuous time is given in Section 2.3. ■

Proof. Note that by (1.2), the reward $r(x,u)$ must be bounded from above. Define the norm $\|v\| = \sup_{x\in E} |v(x)|$. Let $v_1(x)$ and $v_2(x)$ be two functions. Suppose that $u(x)$ satisfies

$$r(x,u(x)) + \mathrm{e}^{-\delta}\,\mathbb{E}[v_1(f(x,u(x),Y))] > \mathcal{V}(v_1)(x) - \varepsilon\,.$$

Then $r(x,u(x))$ is finite and

$$\begin{aligned}\mathcal{V}(v_1)(x) - \mathcal{V}(v_2)(x) &< r(x,u(x)) + \mathrm{e}^{-\delta}\,\mathbb{E}[v_1(f(x,u(x),Y))]\\ &\quad - \{r(x,u(x)) + \mathrm{e}^{-\delta}\,\mathbb{E}[v_2(f(x,u(x),Y))]\} + \varepsilon\\ &= \mathrm{e}^{-\delta}\mathbb{E}[v_1(f(x,u,Y)) - v_2(f(x,u,Y))] + \varepsilon\\ &\le \mathrm{e}^{-\delta}\|v_1 - v_2\| + \varepsilon\,.\end{aligned}$$

Because ε is arbitrary, we have $\mathcal{V}(v_1)(x) - \mathcal{V}(v_2)(x) \le \mathrm{e}^{-\delta}\|v_1 - v_2\|$. Interchanging the rôles of v_1 and v_2, we find that $\|\mathcal{V}(v_1)(x) - \mathcal{V}(v_2)(x)\| \le \mathrm{e}^{-\delta}\|v_1 - v_2\|$. Thus, $\mathcal{V}$ is a contraction. We have already proved in Lemma 1.1 that $\mathcal{V}(V) = V$. If v is a solution to (1.2), then $\|V - v\| = \|\mathcal{V}(V) - \mathcal{V}(v)\| \le \mathrm{e}^{-\delta}\|V - v\|$, and $\|V - v\| = 0$ follows. Finally, for $v_{n+1}(x) = \mathcal{V}(v_n)(x)$, we find that

$$\|v_{n+1} - V\| = \|\mathcal{V}(v_n) - \mathcal{V}(V)\| \le \mathrm{e}^{-\delta}\|v_n - V\| \le \mathrm{e}^{-\delta(n+1)}\|v_0 - V\|\,.$$

Thus, $\|v_n - V\|$ converges to zero. □

Lemma 1.4 provides a possibility to calculate the function V numerically.

An alternative way to find the solution works in the case where both E and $\mathcal{U}$ are finite. This alternative algorithm is often faster than the method of Lemma 1.4, because it solves the problem in finite time.

Let $n = 0$ and choose controls $u_0(x)$ for all $x \in E$. Then

i) Solve the equations

$$V_n(x) = r(x,u_n(x)) + \mathrm{e}^{-\delta}\mathbb{E}[V_n(f(x,u_n(x),Y))] \tag{1.4}$$

in order to find $V_n(x)$.

ii) Choose the largest $u_{n+1}(x)$ (with respect to some ordering of $\mathcal{U}$) maximising

$$r(x,u_{n+1}(x)) + \mathrm{e}^{-\delta}\mathbb{E}[V_n(f(x,u_{n+1}(x),Y))]\,. \tag{1.5}$$

iii) If $u_{n+1}(x) = u_n(x)$ for all $x \in E$, the algorithm terminates; otherwise, increase n by 1 and return to step i).

The next result shows that the procedure works.

Lemma 1.6. *Suppose that E and $\mathcal{U}$ are finite and $\delta > 0$. Then the algorithm described above terminates in finite time m, say. The function $V_m(x)$ is the value function and $u_m(x)$ is an optimal control.*

Proof. Because V_n solves (1.4) and $u_{n+1}(x)$ maximises (1.5), we have

$$\begin{aligned} V_{n+1}(x) &= r(x, u_{n+1}(x)) + \mathrm{e}^{-\delta}\mathbb{E}[V_{n+1}(f(x, u_{n+1}(x), Y))] \,, \\ V_n(x) &\le r(x, u_{n+1}(x)) + \mathrm{e}^{-\delta}\mathbb{E}[V_n(f(x, u_{n+1}(x), Y))] \,. \end{aligned}$$

Taking the difference yields

$$V_{n+1}(x) - V_n(x) \ge \mathrm{e}^{-\delta}\mathbb{E}[V_{n+1}(f(x, u_{n+1}(x), Y)) - V_n(f(x, u_{n+1}(x), Y))] \,.$$

Suppose that $V_{n+1}(x) < V_n(x)$ for some x. We can choose x such that $V_{n+1}(x) - V_n(x) = \inf_{y \in E} V_{n+1}(y) - V_n(y)$. But then

$$\begin{aligned} V_{n+1}(x) - V_n(x) &\ge \mathrm{e}^{-\delta}\mathbb{E}[V_{n+1}(f(x, u_{n+1}(x), Y)) - V_n(f(x, u_{n+1}(x), Y))] \\ &\ge \mathrm{e}^{-\delta}(V_{n+1}(x) - V_n(x)) \end{aligned}$$

would be a contradiction. Thus, $V_n(x)$ is increasing. Because there is only a finite number of controls $u(x)$, there exists m such that $u_m(x) = u_k(x)$ for some $k < m$ for all $x \in E$. Thus, as the solution to (1.4) we also have

$$V_m(x) - V_k(x) = \mathrm{e}^{-\delta}\mathbb{E}[V_m(f(x, u_m(x), Y)) - V_k(f(x, u_m(x), Y))] \,.$$

If we choose x such that $V_m(x) - V_k(x)$ is maximal, we see that $V_k(x) = V_m(x)$. Because $V_n(x)$ is increasing, the algorithm terminates. By the construction of $u_m(x)$, it follows that $V_m(x)$ solves (1.2). By Lemma 1.4, $V_m(x) = V(x)$. Because $r(x, u)$ necessarily is bounded as a function on a finite space, the conditions of Corollary 1.3 are fulfilled and $u_m(x)$ is the optimal strategy. □

In solving the problem numerically, it may happen that the maximiser of (1.5) is not unique. If in this case one chooses the maximal or the minimal u at which the maximum is taken, the algorithm will terminate as soon as the value function is reached.

Bibliographical Remarks

Introductions to discrete-time dynamic programming can be found in such textbooks as [17], [18], [19], [24], [81], [95], [106], or [111]. These textbooks give a more general introduction to the topic than considered in this section.

In our situation we only considered the Markov case. This is because usually a Markov process can be obtained by *Markovization*, that is, by adding more state variables $\{(X_k, J_k)\}$ to the existing process $\{X_k\}$. If the process J is not observable, it is possible to estimate the process by filtering techniques. If one then considers the process $\{(X_k, J'_k)\}$, where J' is the filtered process (for example, the parameters of the distribution of J_k), it is often possible to find the optimal control based on the observable information.

The algorithm in Lemma 1.6 is described by Howard [106].

Discrete-time optimisation in insurance was also considered by Martin-Löf [130].

1.2 Optimal Dividend Strategies in Risk Theory

1.2.1 The Model

Let us consider the following risk model. In a unit interval an insurance company earns some premia and has to pay possible claims. Premia minus payout are denoted by Y_n, where Y_n is an integer. We let $\{Y_n\}$ be an iid sequence, and $p_n = \mathbb{P}[Y_k = n]$ for all $n \in \mathbb{Z}$. As before we work with the natural filtration $\{\mathcal{F}_t = \mathcal{F}_t^Y\}$. We denote by Y a generic random variable and suppose that $\mathbb{E}[|Y|] < \infty$. In order not to deal with a trivial problem we assume that $\mathbb{P}[Y < 0] > 0$; otherwise, the optimal strategy defined below is $U_n = X_n$. At time n the insurer can pay a dividend U_n with $0 \le U_n \le X_n$. The (pre-dividend) surplus process is modelled as $X_0 = x$, and

$$X_{n+1} = X_n - U_n + Y_{n+1} .$$

The process is stopped at the time of ruin, $\tau = \inf\{n : X_n < 0\}$. Ruin is a technical expression and does not necessarily mean that the company is bankrupt, but that the capital reserved for the business was not sufficient; see also Appendix D. It will turn out that under the optimal strategy ruin will happen in finite time almost surely.

The goal is now to maximise the expected discounted dividend payments

$$V^U(x) = \mathbb{E}\Bigl[\sum_{n=0}^{\tau-1} e^{-\delta n} U_n\Bigr] ,$$

where $\delta > 0$. The factor δ has to be considered as an additional discounting. Because the $\{Y_n\}$ are iid, the claim sizes and the premium should be measured in values at time 0, i.e., claim sizes and premia are already discounted. The additional discounting of the dividends can be seen as the investor's preference for dividends today to dividends tomorrow.

In order to fit into the setup of Section 1.1, we let $X_{n+1} = X_n$ if $X_n < 0$, $X_{n+1} = X_n - U_n$ if $U_n > X_n$, and $r(X_n, U_n) = U_n 1\!\!I_{X_n \ge 0} 1\!\!I_{U_n \le X_n}$. Thus, $E = \mathbb{R}$ and $\mathcal{U} = [0, \infty)$. We will soon see that we can work on smaller spaces E and $\mathcal{U}$.

Before we consider the problem in more detail, we prove a useful tool.

Lemma 1.7. *Let $\{U_n\}$ and $\{U_n'\}$ be some dividend strategies. We denote by $W_n = \sum_{k=0}^n U_k$ and $W_n' = \sum_{k=0}^n U_k'$ the accumulated dividend payments. If $W_n \ge W_n'$ for all n, then*

$$\sum_{n=1}^\infty e^{-\delta n} U_n \ge \sum_{n=1}^\infty e^{-\delta n} U_n' .$$

If $\mathbb{P}[U \ne U'] > 0$, then

$$\mathbb{E}\Bigl[\sum_{n=1}^\infty e^{-\delta n} U_n\Bigr] > \mathbb{E}\Bigl[\sum_{n=1}^\infty e^{-\delta n} U_n'\Bigr] .$$

Proof. The discounted dividend payments can be written as

$$\begin{aligned}\sum_{n=0}^\infty e^{-\delta n} U_n &= (1 - e^{-\delta}) \sum_{n=0}^\infty \sum_{k=n}^\infty e^{-\delta k} U_n = (1 - e^{-\delta}) \sum_{k=0}^\infty \sum_{n=0}^k U_n e^{-\delta k} \\ &= (1 - e^{-\delta}) \sum_{k=0}^\infty W_k e^{-\delta k} .\end{aligned}$$

The first inequality follows readily. If $\mathbb{P}[U \ne U'] > 0$, there must be an n such that $\mathbb{P}[W_n > W_n'] > 0$. In particular, the first inequality is strict with strictly positive probability. Taking expected values proves the result. □

The result suggests paying dividends as early as possible. One therefore has the trade-off between paying dividends early and getting ruined early, or paying dividends later and getting ruined later.

We first obtain some upper and lower bounds for $V(x) = \sup_{U \in \mathfrak{U}} V^U(x)$.

Lemma 1.8. i) *The function $V(x)$ is bounded by*

$$x + \frac{\mathbb{E}[Y^+] e^{-\delta}}{1 - p_+ e^{-\delta}} \le V(x) \le x + \frac{\mathbb{E}[Y^+] e^{-\delta}}{1 - e^{-\delta}} , \tag{1.6}$$

where $p_+ = \mathbb{P}[Y \ge 0]$ and $Y^+ = Y \vee 0$ is the positive part of Y.

ii) *$V(x)$ is strictly increasing and $V(x) - V(y) \ge x - y$ for any $x \ge y$.*

iii) *If $\mathbb{P}[Y > 0] = 0$, then $V(x) = x$, and the optimal strategy is $U_0 = x$, resulting in $U_n = 0$ for $n \ge 1$.*

Proof. Consider the following "pseudo strategy" $U_0 = x$ and $U_n = Y_n^+$ for $n \ge 1$. Suppose that we do not stop this process at ruin. Then $X_n - U_n \le 0$ for all n. Moreover, for any strategy $\{U_n'\}$ with $U_k' = 0$ for $k \ge \tau'$, we have for $n < \tau'$

$$X_n' = x + \sum_{k=1}^{n}(Y_k - U_{k-1}') \le x - U_0' + \sum_{k=1}^{n-1}(Y_k^+ - U_k') + Y_n = \sum_{k=0}^{n-1}(U_k - U_k') + Y_n\,.$$

Because $X_n' - Y_n = X_{n-1}' - U_{n-1}' \ge 0$, it follows that $\sum_{k=0}^n U_k' \le \sum_{k=0}^n U_k$. By Lemma 1.7,

$$V^{U'}(x) \le x + \mathbb{E}\Bigl[\sum_{n=1}^{\infty} \mathrm{e}^{-\delta n} Y_n^+\Bigr] = x + \frac{\mathbb{E}[Y^+]\mathrm{e}^{-\delta}}{1-\mathrm{e}^{-\delta}}\,,$$

yielding the upper bound. Using the strategy U, ruin occurs the first time where $Y_n < 0$. Thus, $\mathbb{P}[\tau = n+1] = (1-p_+)p_+^n$. The value of the strategy U is

$$\begin{aligned}
V^U(x) - x &= \mathbb{E}\Bigl[\sum_{n=1}^{\tau-1} \mathrm{e}^{-\delta n} Y_n\Bigr] = (1-p_+)\sum_{n=1}^{\infty} p_+^n \sum_{k=1}^{n} \mathrm{e}^{-\delta k}\mathbb{E}[Y \mid Y \ge 0] \\
&= (1-p_+)\sum_{k=1}^{\infty}\sum_{n=k}^{\infty} p_+^n \mathrm{e}^{-\delta k}\mathbb{E}[Y \mid Y \ge 0] \\
&= \sum_{k=1}^{\infty} p_+^k \mathrm{e}^{-\delta k}\mathbb{E}[Y \mid Y \ge 0] = \frac{\mathbb{E}[Y \mid Y \ge 0] p_+ \mathrm{e}^{-\delta}}{1 - p_+\mathrm{e}^{-\delta}} \\
&= \frac{\mathbb{E}[Y^+]\mathrm{e}^{-\delta}}{1 - p_+\mathrm{e}^{-\delta}}\,.
\end{aligned}$$

This proves the lower bound.

Let U be a strategy used for initial capital y. Apply the strategy U' with $U_0' = U_0 + x - y$ and $U_n' = U_n$ for $n \ge 1$ to the initial capital x. Then $X_n' = X_n$ for all $n \ge 1$, and $V(x) \ge V^{U'}(x) = V^U(y) + x - y$. Taking the supremum over all strategies U yields the second assertion.

Now if $\mathbb{P}[Y > 0] = 0$, then $\mathbb{E}[Y^+] = 0$ and $V(x) = x$ follows. One can readily see that the claimed strategy has the maximal value $V(x) = x$. □

We next prove that one can restrict to integer initial capital and to integer dividend payments. This will simplify our spaces E and $\mathcal{U}$. We denote by $\lfloor x \rfloor = \sup\{n \in \mathbb{Z} : n \le x\}$ the integer part of x. In the rest of this section we only consider strategies with $U_n = 0$ for $n \ge \tau$.

Lemma 1.9. *Let U be a strategy and X be the corresponding surplus process. Then there exists a strategy U' such that the corresponding surplus process X' fulfils $X_n' = \lfloor X_n \rfloor$ for $n \ge 1$ and $V^{U'}(x) \ge V^U(x)$. If $\mathbb{P}[X \ne X'] > 0$, then the strict inequality holds.*

Proof. Define $U'_0 = x - \lfloor x - U_0 \rfloor$ and $U'_n = \lfloor X_n \rfloor - \lfloor X_n - U_n \rfloor$. By the definition, $0 \le U'_n \le \lfloor X_n \rfloor$. For the process X' we obtain $X'_0 = x$, $X'_1 = x + Y_1 - U'_0 = Y_1 + \lfloor x - U_0 \rfloor = \lfloor X_1 \rfloor$ because $Y_1 \in \mathbb{Z}$. By induction,

$$X'_{n+1} = \lfloor X_n \rfloor + Y_{n+1} - U'_n = \lfloor X_n - U_n \rfloor + Y_{n+1} = \lfloor X_{n+1} \rfloor \,.$$

By the definition of ruin, we get that $\tau' = \tau$, i.e., ruin occurs at the same time as for the original strategy. From

$$x + \sum_{k=0}^{n} (Y_{k+1} - U_k) = X_{n+1} \ge \lfloor X_{n+1} \rfloor = x + \sum_{k=0}^{n} (Y_{k+1} - U'_k) \,,$$

we conclude that $\sum_{k=0}^{n} U'_n \ge \sum_{k=0}^{n} U_n$. The assertion now follows from Lemma 1.7, noting that the dividend stream is different if $X \ne X'$. □

If $x \notin \mathbb{N}$, the above result shows that $V(x) = V(\lfloor x \rfloor) + x - \lfloor x \rfloor$ because $x - U'_0 \in \mathbb{N}$. We therefore can restrict to $E = \mathbb{Z}$. If $x \in \mathbb{N}$, it is not optimal to choose a dividend with $U_n \notin \mathbb{N}$. We therefore can also restrict to $\mathcal{U} = \mathbb{N}$.

1.2.2 The Optimal Strategy

We proved in Lemma 1.1 that the function $V(x)$ fulfils the dynamic programming principle. Equation (1.2) reads

$$V(x) = \sup_{0 \le u \le x} \Bigl\{ u + \mathrm{e}^{-\delta} \sum_{j=-(x-u)}^{\infty} p_j V(x - u + j) \Bigr\} \,. \tag{1.7}$$

Here $V : \mathbb{N} \to \mathbb{R}_+$. Because there are only a finite number of values over which the supremum is taken, there exists for each $x \in \mathbb{N}$ a value $u(x)$ such that

$$V(x) = u(x) + \mathrm{e}^{-\delta} \sum_{j=-(x-u(x))}^{\infty} p_j V(x - u(x) + j) \,. \tag{1.8}$$

In order to define the optimal strategy in a unique way, we take the largest u fulfilling (1.8) if $u(x)$ is not uniquely defined.

Theorem 1.10. i) *The strategy $U_n = u(X_n)$ is an optimal strategy.*

ii) *For all $x \in \mathbb{N}$, the equality $u(x - u(x)) = 0$ holds and $u(y) = u(x) - (x - y)$ for all $x - u(x) \le y \le x$.*

iii) *For all $x \in \mathbb{N}$, one has $V(x) = V(x - u(x)) + u(x)$. In particular, $V(x) - V(y) = x - y$ for all $x - u(x) \le y \le x$.*

iv) *The number $x_0 := \sup\{x : u(x) = 0\}$ is finite, i.e., for x large enough a dividend should be paid immediately.*

v) *For all* $x \ge x_0$, $u(x) = x - x_0$, *and* $V(x) = V(x_0) + x - x_0$, *i.e., it is not optimal to have a capital larger than* x_0 *immediately after dividends are paid.*

vi) *Under the optimal dividend strategy ruin occurs almost surely.*

Proof. i) We need to show (1.3). As in the proof of Lemma 1.7, we find for any strategy that

$$\sum_{k=n}^{\infty} e^{-\delta k} U_k = (1 - e^{-\delta}) \sum_{m=n}^{\infty} \sum_{k=n}^{m} U_k e^{-\delta m} .$$

The pseudo-strategy

$$U'_k = \begin{cases} 0, & \text{if } k < n, \\ X'^{+}_n, & \text{if } k = n, \\ Y^+_k, & \text{if } k > n, \end{cases}$$

majorises $\sum_{k=n}^{m} U_k$ for any strategy U. Therefore,

$$\mathbb{E}\Big[\sum_{k=n}^{\infty} e^{-\delta k} U_k\Big] \le \mathbb{E}[(X'_n)^+] e^{-\delta n} + \frac{e^{-\delta(n+1)} \mathbb{E}[Y^+]}{1 - e^{-\delta}} .$$

Because $\mathbb{E}[(X'_n)^+] \le x + n\mathbb{E}[Y^+]$, the assertion follows from Corollary 1.3.

ii) and iii) The assertion is trivial if $y = x$. Let $x - u(x) \le y < x$. For initial capital y, there is the possibility to pay a dividend $u(x) - (x - y)$; thus,

$$\begin{aligned} V(y) &\ge u(x) - (x - y) + e^{-\delta} \sum_{j=-(y-(u(x)-(x-y)))}^{\infty} p_j V(y - (u(x) - (x - y)) + j) \\ &= V(x) - (x - y) . \end{aligned}$$

For initial capital x, paying the dividend $u(y) + x - y$ gives

$$\begin{aligned} V(x) &\ge u(y) + x - y + e^{-\delta} \sum_{j=-\{x-(u(y)+(x-y))\}}^{\infty} p_j V(x - (u(y) + (x - y)) + j) \\ &= V(y) + x - y . \end{aligned}$$

Thus, equality must hold. By our convention to take the largest possible u, we find that $u(x) = u(y) + x - y$. In particular, $V(y) = V(x) - (x - y)$. That $u(x - u(x)) = 0$ and $V(x) = u(x) + V(x - u(x))$ follows for $y = x - u(x)$.

iv) Suppose that $u(x) = 0$. Applying (1.6) and (1.8), we find that

$$\begin{aligned} x + \frac{\mathbb{E}[Y^+] e^{-\delta}}{1 - p_+ e^{-\delta}} &\le V(x) = e^{-\delta} \sum_{j=-x}^{\infty} p_j V(x + j) \\ &\le e^{-\delta} \sum_{j=-x}^{\infty} p_j \Big(x + j + \frac{\mathbb{E}[Y^+] e^{-\delta}}{1 - e^{-\delta}}\Big) \\ &\le e^{-\delta} x + e^{-\delta} \Big(\mathbb{E}[Y^+] + \frac{\mathbb{E}[Y^+] e^{-\delta}}{1 - e^{-\delta}}\Big) = e^{-\delta} x + \frac{\mathbb{E}[Y^+] e^{-\delta}}{1 - e^{-\delta}} . \end{aligned}$$

This yields

$$x \le \frac{\mathbb{E}[Y^+]e^{-2\delta}(1-p_+)}{(1-e^{-\delta})^2(1-p_+e^{-\delta})}\;,$$

i.e., $x_0 < \infty$. Note that $u(0) = 0$, showing that $\{x : u(x) = 0\} \neq \emptyset$. Therefore, $u(x_0) = 0$.

v) From ii) we conclude that $u(x - u(x)) = 0$, and therefore $u(x) \ge x - x_0$ follows. We also obtain $0 = u(x_0) = u(x) - (x - x_0)$. From part iii) we conclude that $V(x) = V(x_0) + u(x) = V(x_0) + x - x_0$.

vi) We have

$$\mathbb{P}[Y_1 < 0, Y_2 < 0, \ldots, Y_{x_0+1} < 0] = (1-p_+)^{x_0+1} > 0\;.$$

Thus, by the strong law of large numbers, $\{Y_{k(x_0+1)+1} < 0, Y_{k(x_0+1)+2} < 0, \ldots, Y_{(k+1)(x_0+1)} < 0\}$ holds for some k. But then ruin must occur. □

The result gives us an alternative way to characterise the optimal strategy $u(x)$. Clearly, $u(0) = 0$. For $x \ge 1$ we obtain

$$u(x) = \sup\{n \in \mathbb{N} : V(x) = V(x-n) + n\}\;.$$

Indeed, by part iii) of the above theorem, $V(x) = V(x-n) + n$ for $n \le u(x)$. If $u(x) = x$, the statement is proved. If $V(x) = V(x - u(x) - 1) + u(x) + 1$, then we could conclude from (1.7) for $V(x)$ and for $V(x - u(x) - 1)$ that $u = u(x - u(x) - 1) + u(x) + 1$ maximises the right-hand side of (1.7) for $V(x)$. But this is not possible, because we assumed that $u(x)$ is the maximal value. Finally, for $2 \le n \le x - u(x)$ we have

$$\begin{aligned} V(x) &> V(x-u(x)-1) + u(x) + 1 \ge (V(x-u(x)-n) + n - 1) + u(x) + 1 \\ &= V(x-u(x)-n) + u(x) + n\;, \end{aligned}$$

by Lemma 1.8.

Let us now consider the numerical algorithm of Lemma 1.4. Choosing

$$x_1 = \left\lfloor \frac{\mathbb{E}[Y^+]e^{-2\delta}(1-p_+)}{(1-e^{-\delta})^2(1-p_+e^{-\delta})} \right\rfloor\;, \tag{1.9}$$

we can restrict to functions $v(x) : \mathbb{N} \to \mathbb{R}_+$ with $v(x_1 + n) = v(x_1) + n$ for all $n \in \mathbb{N}$. The contraction operator then becomes

$$\mathcal{V}(v)(x) = \sup_{0 \le u \le x} u + e^{-\delta} \sum_{j=-(x-u)}^{\infty} p_j v(x - u + j)\;,$$

for $x \le x_1$ and $\mathcal{V}(v)(x) = \mathcal{V}(v)(x_1) + x - x_1$ for $x > x_1$. In the algorithm it is not necessary to consider all possible u. From the proof of Theorem 1.10 parts ii) and iii), we can construct the values $u_v(x)$ maximising the right-hand side of $\mathcal{V}(v)(x)$ by $u_v(0) = 0$, and then only consider the values $u = 0$ and $u = \inf\{n \geq 1 : u_v(x-n) = 0\}$. The supremum on the right-hand side of $\mathcal{V}(v)(x)$ is taken at one of these values. This yields

$$\mathcal{V}(v)(0) = \mathrm{e}^{-\delta} \sum_{j=0}^{\infty} p_j v(j) ,$$

and for $x \leq x_1 - 1$,

$$\mathcal{V}(v)(x+1) = \max\Bigl\{1 + \mathcal{V}(v)(x), \mathrm{e}^{-\delta} \sum_{j=-(x+1)}^{\infty} p_j v(x+j+1)\Bigr\} .$$

For example, let $v_0(x) = x + \mathbb{E}[Y^+]\mathrm{e}^{-\delta}/(1 - p_+\mathrm{e}^{-\delta})$ and $v_{n+1}(x) = \mathcal{V}(v_n)(x)$. We obtain the following

Corollary 1.11. *If $p_x \leq (\mathrm{e}^{\delta} - p_+)(\mathrm{e}^{\delta} - 1)/\mathbb{E}[Y^+]$ for all $x < 0$, then*

$$V(x) = x + \frac{\mathbb{E}[Y^+]\mathrm{e}^{-\delta}}{1 - p_+\mathrm{e}^{-\delta}} ,$$

i.e., the optimal strategy is $u(x) = x$, to pay all capital as dividends.

Proof. We need to show that $v_0(x)$ is a fixed point of $\mathcal{V}$. We get $v_1(0) = v_0(0)$. Suppose that $v_1(x) = v_0(x)$. Then

$$\begin{aligned}
\mathrm{e}^{-\delta} \sum_{j=-(x+1)}^{\infty} p_j v_0(x+j+1) &= \mathrm{e}^{-\delta} \sum_{j=-(x+1)}^{\infty} p_j \Bigl(x + j + 1 + \frac{\mathbb{E}[Y^+]\mathrm{e}^{-\delta}}{1 - p_+\mathrm{e}^{-\delta}}\Bigr) \\
&= \mathrm{e}^{-\delta} p_{-x-1} \frac{\mathbb{E}[Y^+]\mathrm{e}^{-\delta}}{1 - p_+\mathrm{e}^{-\delta}} + \mathrm{e}^{-\delta} \sum_{j=-x}^{\infty} p_j \\
&\quad + \mathrm{e}^{-\delta} \sum_{j=-x}^{\infty} p_j \Bigl(x + j + \frac{\mathbb{E}[Y^+]\mathrm{e}^{-\delta}}{1 - p_+\mathrm{e}^{-\delta}}\Bigr) \\
&\leq \mathrm{e}^{-\delta} \Bigl(p_{-x-1} \frac{\mathbb{E}[Y^+]\mathrm{e}^{-\delta}}{1 - p_+\mathrm{e}^{-\delta}} + \sum_{j=-x}^{\infty} p_j\Bigr) + \mathcal{V}(v_0)(x) \\
&\leq \mathrm{e}^{-\delta} \Bigl(\frac{\mathrm{e}^{-\delta}(\mathrm{e}^{\delta} - p_+)(\mathrm{e}^{\delta} - 1)}{1 - p_+\mathrm{e}^{-\delta}} + 1\Bigr) + \mathcal{V}(v_0)(x) = 1 + \mathcal{V}(v_0)(x) .
\end{aligned}$$

This proves that $v_1(x+1) = 1 + v_1(x) = 1 + v_0(x) = v_0(x+1)$. □

We let $u_n(x) = u_{v_n}(x)$. Then $u_n(x)$ converges to an optimal strategy.

Corollary 1.12. *Let $u_\infty(x) = \limsup_{n\to\infty} u_n(x)$. Then $V^{u_\infty}(x) = V(x)$. Moreover, if $u(x)$ is the unique point at which the maximum in (1.7) is taken, then $\lim_{n\to\infty} u_n(x) = u(x)$.*

Proof. From Lemma 1.4 we know that $\lim_{n\to\infty} v_n(x) = V(x)$. Thus, also

$$\lim_{n\to\infty} u + e^{-\delta} \sum_{j=-(x-u)}^{\infty} p_j v_n(x-u+j) = u + e^{-\delta} \sum_{j=-(x-u)}^{\infty} p_j V(x-u+j) \;.$$

Only considering n_k for which $u_{n_k}(x) = u_\infty(x)$ [this exists because $u_n(x) \in \{0,1,\dots,x\}$] shows that

$$V(x) = \lim_{k\to\infty} v_{n_k+1}(x) = u_\infty + e^{-\delta} \sum_{j=-(x-u_\infty)}^{\infty} p_j V(x-u_\infty+j) \;,$$

where $u_\infty = u_\infty(x)$. Thus, for $u = u(x)$,

$$u_\infty + e^{-\delta} \sum_{j=-(x-u_\infty)}^{\infty} p_j V(x-u_\infty+j) = u + e^{-\delta} \sum_{j=-(x-u)}^{\infty} p_j V(x-u+j) \;.$$

If $u(x)$ is now the unique value maximising (1.7), then $u_\infty(x) = u(x)$. The proof could be repeated with any convergent subsequence of $u_{n_k}(x)$, showing that $u_n(x)$ converges to $u(x)$. □

1.2.3 Premia of Size One

Before we discuss some specific examples, let us discuss in general the case of a premium of size one. In each period the insurer earns a premium of size one and possibly has to pay an integer-valued claim $\tilde{Y}_i$. That is, $Y_i = 1 - \tilde{Y}_i$, and $p_k = 0$ for $k > 1$. Formula (1.7) then reads

$$V(x) = \max\Bigl\{V(x-1)+1, e^{-\delta} \sum_{j=0}^{x+1} p_{j-x} V(j)\Bigr\} \;; \tag{1.10}$$

see Theorem 1.10. Let $n_0 = \inf\{n \in \mathbb{N} : u(n) = 1\}$ be the first point where a dividend is paid. Recall that $1 \le n_0 < \infty$. Then for $0 \le n < n_0$, we obtain

$$V(n+1) = \frac{1}{p_1}\Bigl((e^\delta - p_0)V(n) - \sum_{j=0}^{n-1} p_{j-n} V(j)\Bigr) \;. \tag{1.11}$$

This equation is a linear equation. Let $\rho_0 = 1$, and recursively define

$$\rho_{n+1} = \frac{1}{p_1}\Bigl((e^\delta - p_0)\rho_n - \sum_{j=0}^{n-1} p_{j-n} \rho_j\Bigr) \;.$$

We get

$$\rho_1 = \frac{1}{p_1}(e^\delta - p_0)\rho_0 = \frac{e^\delta - p_0}{p_1} > 1 = \rho_0 \;,$$

because $e^\delta > 1 > p_0 + p_1$. By induction we obtain that ρ_n is strictly increasing:

$$\rho_{n+1} = \frac{1}{p_1}\Bigl((e^\delta - p_0)\rho_n - \sum_{j=0}^{n-1} p_{j-n}\rho_j\Bigr) > \frac{e^\delta - p_0 - \sum_{j=-n}^{-1} p_j}{p_1}\rho_n > \rho_n\;.$$

Comparing with the dynamic programming equation we see that $V(n) = \rho_n V(0)$. $V(0)$ can now be determined through $V(n_0) = V(n_0 - 1) + 1$. That is,

$$V(0) = (\rho_{n_0} - \rho_{n_0-1})^{-1}\;.$$

Starting from zero, we of course choose the strategy that maximises $V(0)$; thus n_0 is the argument such that $\rho_n - \rho_{n-1}$ becomes minimal. From Theorem 1.10 part iv), we know that n_0 is finite.

Once having determined n_0, one can determine $n_1 = \inf\{n > n_0 : u(n) > 0\}$. Instead of a multiple of $V(0)$ in Equation (1.11), we obtain $V(n) = q_n V(n_0 + 1) + k_n$. From $V(n_1) = V(n_1 - 1) + 1$ it is possible to determine $V(n_0 + 1)$. Doing this for each possible n_1, $V(n_0 + 1)$ and n_1 can be determined by choosing the values that maximise $V(n_0 + 1)$.

Example 1.13. De Finetti [59] considered the example where $p_1 = p$ and $p_{-1} = 1 - p$. Then we have the equations

$$V(n+1) = \frac{e^\delta}{p}V(n) - \frac{1-p}{p}V(n-1)\;,$$

if $n \ge 1$, and

$$V(1) = \frac{e^\delta}{p}V(0)\;.$$

Let us first determine conditions under which $x_0 = 0$. If $x_0 = 0$, then

$$1 + V(0) = V(1) = \frac{e^\delta}{p}V(0)\;.$$

It follows that

$$V(n) = n + \frac{p}{e^\delta - p}\;.$$

We have now to verify that for $n \ge 1$

$$\begin{aligned} n + \frac{p}{e^\delta - p} &= V(n) \ge e^{-\delta}\Bigl[p\Bigl(n + 1 + \frac{p}{e^\delta - p}\Bigr) + (1-p)\Bigl(n - 1 + \frac{p}{e^\delta - p}\Bigr)\Bigr] \\ &= e^{-\delta}\Bigl(n + \frac{p}{e^\delta - p} + 2p - 1\Bigr)\;. \end{aligned}$$

This is equivalent to

$$(e^\delta - 1)\Bigl(n + \frac{p}{e^\delta - p}\Bigr) \ge 2p - 1\;.$$

The inequality is fulfilled whenever it is fulfilled for $n = 1$, i.e.,

$$(\mathrm{e}^\delta - 1)\frac{\mathrm{e}^\delta}{\mathrm{e}^\delta - p} \ge 2p - 1 .$$

Multiplying by the denominator yields the equation

$$\mathrm{e}^{2\delta} - 2p\mathrm{e}^\delta + p^2 \ge p(1-p) .$$

Hence,

$$\mathrm{e}^\delta \ge p + \sqrt{p(1-p)} .$$

It is not surprising that the inequality is always fulfilled if $p \le \frac{1}{2}$.

Let us now assume that $\mathrm{e}^\delta < p + \sqrt{p(1-p)}$. In particular, $\mathrm{e}^\delta < 2p$. We want to determine n_0. We have $\rho_0 = 1$ and $\rho_1 = \mathrm{e}^\delta / p$. We recursively find that

$$\rho_{n+1} = \frac{\mathrm{e}^\delta}{p}\rho_n - \frac{1-p}{p}\rho_{n-1} .$$

Letting $\rho_n = \lambda^n$, this is

$$\lambda^2 = \frac{\mathrm{e}^\delta}{p}\lambda - \frac{1-p}{p} ,$$

or

$$\lambda_{1/2} = \frac{\mathrm{e}^\delta \pm \sqrt{\mathrm{e}^{2\delta} - 4p(1-p)}}{2p} = \frac{\mathrm{e}^\delta \pm \sqrt{\mathrm{e}^{2\delta} - 1 + (2p-1)^2}}{2p} .$$

We see that $\lambda_1 > 1 > \lambda_2$. The solution is then $\rho_n = A\lambda_1^n + B\lambda_2^n$. From the initial conditions we conclude that

$$\rho_n = \frac{\lambda_1^{n+1} - \lambda_2^{n+1}}{\lambda_1 - \lambda_2} .$$

We have to minimise

$$(\lambda_1 - \lambda_2)(\rho_n - \rho_{n-1}) = (\lambda_1 - 1)\lambda_1^n + (1 - \lambda_2)\lambda_2^n .$$

The minimum is determined through

$$(\lambda_1 - 1)\lambda_1^n \log \lambda_1 + (1 - \lambda_2)\lambda_2^n \log \lambda_2 = 0 .$$

This has the solution

$$n = \frac{\log[\{-(1-\lambda_2)\log \lambda_2\}/\{(\lambda_1 - 1)\log \lambda_1\}]}{\log(\lambda_1/\lambda_2)} .$$

If $n \notin \mathbb{N}$, then either $n_0 = \lfloor n \rfloor$ or $n_0 = \lfloor n+1 \rfloor$.

Suppose that we have now determined n_0. Then $V(n_0 - 1) = V(n_0) - 1$ and $V(n_0 + 1) \ge V(n_0) + 1$. Thus,

$$\mathrm{e}^\delta V(n_0) \ge pV(n_0+1) + (1-p)V(n_0-1) \ge p(V(n_0)+1) + (1-p)(V(n_0)-1) .$$

It follows for any $n \ge 1$ that

$$e^{\delta} V(n_0 + n) \ge e^{\delta}(V(n_0) + n) > p(V(n_0) + n + 1) + (1 - p)(V(n_0) + n - 1) .$$

The right-hand side multiplied by $e^{-\delta}$ is the value if we do not pay a dividend. Because the value of $V(n_0 + n)$ is strictly larger, this means that a dividend has to be paid. By induction, this implies that

$$V(n) = V(n_0) + n - n_0$$

for all $n \ge 0$, because the function $V(n)$ defined in this way fulfils (1.10). The optimal strategy is thus a barrier strategy. ■

Example 1.14. We now consider geometrically distributed claim sizes. We let $p_k = (1 - p)p^{1-k}$ for $k \le 1$. We want to solve

$$\rho_{n+1} = \Bigl(\frac{e^{\delta}}{1 - p} - p\Bigr)\rho_n - \sum_{j=0}^{n-1} p^{n+1-j} \rho_j .$$

Letting $q_n = \rho_n p^{-n}$, we have

$$q_{n+1} = \Bigl(\frac{e^{\delta}}{p(1 - p)} - 1\Bigr) q_n - \sum_{j=0}^{n-1} q_j = \frac{e^{\delta}}{p(1 - p)} q_n - \sum_{j=0}^{n} q_j .$$

Let $\ell_n = \sum_{j=0}^{n-1} q_j$. We can express the equation in matrix form as

$$\begin{pmatrix} q_{n+1} \\ \ell_{n+1} \end{pmatrix} = \begin{pmatrix} (\frac{e^{\delta}}{p(1-p)} - 1) & -1 \\ 1 & 1 \end{pmatrix} \begin{pmatrix} q_n \\ \ell_n \end{pmatrix} .$$

The initial vector is $(1, 0)^{\top}$. Let us calculate the eigenvalues. We have to solve

$$\Bigl(\frac{e^{\delta}}{p(1 - p)} - 1 - \lambda\Bigr)(1 - \lambda) + 1 = 0 .$$

This is equivalent to

$$\lambda^2 - \lambda \frac{e^{\delta}}{p(1 - p)} + \frac{e^{\delta}}{p(1 - p)} = 0 .$$

The eigenvalues are

$$\lambda_{1/2} = \frac{e^{\delta} \pm \sqrt{e^{2\delta} - 4p(1 - p)e^{\delta}}}{2p(1 - p)} .$$

Note that $\lambda_1 > p^{-1} > \lambda_2 > 1$. Thus, we have $q_n = A\lambda_1^n + B\lambda_2^n$. From the initial conditions $q_0 = 1$ and $q_1 = \lambda_1 + \lambda_2 - 1$, we see that

$$q_n = \frac{(\lambda_1-1)\lambda_1^n - (\lambda_2-1)\lambda_2^n}{\lambda_1-\lambda_2}\,, \qquad \rho_n = \frac{(\lambda_1-1)(\lambda_1 p)^n - (\lambda_2-1)(\lambda_2 p)^n}{\lambda_1-\lambda_2}\,.$$

We have to minimise

$$(\lambda_1-\lambda_2)(\rho_n-\rho_{n-1}) = (\lambda_1-1)(\lambda_1 p-1)(\lambda_1 p)^{n-1} + (\lambda_2-1)(1-\lambda_2 p)(\lambda_2 p)^{n-1}\,.$$

Thus, we find that

$$n = 1 + \frac{\log[\{-(\lambda_2-1)(1-\lambda_2 p)\log(\lambda_2 p)\}/\{(\lambda_1-1)(\lambda_1 p-1)\log(\lambda_1 p)\}]}{\log(\lambda_1/\lambda_2)}\,.$$

We have either $n_0 = \lfloor n \rfloor$ or $n_0 = \lfloor n+1 \rfloor$. Thus, we have found $V(x)$ for $x \le n_0$. ∎

Bibliographical Remarks

The model considered in this section was introduced by de Finetti [59]. His motivation was to measure an insurance portfolio by the value of future profits (as done in economics) instead of ruin probabilities (as done in actuarial science). He considered the model of Example 1.13. Gerber [72] found Equation (1.7), proved the bounds (1.6), and the finiteness of the set $\{x : u(x) = 0\}$. Dickson and Waters [47] considered premia of size one and studied the mth moment of the discounted dividends for barrier strategies, that is, $u(x) = 1\!\!\mathrm{I}_{x=b+1}$ for some $b \in \mathbb{N}$. Frisque [63] considered the expected utility of the future dividends. A general introduction to ruin theory can be found in [152].

1.3 Minimising Ruin Probabilities

1.3.1 Optimal Reinsurance

We now consider the (discrete-time) surplus process connected to some insurance portfolio. The insurer earns some premium. With the surplus, reinsurance is bought, and a premium has to be paid. The insurer can control how much reinsurance is bought. Our goal will be to maximise the survival probability, or, equivalently, to minimise the ruin probability. In order for the problem not to become trivial, we will have to assume that the cedent (first insurer) would have to pay more for full reinsurance than the premium he receives.

Let $Y_i \ge 0$ be the aggregate claim in period i and let $G(y)$ denote its distribution function. The sequence is assumed to be iid. We work again with the filtration $\{\mathcal{F}_t = \mathcal{F}_t^Y\}$. The insurer can at each time i choose the reinsurance for the next period. The set of possible reinsurance treaties is a compact connected subset $\mathcal{U} \subset \mathbb{R}^d$. If the reinsurance treaty $b \in \mathcal{U}$ is chosen, the insurer has to pay $s(Y_i, b)$ for the aggregate claim Y_i in the ith period, and the rest is paid by the reinsurer. Here $s : (0, \infty) \times \mathcal{U} \to [0, \infty)$ is some function with properties we will assume below. We only allow reinsurance treaties with $0 \le s(y, b) \le y$, i.e., the reinsurer pays at most the whole claim size. For this protection the insurer has to pay a reinsurance premium. Let $c(b) \in \mathbb{R}$ denote the premium left for the insurer if reinsurance b is chosen, i.e., the original premium minus the reinsurance premium. The reinsurer can at any time change the reinsurance for the next period, i.e., he chooses an adapted strategy $\{b_n\}$. The income for the next period is then $c(b_n)$. We will treat more specific examples ahead. For popular reinsurance forms, see Section D.5.

We assume that more reinsurance is more expensive $[s(y, b) \ge s(y, b')$ for all y implies that $c(b) \ge c(b')]$ and that full reinsurance leads to a strictly negative income ($c(b_{\text{fr}}) < 0$). We also assume that $c(b)$ and $s(y, b)$ are continuous in b. The income in case of no reinsurance is denoted by c. We then have that $c(b) \le c$. We also assume that $s(y, b)$ is increasing in y. In this case the generalised inverse $\rho(z, b) := \sup\{y : s(y, b) \le z\}$ is well defined. Note that $\rho(z, b)$ is increasing and right-continuous in z. We also assume that $\mathbb{P}[s(Y, b) > c(b)] > 0$ for all $b \in \mathcal{U}$. Otherwise, ruin can be prevented by reinsurance and the problem considered in this section becomes trivial. We also assume the net profit condition $\mathbb{E}[c(b) - s(Y, b)] > 0$ for some b. Otherwise, ruin cannot be prevented because the surplus would be decreasing in time for all reinsurance treaties.

We now fit the optimisation problem into the setup of Section 1.1. Let the initial capital be x. Then the surplus process is $X_0 = x$ and

$$X_{n+1} = X_n + c(b_n) - s(Y_{n+1}, b_n) ,$$

as long as $X_n \ge 0$. If $b \in \mathcal{U}$ is now a reinsurance treaty fulfilling the net profit condition $\mathbb{E}[c(b) - s(Y, b)] > 0$, then by the law of large numbers, $n^{-1} \sum_{k=1}^{n} c(b) - s(Y_{k+1}, b) \to \mathbb{E}[c(b) - s(Y, b)]$. This implies that for the constant strategy $b_n = b$ the process X_n tends to infinity. In particular, $\inf_n X_n > -\infty$. Hence, there is an initial capital x_0 such that $\mathbb{P}[\inf_n X_n \ge 0 \mid X_0 = x_0] > 0$. Because there is a strictly positive probability that from initial capital zero the set $[x_0, \infty)$ is reached before the set $(-\infty, 0)$, we get also that $\mathbb{P}[\inf_n X_n \ge 0 \mid X_0 = 0] > 0$. Hence, we have a strategy such that ruin is not certain.

Following the idea of Schäl [154] we introduce a cemetery state Δ. If $X_n < 0$ or $X_n = \Delta$, we let $X_{n+1} = \Delta$. This allows us to formulate the problem in the setup of Section 1.1.

We let $\delta = 0$ and choose the reward function

$$r(X_n, b_n) = \begin{cases} 0, & \text{if } X_n \ge 0 \text{ or } X_n = \Delta, \\ -1, & \text{if } X_n < 0. \end{cases}$$

In this way the cost is paid at most once. The value of a reinsurance strategy is $V^b(x) = -\mathbb{P}[X_n = \Delta \text{ for some } n]$, and the value function is

$V(x) = \sup_b V^b(x)$, where we take the supremum over all adapted reinsurance strategies b. Clearly, $V(x) \in (-1,0)$; hence, $V: \mathbb{R}_+ \to (-1,0)$. Equation (1.2) then reads for $x \ge 0$

$$\begin{aligned} V(x) &= \sup_{b\in\mathcal{U}} \int_0^\infty V(x+c(b)-s(y,b))\,\mathrm{d}G(y) \\ &= \sup_{b\in\mathcal{U}} \int_0^{\rho(x+c(b),b)} V(x+c(b)-s(y,b))\,\mathrm{d}G(y) - (1-G(\rho(x+c(b),b)))\,, \end{aligned}$$

where we used that $V(x) = -1$ for $x < 0$.

The difficulty with solving the equation is that it does not have a unique solution within the set of real functions. We have to pick the solution with $\lim_{x\to\infty} V(x) = 0$. It turns out to be simpler to consider the survival function $\delta(x) = 1 + V(x)$. Then

$$\delta(x) = \sup_{b\in\mathcal{U}} \int_0^{\rho(x+c(b),b)} \delta(x+c(b)-s(y,b))\,\mathrm{d}G(y)\,.$$

Any multiple of $\delta(x)$ also solves the equation. Let us therefore consider

$$f(x) = \sup_{b\in\mathcal{U}} \int_0^\infty f(x+c(b)-s(y,b))\,\mathrm{d}G(y) \tag{1.12}$$

with $f(0) = 1$, imposing $f(x) = 0$ for $x < 0$.

For the following results we will need that either $\tau < \infty$ or $X_n \to \infty$. For a proof we need the following lemma, taken from [140, p. 267].

Lemma 1.15. *Let $S_n = \sum_{k=1}^n W_k - Z_k$ with $0 \le W_k \le w$ for some $w < \infty$ and $Z_k \ge 0$. If $\{S_n\}$ is a submartingale, then*

$$\mathbb{P}\Bigl[\sum_{k=1}^\infty Z_k = \infty, \sum_{k=1}^\infty W_k < \infty\Bigr] = 0\,.$$

Proof. Let $a \in (0,\infty)$ and $N = \inf\{n : S_n > a\}$. Then $\{S_{n\wedge N}\}$ is a submartingale that is bounded from above by $a + w$. Thus, $S_{n\wedge N}$ converges to an integrable random variable S_N. In particular, if $\sup_n S_n \le a$, then $\inf_n S_n > -\infty$. Because this holds for all $a > 0$, we have that $\sup_n S_n < \infty$ implies that $\inf_n S_n > -\infty$. If $\sup_n S_n = \infty$, then $\sum_{k=1}^\infty W_k = \infty$. If $\sup_n S_n < \infty$ and $\sum_{k=1}^\infty Z_k = \infty$, then, because $\inf_n S_n > -\infty$, we also have $\sum_{k=1}^\infty W_k = \infty$. □

Lemma 1.16. *For any strategy either ruin occurs or the capital tends to infinity.*

Proof. The function $b \mapsto \mathbb{P}[s(Y,b) > c(b)]$ is lower semi-continuous. Thus, there is b_0 where the infimum is taken. By our assumption we have $\mathbb{P}[s(Y,b) > c(b)] \ge 2\delta$ for some $\delta > 0$. For each b let

$$\varepsilon(b) = \sup\{\varepsilon : \mathbb{P}[s(Y,b) > c(b) + \varepsilon] \ge \delta\} .$$

By continuity and compactness we have that $\varepsilon(b)$ is bounded away from zero. Thus, there is $\varepsilon > 0$ such that $\mathbb{P}[s(Y,b) > c(b) + \varepsilon] \ge \delta$ for all b.

Choose $a > 0$. Let $W_n = \mathbb{I}_{X_n \le a, s(Y_{n+1}, b_n) > c(b_n) + \varepsilon}$ and $Z_n = \delta \mathbb{I}_{X_n \le a}$. Then $S_n = \sum_{k=1}^n W_k - Z_k$ fulfils the assumptions of Lemma 1.15. Thus, $X_n \le a$ infinitely often implies that $X_{n+1} \le a - \varepsilon$ infinitely often. Therefore, if $\liminf X_n \le a$, then $X_n \le a + \varepsilon/2$ infinitely often. This implies that $X_n \le a - \varepsilon/2$ infinitely often, or $\liminf X_n \le a - \varepsilon/2$. By induction, $\liminf X_n \le -\varepsilon/2$, so ruin occurs almost surely if $\liminf X_n < \infty$. □

Clearly, the solution we are looking for is increasing. If we use the same strategy for initial capital x and initial capital $x+h$, then ruin cannot occur for initial capital $x+h$ unless it occurs for initial capital x. We can say something about the uniqueness of the solution if we restrict to increasing functions.

Theorem 1.17. *Suppose that $f(x)$ is an increasing solution to* (1.12) *with $f(0) = 1$. Then $f(x)$ is bounded and $f(x) = \delta(x)/\delta(0)$.*

Proof. Denote by $f(\infty) = \lim_{x\to\infty} f(x) \in [1, \infty]$. Let $\{b_n\}$ be an arbitrary strategy. We prove that $\{f(X_{\tau\wedge n})\}$ is a supermartingale. We have

$$\begin{aligned}
\mathbb{E}[f(X_{\tau\wedge(n+1)}) \mid \mathcal{F}_n] &= \int_0^\infty f(X_n + c(b_n) - s(y, b_n)) \,\mathrm{d}G(y) \mathbb{I}_{\tau>n} \\
&\le \sup_{b\in\mathcal{U}} \int_0^\infty f(X_n + c(b) - s(y,b)) \,\mathrm{d}G(y) \mathbb{I}_{\tau>n} \\
&= f(X_n)\mathbb{I}_{\tau>n} = f(X_{\tau\wedge n}) ,
\end{aligned}$$

where we used (1.12). Then $\{f(X_{\tau\wedge n})\}$ is a positive supermartingale and $\lim_{n\to\infty} f(X_{\tau\wedge n})$ exists by the martingale convergence theorem. This limit must be $f(\infty)$ or zero by Lemma 1.16. For an arbitrary strategy $\{b_t\}$ we let $\delta^b(x)$ be the survival probability that ruin does not occur. Because it is possible to choose a strategy such that $\delta^b(x) > 0$ and $\lim_{n\to\infty} f(X_{\tau\wedge n})$ must be integrable, we have that $f(\infty) < \infty$. By bounded convergence, $f(x) \ge f(\infty)\delta^b(x)$. Choose $\varepsilon > 0$. We now choose a strategy $\{b_n\}$ such that

$$f(X_n) < \int_0^\infty f(X_n + c(b_n) - s(y, b_n)) \,\mathrm{d}G(y) + \frac{\varepsilon}{(n+1)^2} .$$

Then $\{f(X_{\tau\wedge n}) + \sum_{k=0}^{n-1} \varepsilon/(k+1)^2\}$ is a bounded submartingale. Letting $n \to \infty$ shows that $f(x) < f(\infty)\delta(x) + \sum_{k=0}^\infty \varepsilon/(k+1)^2$. Thus, $f(x) = \delta(x) f(\infty)$. □

We next show how the solution can be calculated. Let $f_0(x) = \mathbb{I}_{x\ge 0}$ and define recursively

$$f_{n+1}(x) = \Bigl(\sup_{b\in\mathcal{U}} \int_0^\infty f_n(x + c(b) - s(y,b)) \,\mathrm{d}G(y)\Bigr) \mathbb{I}_{x\ge 0} .$$

Then the following proposition holds.

Proposition 1.18. *The functions $f_n(x)$ converge to $\delta(x)$.*

Proof. By induction it follows that $0 \le f_n(x) \le 1$. We next show by induction that the functions $f_n(x)$ are increasing in x. Let $h > 0$ and $b \in \mathcal{U}$ be arbitrary. Then because $f_n(x)$ is increasing,

$$\begin{aligned}\int_0^\infty & f_n(x + c(b) - s(y,b))\,\mathrm{d}G(y) \\ &\le \int_0^\infty f_n(x + h + c(b) - s(y,b))\,\mathrm{d}G(y) \le f_{n+1}(x+h)\,.\end{aligned}$$

Taking the supremum over all b gives $f_{n+1}(x) \le f_{n+1}(x+h)$. We show that $\{f_n(x) : n \in \mathbb{N}\}$ is monotone in n. Clearly, $f_0(x) = 1 \ge f_1(x)$. Suppose that $f_{n-1}(x) \ge f_n(x)$ for all x. Let $\varepsilon > 0$ and fix x. Denote by b an argument such that

$$\int_0^\infty f_n(x + c(b) - s(y,b))\,\mathrm{d}G(y) > f_{n+1}(x) - \varepsilon\,.$$

Then

$$\begin{aligned}f_n(x) &- f_{n+1}(x) \\ &\ge \int_0^\infty (f_{n-1}(x + c(b) - s(y,b)) - f_n(x + c(b) - s(y,b)))\,\mathrm{d}G(y) - \varepsilon\,.\end{aligned}$$

By our assumption $f_n(x) - f_{n+1}(x) > -\varepsilon$. Because ε was arbitrary, we get that $\{f_n(x)\}$ is a decreasing sequence in n. In particular, it converges pointwise to a function $f(x)$. By monotone convergence the limit has to fulfil (1.12). Moreover, $f(x)$ is increasing because $f_n(x)$ is increasing for each n. By Theorem 1.17 we have $f(x) = \delta(x)f(\infty)$. In order to show that $f(\infty) = 1$, we show that $f_n(x) \ge \delta(x)$. Clearly, $f_0(x) = 1 \ge \delta(x)$. By induction it follows that $f_n(x) \ge \delta(x)$ because $\delta(x)$ solves (1.12). Thus, we also have $f(x) \ge \delta(x)$. Because $1 \ge f(\infty) \ge \delta(\infty) = 1$, the result is proved. □

1.3.2 Optimal Investment

Let $\{R_n\}$ be the return of a risky portfolio in period n, and we suppose that $\{R_n\}$ is iid. That is, if we invest one monetary unit at time $n-1$, we get back $1 + R_n$ units at time n. As before, $\{Y_n\}$ is the aggregate claim in period n. We let $\{R_n\}$ and $\{Y_n\}$ be independent. The insurer can now choose U_n, the amount that is invested at time n into the risky portfolio. The surplus process then becomes

$$X_{n+1} = X_n + c + U_n R_{n+1} - Y_{n+1}$$

if $X_n \ge 0$. The reward function and the process for $X_n < 0$ are chosen as in Section 1.3.1. In order for the problem not to have the trivial solution $U_n = 0$, we assume that $\mathbb{P}[Y > c] > 0$. The filtration we work with is now $\{\mathcal{F}_t = \mathcal{F}_t^{(Y,R)}\}$, the natural filtration of the process $\{(Y_n, R_n)\}$.

We assume that the market is arbitrage-free, i.e., $\mathbb{P}[R_n \ge 0] \in (0,1)$, and that $\mathbb{E}[R_n] > 0$. In particular, the distribution function $H(z)$, say, of R_n is nondegenerate. If the market were not arbitrage-free, it would be possible to make a profit without risk. In order for an investor to be willing to take over the risk, he should on average get a higher return than with a risk-free investment. Note that no net profit condition is needed. Choose $u > 0$ such that $c + u\mathbb{E}[R_n] > \mathbb{E}[Y_n]$. Then the process with the fixed strategy $U_n = u$ has a strictly positive survival probability.

The value function then fulfils (1.2):

$$V(x) = \sup_{u\in\mathbb{R}} \int_0^\infty \int_{-\infty}^\infty V(x + c + ur - y)\,\mathrm{d}H(r)\,\mathrm{d}G(y)\;,$$

with the convention that $V(x) = -1$ for $x < 0$. We again consider $\delta(x) = 1 + V(x)$ and $f(x) = \delta(x)/\delta(0)$. Then $f(x) = 0$ for $x < 0$ and for $x \ge 0$,

$$\begin{aligned} f(x) &= \sup_{u\in\mathbb{R}} \int_0^\infty \int_{-\infty}^\infty f(x + c + ur - y)\,\mathrm{d}H(r)\,\mathrm{d}G(y) \\ &= \sup_{u\in\mathbb{R}} \int_{-\infty}^\infty \int_0^{(x+c+ur)^+} f(x + c + ur - y)\,\mathrm{d}G(y)\,\mathrm{d}H(r)\;. \end{aligned} \tag{1.13}$$

We now prove the results from Section 1.3.1 for the present case.

Lemma 1.19. *For any strategy either ruin occurs or $X_n \to \infty$.*

Proof. There is $\varepsilon > 0$ such that $\mathbb{P}[Y > c + \varepsilon] > 0$. Let $\delta = \min\{\mathbb{P}[Y > c + \varepsilon, R \ge 0], \mathbb{P}[Y > c + \varepsilon, R \le 0]\}$. Choose $a > 0$. We let $W_n = 1\!\!1_{X_n \le a, Y_{n+1} > c+\varepsilon, U_{n+1}R_{n+1} \le 0}$ and $Z_n = \delta 1\!\!1_{X_n \le a}$. Now the proof follows similarly to the proof of Lemma 1.16. □

We cannot conclude as in Section 1.3.1 that a solution is positive, because there is no strategy such that X remains bounded. But again an increasing solution is unique.

Theorem 1.20. *Suppose that $f(x)$ is an increasing solution to* (1.13) *with $f(0) = 1$. Then $f(x)$ is bounded and $f(x) = \delta(x)/\delta(0)$.*

Proof. Choose u such that $c + u\mathbb{E}[R_n] > \mathbb{E}[Y_n]$. Consider first the constant strategy $U_n = u$. Then $\{X_n\}$ converges to infinity or ruin occurs. Because $\{f(X_n)\}$ is a supermartingale, it converges to an integrable random variable. The rest of the proof follows as in the proof of Theorem 1.17. □

Again let $f_0(x) = 1$ and

$$f_{n+1}(x) = \sup_{u\in\mathbb{R}} \int_{-\infty}^\infty \int_0^{(x+c+ur)^+} f_n(x + c + ur - y)\,\mathrm{d}G(y)\,\mathrm{d}H(r)\;.$$

The convergence result can be proved similarly as Proposition 1.18.

Proposition 1.21. *The functions $f_n(x)$ converge to $\delta(x)$.* □

Bibliographical Remarks

Schäl [154] introduced a similar model that considered optimal reinsurance and investment. The existence of an optimal strategy is proved from general results on dynamic programming in infinite time. In particular, the case of exponentially distributed claim sizes is discussed. A second problem considered is a model in finite time where exponential utility is maximised.

2

Stochastic Control in Continuous Time

In this chapter we consider processes in continuous time, i.e., $I = \mathbb{R}_+$. In many cases, such as the problem considered in Section 2.4, the continuous-time problem can be approximated by models in discrete time. One therefore expects also that a limit of the dynamic programming equation (1.1) or (1.2) will converge to an equation the value function has to fulfil. We will discuss this point heuristically in Section 2.1. For the solution, however, this connection to discrete-time dynamic programming will not play any rôle. We will relay with our proofs on martingale techniques. The connection to martingales will be discussed in Section 2.1 as well.

Two main problems will arise when solving optimisation problems in continuous time. The equation connected to the problem, the so-called *Hamilton–Jacobi–Bellman equation*, can usually only be motivated but not proved directly. Hence, we cannot be sure that a solution to the equation really is the desired value function. One always needs to prove a so-called *verification theorem.* There are many examples in the literature where a verification theorem is missing and the solution found is not the value function. For some problems it may even occur that the natural equation is not the correct Hamilton–Jacobi–Bellman equation; see the hedging problem under gamma constraints in [182].

A second necessary step is to prove that the value function solves the Hamilton–Jacobi–Bellman equation. The simplest case arises when one can guess the value function or find the unique solution to the Hamilton–Jacobi–Bellman equation. Then nothing more than the verification theorem has to be proved. If one has already proven that any solution to the equation is the value function, then it is enough to show that a solution exists. A usually more complicated procedure could be to show directly that the value function solves the Hamilton–Jacobi–Bellman equation. The latter is necessary if, for example, the solution to the equation is not unique and one cannot guess the solution, see, for example, Section 2.4.

The methods presented were developed for diffusion processes. In some sense the diffusion case is simpler because the equations are differential equations and not integro-differential equations. In many cases the supremum can be calculated (in terms of the unknown function) and it remains to solve an ordinary (nonlinear) differential equation. If the processes involved allow jumps, there will be an integral part in the equation. This often makes it impossible to express the argument at which the supremum is taken in terms of the derivatives. We therefore in general cannot find an equivalent ordinary integro-differential equation. That makes the proofs more complicated. On the other hand, if the process does not contain a diffusion part, we will just get integro-differential equations of first order, that are easier to discuss than second-order equations. Another simplification is that the martingale arguments work for a so-called weak solution. That is, the solution is absolutely continuous and solves the equation at all points where the derivative exists.

Another problem connected to problems containing a diffusion part is that the value function must be twice continuously differentiable. It may happen that this is not the case. For example, the value function may be absolutely continuous but not twice continuously differentiable. Then Itô's formula can not always be applied. A nice way around this problem then is the concept of *viscosity solutions*. We will not use the concept in this book, because either it is enough to have a weak solution (first-order equations) or the value function is a strong solution. But because the concept of viscosity solutions is important, we will explain the method in Section 2.5.3. The reader should, however, be aware that additional problems may arise. For our problems the verification theorem leads naturally to an optimal strategy. For viscosity solutions it may happen that the natural "optimal" strategy is not optimal. In this case no optimal strategy may exist. That is, one has to approximate the optimal value but will not be able to reach it.

In this chapter we will consider the same two problems as in discrete time. The reader might note some similarities between the discrete-time and the continuous-time models. Actually, considering the discrete-time model first can lead to ideas about what properties the continuous-time solution has. We will consider both diffusion approximations and classical risk models. We will note that the optimal strategies look different for diffusion approximations and for classical risk models. This is in order to warn the reader that the use of diffusion approximations may fail in insurance. The reason is that often the optimal strategies converge very slowly to the optimal strategy in the diffusion problem.

2.1 The Hamilton–Jacobi–Bellman Approach

This section is mainly to help in understanding the approach we will use in the following examples and to show the connection to Markov process theory.

In order to prove the results of this section, one has to make quite strong assumptions. It is often easier to prove existence and verification theorems in specific situations. Therefore in this section we will discuss the *intuition* behind the approach and not give a mathematically rigorous treatment of the topic. Upon first reading of this chapter the reader may skip this section and proceed directly to the problems considered. We will prove the results of the other sections without using this introduction. The reader should keep in mind that the treatment in this section is *heuristic* only and not rigorous.

Let X^U be a stochastic process for which the law is determined by a control process U with state space $\mathcal{U}$. Because the choice of U_t cannot be made with the knowledge of future events, we restrict U to be adapted. We just take a filtration $\{\mathcal{F}_t\}$ to be given.

Remark 2.1. In the literature one often deals with previsible controls. This implies that the control process has to be left-continuous, and not cadlag, as we assume in this book. The reason is that the control should simultaneously affect the jump dynamics and be dependent on the post-jump value of the process. In order to obtain a proper formulation of the problem, we usually use U_{t-} in the definition of the processes, or take the left limit $\int_0^{T-}$ in the integrals in the definition of the value function. This simplifies the presentation of the material. However, the drawback is that if U leads to a jump of the controlled process, one does not observe the pre-control but the post-control process. ■

The set of *admissible* strategies is denoted by $\mathfrak{U}$. We consider here the situation where for a constant strategy $U_t = u$ the process X^u is a Markov process with generator $\mathfrak{A}_u$. Note that in order to allow non-Markovian controls, we cannot assume that X^U is a Markov process.

We now modify the approach of Section 1.1 to continuous time. The value connected to the strategy U for the value $X_t^U = x$ at time $t < T$ is

$$V^U(t,x) = \mathbb{E}\Bigl[\int_t^T e^{-\delta(s-t)} r(X_s^U, U_s)\, ds + e^{-\delta(T-t)} r_T(X_T^U) \Bigm| \mathcal{F}_t\Bigr] .$$

T is a stopping time. To simplify the presentation of this discussion, we consider value functions not depending on t in order to skip the partial derivative with respect to t,

$$V^U(x) = \mathbb{E}\Bigl[\int_t^T e^{-\delta(s-t)} r(X_s^U, U_s)\, ds + e^{-\delta(T-t)} r_T(X_T^U) \Bigm| X_t = x\Bigr] .$$

For example, deterministic times T are not allowed for the moment. First entrance times such as the time of ruin are stopping times yielding value functions that are not time-dependent. Here the law of the process $\{X_s\}$ should depend on $\mathcal{F}_t$ via X_t and $\{U_s : 0 \le s \le t\}$ only. Time-dependent value functions could then be treated via the process $\{(t, X_t)\}$. This trick is quite common in Markov process theory where in this way a homogeneous Markov process is obtained.

There are a continuous reward at rate $r(X_t, U_t)$ depending on the present state and the chosen control and a final reward r_T depending on the final state only. We could also make the final reward dependent on the control at time T. This will, however, in most examples not be the case. In this section we do not allow "discontinuous" rewards such as a lump sum payment. In order to explain the approach, it seems simpler to consider the present setting. However, only small changes are needed also to cover the case where rewards are earned at deterministic times or at times where $\{X_t\}$ jumps.

That the discounting rate δ is constant is also just a simplification of the presentation. Only slight changes are needed to treat a discount at rate $\delta(X_t^U, U_t)$. The Hamilton–Jacobi–Bellman equation obtained below will just become more complicated, that is δ has to be replaced by $\delta(x, u)$. Moreover, the functions r and δ could, in addition, depend on time. In this case the value function $V(x)$ should be replaced by $V(x, t)$, and a partial derivative with respect to time will appear in the Hamilton–Jacobi–Bellman equation.

The goal is now to maximise the value, i.e., we look for the function $V(x) = \sup_{U \in \mathfrak{U}} V^U(x)$. We first prove the dynamic programming principle, i.e., an equation corresponding to the results of Lemma 1.1.

Take an arbitrary strategy U in the interval $[0, T \wedge t]$. From time $T \wedge t$ choose a strategy $\tilde{U}$ such that $V(X_{T\wedge t}^U) < V^{\tilde{U}}(X_{T\wedge t}^U) + \varepsilon$. We suppose that this can be done in a measurable way. Then, conditioning on $\mathcal{F}_t$,

$$\begin{aligned}
V(x) \ge V^U(x) &= \mathbb{E}\Bigl[\int_0^{T\wedge t} e^{-\delta s} r(X_s^U, U_s)\,ds + \int_{T\wedge t}^T e^{-\delta s} r(X_s^{\tilde{U}}, \tilde{U}_{s-t})\,ds \\
&\qquad + e^{-\delta T} r_T(X_T^{\tilde{U}})\Bigr] \\
&= \mathbb{E}\Bigl[\int_0^{T\wedge t} e^{-\delta s} r(X_s^U, U_s)\,ds + e^{-\delta(T\wedge t)} \\
&\qquad \times \mathbb{E}\Bigl[\int_0^{(T-t)^+} e^{-\delta s} r(X_{t+s}^{\tilde{U}}, \tilde{U}_s)\,ds + e^{-\delta(T-t)^+} r_T(X_T^{\tilde{U}}) \Bigm| \mathcal{F}_t\Bigr]\Bigr] \\
&= \mathbb{E}\Bigl[\int_0^{T\wedge t} e^{-\delta s} r(X_s^U, U_s)\,ds + e^{-\delta(T\wedge t)} V^{\tilde{U}}(X_{T\wedge t}^U)\Bigr] \\
&> \mathbb{E}\Bigl[\int_0^{T\wedge t} e^{-\delta s} r(X_s^U, U_s)\,ds + e^{-\delta(T\wedge t)} V(X_{T\wedge t}^U)\Bigr] - \varepsilon\,.
\end{aligned}$$

Because the right-hand side does not directly depend on $\tilde{U}$, the (weak) inequality must also hold for $\varepsilon = 0$. Taking the supremum over U shows that

$$V(x) \ge \sup_U \mathbb{E}\Bigl[\int_0^{T\wedge t} e^{-\delta s} r(X_s^U, U_s)\,ds + e^{-\delta(T\wedge t)} V(X_{T\wedge t}^U)\Bigr]\,.$$

Starting with an arbitrary strategy yields

$$
\begin{aligned}
V^U(x) &= \mathbb{E}\Bigl[\int_0^{T\wedge t} \mathrm{e}^{-\delta s} r(X_s^U, U_s)\,\mathrm{d}s + \mathrm{e}^{-\delta(T\wedge t)} V^U(X_{T\wedge t}^U)\Bigr] \\
&\le \mathbb{E}\Bigl[\int_0^{T\wedge t} \mathrm{e}^{-\delta s} r(X_s^U, U_s)\,\mathrm{d}s + \mathrm{e}^{-\delta(T\wedge t)} V(X_{T\wedge t}^U)\Bigr] \\
&\le \sup_{\bar U} \mathbb{E}\Bigl[\int_0^{T\wedge t} \mathrm{e}^{-\delta s} r(X_s^{\bar U}, \bar U_s)\,\mathrm{d}s + \mathrm{e}^{-\delta(T\wedge t)} V(X_{T\wedge t}^{\bar U})\Bigr] ,
\end{aligned}
$$

where $\tilde U_s = U_{t+s}$. Taking the supremum over U yields

$$
V(x) \le \sup_U \mathbb{E}\Bigl[\int_0^{T\wedge t} \mathrm{e}^{-\delta s} r(X_s^U, U_s)\,\mathrm{d}s + \mathrm{e}^{-\delta(T\wedge t)} V(X_{T\wedge t}^U)\Bigr] .
$$

Thus, we found the *dynamic programming principle*

$$
V(x) = \sup_{U\in\mathfrak{U}} \mathbb{E}\Bigl[\int_0^{T\wedge t} \mathrm{e}^{-\delta s} r(X_s^U, U_s)\,\mathrm{d}s + \mathrm{e}^{-\delta(T\wedge t)} V(X_{T\wedge t}^U)\Bigr] . \tag{2.1}
$$

In order to get some information on U, one should consider (2.1) for small t. However, letting $t \downarrow 0$ just yields $V(x) = V(x)$. Rearranging the terms and dividing by t gives

$$
\begin{aligned}
\sup_U \mathbb{E}\Bigl[\frac{1}{t}\int_0^{T\wedge t} \mathrm{e}^{-\delta s} r(X_s^U, U_s)\,\mathrm{d}s &+ \mathrm{e}^{-\delta(T\wedge t)} \frac{V(X_{T\wedge t}^U) - V(x)}{t} \\
&- \frac{1 - \mathrm{e}^{-\delta(T\wedge t)}}{t} V(x)\Bigr] = 0 .
\end{aligned}
$$

Letting $t \downarrow 0$, the first term converges (under some conditions allowing one to interchange the limit, supremum, and integration) to $r(x, U_0)$. Note that U_0 exists because U is cadlag. The last term converges to $\delta V(x)$. The middle term is just the definition of the infinitesimal generator. Hence, if $V(x)$ is in the domain of the generator of the process $\{X_{T\wedge t}\}$, then the middle term converges to $\mathfrak{A}_{U_0} V(x)$, where $\mathfrak{A}_u$ denotes the generator of the process controlled by the constant control u. For an introduction to the generator of a Markov process, see Appendix B.

We therefore would expect that $V(x)$ fulfils the *Hamilton–Jacobi–Bellman equation*

$$
\sup_{u\in\mathcal{U}} \{\mathfrak{A}_u V(x) - \delta V(x) + r(x, u)\} = 0 . \tag{2.2}
$$

Because we have interchanged the limit and supremum, our derivation does not hold formally. But in this way we can see the connection between dynamic programming in discrete time and in continuous time. The difference is mainly that we have to consider infinitesimal changes instead of changes between two time points. In this way the differential operator $\mathfrak{A}_u$ appears.

The name Bellman comes from Equation (1.1). The name Hamilton–Jacobi comes from physics. If we consider a diffusion process the corresponding Hamilton–Jacobi–Bellman equation looks similar to the Hamilton–Jacobi equation from physics.

We now give another intuitive derivation of (2.2). If we condition on $\mathcal{F}_t$, we obtain on $\{T > t\}$

$$\begin{aligned}\mathbb{E}\Bigl[\int_0^T e^{-\delta t} r(X_t^U, U_t)\,dt + e^{-\delta T} r_T(X_T^U) \Bigm| \mathcal{F}_t\Bigr] &= \int_0^t e^{-\delta s} r(X_s^U, U_s)\,ds \\ &\quad + e^{-\delta t}\mathbb{E}\Bigl[\int_t^T e^{-\delta(s-t)} r(X_s^U, U_s)\,ds + e^{-\delta(T-t)} r_T(X_T^U) \Bigm| \mathcal{F}_t\Bigr]\,.\end{aligned}$$

The conditional expected value on the right-hand side is just the remaining discounted value of the strategy U. The first term is the discounted value of the strategy until time t and independent of the strategy chosen after time t. Trying to maximise the expected value of the strategy, it does not matter how the capital X_t^U was obtained. The optimal value can be approximated by just deciding on the strategy once one knows the capital X_t^U. We therefore can assume that the strategy U after time t depends on $\mathcal{F}_t$ via X_t only. Thus, the expected value on the right-hand side is $V^U(X_t^U)$. This implies that

$$\Bigl\{\int_0^{T\wedge t} e^{-\delta s} r(X_s^U, U_s)\,ds + e^{-\delta T} r_T(X_T^U)\mathbb{1}_{T\le t} + e^{-\delta t} V^U(X_t^U)\mathbb{1}_{T>t}\Bigr\} \tag{2.3}$$

is a martingale.

In order to apply Markov process theory, consider the process (X_t^U, t). If U_t was of the form $u(X_t^U)$ and $f(x,t) = V^U(x)e^{-\delta t}$ was in the domain of the generator, then the process

$$\Bigl\{V^U(X_{T\wedge t})e^{-\delta(T\wedge t)} - \int_0^{T\wedge t} [\mathfrak{A}_{U_s} V^U(X_s) - \delta V^U(X_s)]e^{-\delta s}\,ds\Bigr\} \tag{2.4}$$

would be a martingale. We will see that for our examples (2.4) is a martingale for general adapted controls $\{U_t\}$ under quite mild regularity conditions. In particular, in the case of diffusions Itô's formula yields the martingale property. We work out the details in the specific examples later.

Taking the difference between the martingales (2.3) and (2.4) yields a continuous martingale of bounded variation. By Lemma A.9 this martingale must be constant. Thus, we obtain the equation

$$\mathfrak{A}_{U_t} V^U(X_t^U) - \delta V^U(X_t^U) + r(X_t^U, U_t) = 0\;,$$

with boundary condition $V(X_T^U) = r_T(X_T^U)$. Note that if the function $V(x)$ is time-dependent, the process X contains the time t as a component. Therefore, $\mathfrak{A}_u$ may contain a partial derivative with respect to time.

We obtained before the dynamic programming equality

$$V(x) = \sup_{U\in\mathfrak{U}} \mathbb{E}\Bigl[\int_0^{T\wedge t} e^{-\delta s} r(X_s^U, U_s)\,ds + e^{-\delta(T\wedge t)} V(X_{T\wedge t}^U)\Bigr]\,.$$

The same considerations as for the martingale (2.4) establish the equation

$$V(x) = \sup_{U\in\mathfrak{U}} \mathbb{E}\Bigl[-\int_0^{T\wedge t} (\mathfrak{A}_{U_s} V(X_s^U) - \delta V(X_s^U))\mathrm{e}^{-\delta s}\,\mathrm{d}s + \mathrm{e}^{-\delta(T\wedge t)} V(X_{T\wedge t}^U)\Bigr] . \tag{2.5}$$

If there is an optimal strategy, we can replace the supremum by this optimal strategy. Comparing the strategies, we get

$$\mathfrak{A}_{u^*} V(x) - \delta V(x) + r(x, u^*) = 0 ,$$

where u^* is the optimal strategy close to time zero.

We found an equation for V, but we do not know u^* yet. Consider now an arbitrary strategy U and the process $\{V(X_t^U)\}$. Then, as above,

$$\Bigl\{V(X_t^U)\mathrm{e}^{-\delta t} - \int_0^t [\mathfrak{A}_{U_s} V(X_s^U) - \delta V(X_s^U)]\mathrm{e}^{-\delta s}\,\mathrm{d}s\Bigr\}$$

is a martingale. Thus,

$$V(x) = \mathbb{E}\Bigl[V(X_t^U)\mathrm{e}^{-\delta t} - \int_0^t [\mathfrak{A}_{U_s} V(X_s^U) - \delta V(X_s^U)]\mathrm{e}^{-\delta s}\,\mathrm{d}s\Bigr] .$$

Comparing with (2.3) and noting that $V(x) \ge V^U(x)$, we obtain

$$\mathfrak{A}_u V(x) - \delta V(x) + r(x, u) \le 0 ,$$

where u is the limit from the right of U_t as $t \to 0$. Thus, we have motivated the Hamilton–Jacobi–Bellman equation (2.2). The optimal strategy should therefore be of the form $u^*(X_t)$, where $u^*(x)$ is the argument maximising the left-hand side of (2.2).

The development relied on several regularity conditions. For a concrete situation one will have to verify that $V(x)$ solves these regularity conditions. One step is called a *verification theorem.* Suppose that there is a solution to (2.2) with boundary condition $V(X_T) = r_T(X_T)$ such that the martingale assumptions below hold. Suppose further that the function $u^*(x)$ is measurable. Let $\{X_t^*\}$ be the process for the strategy $U_t^* = u^*(X_t^*)$. Then it will turn out that

$$\Bigl\{V(X_{T\wedge t}^*)\mathrm{e}^{-\delta(T\wedge t)} - \int_0^{T\wedge t} [\mathfrak{A}_{U_s^*} V(X_s^*) - \delta V(X_s^*)]\mathrm{e}^{-\delta s}\,\mathrm{d}s\Bigr\}$$

is a martingale. From (2.2) it follows that

$$\Bigl\{V(X_{T\wedge t}^*)\mathrm{e}^{-\delta(T\wedge t)} + \int_0^{T\wedge t} r(X_s^*, U_s^*)\mathrm{e}^{-\delta s}\,\mathrm{d}s\Bigr\}$$

is a martingale. Thus,

$$V(x) = \mathbb{E}\Bigl[V(X^*_{T\wedge t})\mathrm{e}^{-\delta(T\wedge t)} + \int_0^{T\wedge t} r(X^*_s, U^*_s)\mathrm{e}^{-\delta s}\,\mathrm{d}s\Bigr]\;.$$

Suppose that $\lim_{t\to\infty}$ and the expectation can be interchanged. Then it follows that $V(x) = V^{U^*}(x)$, provided that $V(X^*_{T\wedge t})\mathrm{e}^{-\delta(T\wedge t)} \to 0$ as $t \to \infty$, or that $V(X^*_T)\mathrm{e}^{-\delta T} = r_T(X^*_T)\mathrm{e}^{-\delta T}$ is well defined. We need to show that this is a maximum. Let U be an arbitrary strategy. Then, provided that the martingale property holds, (2.5) and (2.2) imply that

$$V(x) \ge \mathbb{E}\Bigl[\int_0^{T\wedge t} r(X^U_s, U_s)\mathrm{e}^{-\delta s}\,\mathrm{d}s + \mathrm{e}^{-\delta(T\wedge t)}V(X^U_{T\wedge t})\Bigr]\;.$$

If here the limit and expectation can be interchanged, we obtain $V(x) \ge V^U(x)$, provided that $\mathrm{e}^{-\delta(T\wedge t)}V(X^U_{T\wedge t}) \to 0$ as $t \to \infty$. Thus, $V(x)$ is the desired function and $\{U^*_t\}$ is an optimal strategy.

It will remain to show that $V(x)$ really solves (2.2). One way is to show that there is a solution to (2.2) with the right regularity properties. By the verification theorem, this solution will be $V(x)$.

Bibliographical Remarks

There are many textbooks on the Hamilton–Jacobi–Bellman method. Possible references are Davis [40], Fleming and Soner [61], Korn and Korn [121], Øksendal and Sulem [144], Touzi [182], or Yong and Zhou [187]. Most of these references use viscosity solutions because it is often difficult to verify that the solution is twice continuously differentiable.

2.2 Minimising Ruin Probabilities for a Diffusion Approximation

We start by considering a diffusion approximation instead of a classical Cramér–Lundberg model. This simplifies the equations because we get ordinary differential equations. Because we are able to solve these equations explicitly, we can easily find the value function and the optimal controls.

2.2.1 Optimal Reinsurance

A reader not familiar with the construction of diffusion approximations can just proceed to (2.6) for the definition of the surplus process. We motivate the model (2.6) in the following way. Consider a diffusion approximation to a classical risk process where the claims are reinsured by proportional reinsurance with retention level $b \in [0, 1]$. We write for the premium income $c = (1+\eta)\lambda\mu$, where we assume that $\eta > 0$. For a claim Y_i the cedent pays bY_i, and the reinsurer pays $(1-b)Y_i$. The reinsurer uses an expected value principle for the calculation of the reinsurance premium. Thus, the premium rate for the reinsurance is $(1+\theta)\lambda\mathbb{E}[(1-b)Y_i] = (1+\theta)(1-b)\lambda\mu$. That is, the premium rate remaining for the insurer is $c(b) = (b(1+\theta)-(\theta-\eta))\lambda\mu$. The diffusion approximation from Section D.3 then reads

$$x + (b\theta - (\theta-\eta))\lambda\mu t + b\sqrt{\lambda\mu_2}W_t\;,$$

where W is a standard Brownian motion. The parameters are as in Appendix D. To simplify the notation, we choose $\lambda\mu = 1$ (change of the time unit) and let $\sigma = \sqrt{\lambda\mu_2}$ (change of the monetary unit). The insurer can now choose a strategy $b = \{b_t\}$, i.e., the retention level can be changed continuously. Then the surplus process becomes

$$X_t^b = x + \int_0^t (b_s\theta - (\theta-\eta))\,\mathrm{d}s + \sigma\int_0^t b_s\,\mathrm{d}W_s\,. \tag{2.6}$$

The set of admissible strategies is the set of all cadlag adapted processes $\{b_t\}$ with values in $[0,1]$. The filtration $\{\mathcal{F}_t\}$ is the filtration generated by the Brownian motion.

The time of ruin is $\tau^b = \inf\{t \ge 0 : X_t^b < 0\}$, the ruin probability is $\psi^b(x) = \mathbb{P}[\tau^b < \infty]$, and the survival probability is $\delta^b(x) = 1 - \psi^b(x)$. Our goal is to minimise the ruin probability or, equivalently, to maximise the survival probability $\delta(x) = \sup_{b\in\mathfrak{U}} \delta^b(x)$. So that the problem does not have the trivial solution $b_t = 0$, we assume that $\theta > \eta$.

Lemma 2.2. *The function $\delta(x)$ fulfils $\delta(0) = 0$ and*

$$(1 - \mathrm{e}^{-2\eta x/\sigma^2}) \le \delta(x) \le 1$$

for all $x \ge 0$.

Proof. Let $x = 0$. Let b be an admissible strategy. Then $b_0 = \lim_{t\downarrow 0} b_t$ exists. If $b_0 \ne 0$, ruin occurs immediately because the Brownian motion crosses the t-axis infinitely often in any small interval. If $b_0 = 0$, the process has a negative drift and therefore also $\tau^b = 0$. Thus, $\delta^b(0) = 0$ for all b, and $\delta(0) = 0$ follows. Choosing the strategy $b = 1$ (no reinsurance) gives the ruin probability $\psi^1(x) = \exp\{-2\eta x/\sigma^2\}$; see (A.2). This yields the lower bound. That $\delta(x) \le 1$ is clear because $\delta^b(x)$ is a probability. □

Lemma 2.3. *The function $\delta(x)$ is strictly increasing on the set $\{x : \delta(x) < 1\}$.*

Proof. Let $y > x$ and let b be a strategy for initial capital x. Let τ_x^b be the time of ruin with initial capital x. For initial capital y we use the following strategy: $\tilde{b}_t = b_t$ for $t \le \tau_x^b$, and $\tilde{b}_t = 1$ for $t > \tau_x^b$. This shows that

$$\delta^{\tilde{b}}(y) = \delta^b(x) + (1 - \delta^b(x))\delta^1(y - x) = \delta^b(x) + (1 - \delta^b(x))(1 - \mathrm{e}^{-2\eta(y-x)/\sigma^2}) .$$

Taking the supremum over all strategies b yields

$$\delta(y) \ge \delta(x) + (1 - \delta(x))(1 - \mathrm{e}^{-2\eta(y-x)/\sigma^2}) .$$

If $\delta(x) < 1$, this implies that $\delta(y) > \delta(x)$. Thus, $\delta(x)$ is strictly increasing on the set $\{x : \delta(x) < 1\}$. □

We now motivate the Hamilton–Jacobi–Bellman equation. Let $(0, h]$ be a small interval, and suppose that we have for each capital $x_h > 0$ at time h a strategy b^ε such that $\delta^{b^\varepsilon}(x_h) > \delta(x_h) - \varepsilon$. We let $b_t = b \in [0, 1]$ for $t \le h$. Then

$$\delta(x) \ge \delta^b(x) = \mathbb{E}[\delta^{b^\varepsilon}(X_h^b)\mathbb{1}_{\tau^b > h}] = \mathbb{E}[\delta^{b^\varepsilon}(X_{\tau^b \wedge h}^b)] \ge \mathbb{E}[\delta(X_{\tau^b \wedge h}^b)] - \varepsilon .$$

Because ε is arbitrary, we can choose $\varepsilon = 0$. By Itô's formula, provided that $\delta(x)$ is twice continuously differentiable,

$$\delta(X_{\tau^b \wedge h}^b) = \delta(x) + \int_0^{\tau^b \wedge h} \Bigl\{(b\theta - (\theta - \eta))\delta'(X_s^b) + \frac{\sigma^2 b^2}{2}\delta''(X_s^b)\Bigr\} \,\mathrm{d}s + \int_0^{\tau^b \wedge h} b\sigma\delta'(X_s^b) \,\mathrm{d}W_s .$$

Suppose that the stochastic integral is a true martingale. Plugging the above formula into the expected value yields

$$\mathbb{E}\Bigl[\int_0^{\tau^b \wedge h} \Bigl\{(b\theta - (\theta - \eta))\delta'(X_s^b) + \frac{\sigma^2 b^2}{2}\delta''(X_s^b)\Bigr\} \,\mathrm{d}s\Bigr] \le 0 .$$

Dividing by h and letting h to zero yields, provided that limit and expectation can be interchanged,

$$(b\theta - (\theta - \eta))\delta'(x) + \frac{\sigma^2 b^2}{2}\delta''(x) \le 0 .$$

This equation must hold for all $b \in [0, 1]$, i.e.,

$$\sup_{b \in [0,1]} \Bigl\{(b\theta - (\theta - \eta))\delta'(x) + \frac{\sigma^2 b^2}{2}\delta''(x)\Bigr\} \le 0 .$$

Suppose that there is an optimal strategy b such that $\lim_{t \downarrow 0} b_t = b_0$. Then, as above,

$$\mathbb{E}\Bigl[\int_0^{\tau^b \wedge h} \Bigl\{(b_s\theta - (\theta - \eta))\delta'(X_s^b) + \frac{\sigma^2 b_s^2}{2}\delta''(X_s^b)\Bigr\} \,\mathrm{d}s\Bigr] = 0 .$$

Dividing by h and letting $h \to 0$ yields

$$(b_0\theta - (\theta - \eta))\delta'(x) + \frac{\sigma^2 b_0^2}{2}\delta''(x) = 0 \,.$$

We therefore now consider the *Hamilton–Jacobi–Bellman equation*

$$\sup_{b\in[0,1]} \left\{ (b\theta - (\theta - \eta)) f'(x) + \frac{\sigma^2 b^2}{2} f''(x) \right\} = 0 \tag{2.7}$$

with the boundary condition $f(0) = 0$. The solution we are looking for is strictly increasing. We now denote the function by f because we do not know whether $\delta(x)$ satisfies the conditions used in the derivation above.

Equation (2.7) admits the following solution.

Lemma 2.4. *The function*

$$f(x) = 1 - e^{-\kappa x} \tag{2.8}$$

solves (2.7) *where*

$$\kappa = \begin{cases} \theta^2/(2\sigma^2(\theta - \eta)), & \text{if } \eta < \theta < 2\eta, \\ 2\eta/\sigma^2, & \text{if } \theta \ge 2\eta. \end{cases}$$

The value $b^*(x)$ that maximises the left-hand side of (2.7) *is constant and*

$$b^* = 2(1 - \eta/\theta) \wedge 1 \,. \tag{2.9}$$

Proof. Of course, one can verify directly that $f(x)$ solves (2.7). But in order to illustrate the method, we will solve (2.7) analytically. The equation is quadratic in b. If $f''(x) > 0$, then the minimum is attained for $b = -\theta f'(x)/(\sigma^2 f''(x)) < 0$. The maximum in $[0, 1]$ is therefore attained at $b = 1$. This yields the solution $f'(x) = C_1 e^{-2\eta x/\sigma^2} > 0$, that is, $f''(x) < 0$, which is a contradiction. $f''(x) = 0$ on some interval implies that $b = 1$ and therefore $f'(x) = 0$. By Lemma 2.3 the solution should be $f(x) = 1$ on this interval. Because $f(0) = 0$, we conclude that $f''(x) < 0$ almost everywhere for x small enough and the supremum is attained at

$$b = -\frac{\theta f'(x)}{\sigma^2 f''(x)} \,.$$

If b is larger than 1, then $b^*(x) = 1$ and the solution to (2.7) is $f'(x) = C_1 e^{-2\eta x/\sigma^2} > 0$. If $b \in (0, 1)$, we obtain

$$-\frac{\theta^2 f'(x)^2}{2\sigma^2 f''(x)} - (\theta - \eta) f'(x) = 0 \,.$$

Because the solution we are looking for satisfies $f'(x) > 0$, we can divide by $f'(x)$ and obtain $f'(x) = C_2 e^{-\theta^2 x/(2\sigma^2(\theta-\eta))} > 0$. Because $f'(x) \neq 0$ for all x, none of these solutions can be combined with $f(x) = 1$ without losing differentiability. Because $f(0) = 0$ for the solution we are looking for, we must have $f'(x) > 0$ for all x. Note that $\theta^2/(2\sigma^2(\theta - \eta)) \ge 2\eta/\sigma^2$ and that equality holds if and only if $\theta = 2\eta$. In the latter case we find that $b = 1$, so the two solutions coincide. Assume that there is a point x_0 where the solution $f(x)$ changes from one of the solutions $f_1(x)$ to another of the solutions $f_2(x)$. Because we want $f(x)$ to be twice continuously differentiable, we get the two equations $f_1'(x_0) = f_2'(x_0)$ and $f_1''(x_0) = f_2''(x_0)$. These two equations can never be fulfilled simultaneously, except if $\theta = 2\eta$; thus, the desired solution is one of the functions obtained above. From $f(0) = 0$ and $\lim_{x\to\infty} f(x) = 1$, the possible solutions are (2.8) with one of the two possible values of κ given above. The maximum in (2.7) is attained for $b = \theta/(\sigma^2\kappa)$, which is smaller than 1 if and only if $\theta < 2\eta$. This proves the lemma. □

Because we have an explicit solution to (2.7), the *verification theorem* is easy to prove.

Theorem 2.5. *We have $\delta(x) = f(x)$, where $f(x)$ is given by* (2.8). *The constant strategy $b_t = b^*$ with b^* given by* (2.9) *is an optimal reinsurance strategy.*

Remark 2.6. It is not surprising that $\psi(x)$ is an exponential function and that the optimal strategy is constant. Starting with initial capital $x_1 + x_2$, ruin can occur only if first the surplus x_1 is reached. Due to the strong Markov property, the ruin probability at the point where x_1 is reached is $\psi(x_1)$. Before x_1 is reached, it is therefore optimal to use a strategy that minimises the probability of reaching x_1. Thus, $\psi(x_1 + x_2) = \psi(x_1)\psi(x_2)$, yielding the exponential form. Let $n \in \mathbb{N}$ and consider the interval $[k/n, (k+1)/n]$. The optimal strategy in this interval has to be the same as in the interval $[0, 1/n]$, because we want to minimise the probability of reaching k/n. A strategy that is the same in all such intervals must be constant. ■

Proof. That $\delta^{b^*}(x) = f(x)$ follows from (A.2) in Appendix A. Let b be an arbitrary strategy. By Itô's formula,

$$\begin{aligned} f(X^b_{\tau\wedge t}) &= f(x) + \int_0^{\tau\wedge t} (b_s\theta - (\theta - \eta))f'(X^b_s) + \frac{b_s^2\sigma^2}{2} f''(X^b_s)\,\mathrm{d}s \\ &\quad + \int_0^{\tau\wedge t} b_s\sigma f'(X^b_s)\,\mathrm{d}W_s \\ &\le f(x) + \int_0^{\tau\wedge t} b_s\sigma f'(X^b_s)\,\mathrm{d}W_s \end{aligned}$$

because $f(x)$ solves (2.7). Since $f'(x)$ is bounded by κ, the stochastic integral is a martingale. Taking expectations yields

$$\mathbb{E}[f(X^b_{\tau\wedge t})] \le f(x) . \tag{2.10}$$

In particular, we find that $\{f(X^b_{\tau\wedge t})\}$ is a supermartingale. By Proposition A.1, $\{f(X^b_{\tau\wedge t})\}$ converges as $t \to \infty$. If, on $\{\tau = \infty\}$, $\lim_{t\to\infty} f(X^b_t) \in (0,1)$, then $X^b_\infty = \lim_{t\to\infty} X^b_t \in (0,\infty)$. Fix $x_0 > 0$ and suppose that $\mathbb{P}[X^b_\infty \le x_0] > 0$. The expression $\int_t^{t+1} b_s \,\mathrm{d}W_s$ has mean value 0 and nonzero variance unless $b_s = 0$. Thus, $\mathbb{P}[\int_t^{t+1} b_s \,\mathrm{d}W_s < 0] > 0$. Suppose that $\int_t^{t+1} \mathbb{I}_{2b_s \ge (1-\eta/\theta)} \,\mathrm{d}s < (1-\eta/\theta)/4$. Then

$$\int_t^{t+1} \theta b_s - (\theta - \eta) \,\mathrm{d}s \le \frac{1-\eta/\theta}{4}\eta - \frac{3+\eta/\theta}{4}\frac{\theta-\eta}{2} = -\frac{(\theta-\eta)(3\theta-\eta)}{8\theta}$$
$$< -\frac{\theta-\eta}{4}\,.$$

Thus, $X^b_{t+1} - X^b_t \le -(\theta-\eta)/4$ with strictly positive probability. If instead we have $\int_t^{t+1} \mathbb{I}_{2b_s \ge (1-\eta/\theta)} \,\mathrm{d}s \ge (1-\eta/\theta)/4$, then $\int_t^{t+1} b_s^2 \,\mathrm{d}s \ge (1-\eta/\theta)^2/8$. Let $T = \inf\{s \ge t : \int_t^s b_v^2 \,\mathrm{d}v > (1-\eta/\theta)^2/8\}$. Then $\int_t^T \sigma b_v \,\mathrm{d}W_v$ is normally distributed with mean value 0 and variance $\sigma^2(1-\eta/\theta)^2/8$. In particular, $\sup_{s\in(t,t+1)}\{X^b_s - X^b_t\} \le -(\theta-\eta)/4$ holds with strictly positive probability also on this set. Thus, there exists $\delta > 0$ such that $\mathbb{P}[\sup_{s\in(t,t+1)}\{X^b_s - X^b_t\} \le -(\theta-\eta)/4 \mid \mathcal{F}_t] \ge \delta$. Now let $B_k = \{\sup_{s\in(k,k+1)}\{X^b_s - X^b_k\} \le -(\theta-\eta)/4\}$ and $A_k = \mathbb{I}_{B_k}$. Then $\{\sum_{k=1}^n (A_k - \delta)\}$ is a submartingale. From Lemma 1.15 we conclude that $\{A_k = 1\}$ holds infinitely often. In particular, X^b_t cannot converge. We thus find that $\lim_{t\to\infty} X^b_t = \infty$ on $\{\tau = \infty\}$. From (2.10) we can conclude by bounded convergence that $\delta^b(x) \le f(x)$. □

2.2.2 Optimal Investment

Let us start with the diffusion approximation to the surplus process

$$X^0_t = x + \eta t + \sigma_{\mathrm{S}} W^{\mathrm{S}}_t\,,$$

where $\eta, \sigma_{\mathrm{S}} > 0$. The insurer now has the possibility of investing in a risky asset, modelled as a Black–Scholes model

$$Z_t = \exp\Bigl\{\Bigl(m - \frac{\sigma_{\mathrm{I}}^2}{2}\Bigr)t + \sigma_{\mathrm{I}} W^{\mathrm{I}}_t\Bigr\}\,,$$

where $m, \sigma_{\mathrm{I}} > 0$, or equivalently,

$$\mathrm{d}Z_t = mZ_t \,\mathrm{d}t + \sigma_{\mathrm{I}} Z_t \,\mathrm{d}W^{\mathrm{I}}_t\,, \qquad Z_0 = 1\,.$$

The Brownian motions W^{S} and W^{I} are supposed to be independent. The insurer can choose the amount A_t invested at time t. Given the investment strategy A, the surplus process fulfils

$$\mathrm{d}X^A_t = (\eta + A_t m) \,\mathrm{d}t + \sigma_{\mathrm{S}} \,\mathrm{d}W^{\mathrm{S}}_t + \sigma_{\mathrm{I}} A_t \,\mathrm{d}W^{\mathrm{I}}_t\,, \qquad X^A_0 = x\,. \tag{2.11}$$

The set of admissible strategies are all adapted cadlag strategies A such that (2.11) admits a unique solution. We use the filtration $\{\mathcal{F}_t\}$ generated by the two-dimensional Brownian motion $\{(W_t^{\mathrm{S}}, W_t^{\mathrm{I}})\}$.

Our goal is to minimise the ruin probability

$$\psi^A(x) = \mathbb{P}[\inf_{t \ge 0} X_t^A < 0] ,$$

or alternatively to maximise $\delta^A(x) = 1 - \psi^A(x)$. The value function becomes $\delta(x) = \sup_A \delta^A(x)$. Note that $\delta(0) = 0$.

Let us start by considering a strategy

$$A_t = \begin{cases} A, & \text{if } t \le h \wedge \tau, \\ A_{t-h}^{\varepsilon}, & \text{if } t > h. \end{cases}$$

Here $\tau = \inf\{t \ge 0 : X_t^A < 0\}$ is the time of ruin and A^ε is a strategy such that $\delta(X_h^A) < \delta^{A^\varepsilon}(X_h^A) + \varepsilon$. By the Markov property,

$$\delta(x) \ge \delta^A(x) = \mathbb{E}[\delta^{A^\varepsilon}(X_h^A)\mathbb{1}_{\tau > h}] \ge \mathbb{E}[\delta(X_h^A)\mathbb{1}_{\tau > h}] - \varepsilon = \mathbb{E}[\delta(X_{\tau \wedge h}^A)] - \varepsilon .$$

Because ε is arbitrary, we can let ε tend to zero. By Itô's formula, under the assumption that $\delta(u)$ is twice continuously differentiable, we find that

$$\begin{aligned} \delta(X_{\tau\wedge h}^A) = \delta(x) &+ \int_0^{\tau \wedge h} (\eta + Am)\delta'(X_t^A) + \tfrac{1}{2}(\sigma_{\mathrm{S}}^2 + \sigma_{\mathrm{I}}^2 A^2)\delta''(X_t^A)\,\mathrm{d}t \\ &+ \int_0^{\tau\wedge h} \sigma_{\mathrm{S}}\delta'(X_t^A)\,\mathrm{d}W_t^{\mathrm{S}} + \int_0^{\tau \wedge h} \sigma_{\mathrm{I}} A\delta'(X_t^A)\,\mathrm{d}W_t^{\mathrm{I}} . \end{aligned}$$

Taking expected values and assuming that the integrals with respect to Brownian motion are martingales, we obtain

$$\mathbb{E}\Bigl[\int_0^{\tau\wedge h} (\eta + Am)\delta'(X_t^A) + \tfrac{1}{2}(\sigma_{\mathrm{S}}^2 + \sigma_{\mathrm{I}}^2 A^2)\delta''(X_t^A)\,\mathrm{d}t\Bigr] \le 0 .$$

Dividing by h and letting $h \to 0$ yields

$$(\eta + Am)\delta'(x) + \tfrac{1}{2}(\sigma_{\mathrm{S}}^2 + \sigma_{\mathrm{I}}^2 A^2)\delta''(x) \le 0 .$$

This motivates the Hamilton–Jacobi–Bellman equation

$$\sup_{A \in \mathbb{R}} \{(\eta + Am)f'(x) + \tfrac{1}{2}(\sigma_{\mathrm{S}}^2 + \sigma_{\mathrm{I}}^2 A^2)f''(x)\} = 0 . \tag{2.12}$$

The boundary conditions are $f(0) = 0$ and $f(\infty) = 1$. We are again interested in a strictly increasing, twice continuously differentiable solution to (2.12).

If $f''(x) > 0$, the supremum is infinite, and $f(x)$ would not be a solution. If $f''(x) = 0$ on some interval, then the supremum is again infinite unless

$f'(x) = 0$. This contradicts the assumption that $f(x)$ is strictly increasing. Therefore, $f''(x) < 0$. The supremum is taken at

$$A(x) = -\frac{mf'(x)}{\sigma_{\rm I}^2 f''(x)} \,.$$

The equation to solve is then

$$2\eta\sigma_I^2 f'(x)f''(x) + \sigma_{\rm S}^2\sigma_{\rm I}^2 f''(x)^2 - m^2 f'(x)^2 = 0 \,.$$

Letting $f'(x) = {\rm e}^{-g(x)}$ for some function $g(x)$, we find that

$$-2\eta\sigma_I^2 g'(x) + \sigma_{\rm S}^2\sigma_{\rm I}^2 g'(x)^2 - m^2 = 0 \,.$$

Because we look for a continuously differentiable $f(x)$, the function $g'(x)$ must be constant:

$$\begin{aligned} R := g'(x) &= \frac{1}{\sigma_{\rm S}^2\sigma_{\rm I}^2}\Bigl(\eta\sigma_I^2 \pm \sqrt{\eta^2\sigma_I^4 + m^2\sigma_{\rm S}^2\sigma_{\rm I}^2}\Bigr) \\ &= \frac{1}{\sigma_{\rm S}^2\sigma_{\rm I}}\Bigl(\eta\sigma_I + \sqrt{\eta^2\sigma_I^2 + m^2\sigma_{\rm S}^2}\Bigr) \,. \end{aligned}$$

The plus sign follows from $f''(x) < 0$. The desired solution is therefore of the form $f(x) = 1 - {\rm e}^{-Rx}$. The optimal strategy becomes constant $A^* = m/(\sigma_{\rm I}^2 R)$.

We next prove the verification theorem.

Theorem 2.7. *The value function is* $\delta(x) = 1 - {\rm e}^{-Rx}$ *with*

$$R = \frac{1}{\sigma_{\rm S}^2\sigma_{\rm I}}\Bigl(\eta\sigma_I + \sqrt{\eta^2\sigma_I^2 + m^2\sigma_{\rm S}^2}\Bigr) \,.$$

Moreover, the strategy

$$A_t^* = A^* = \frac{m}{\sigma_{\rm I}^2 R} = \frac{1}{m\sigma_{\rm I}}\Bigl(\sqrt{\eta^2\sigma_I^2 + m^2\sigma_{\rm S}^2} - \eta\sigma_I\Bigr)$$

is optimal.

Proof. Note that $f(x)$ is a solution to (2.12). Denote by $\{X_t^*\}$ the surplus process by following the strategy A^*, and let $\{A_t\}$ be an arbitrary strategy. By Itô's formula and (2.12),

$$\begin{aligned} f(X_{\tau\wedge t}^A) &= f(x) + \int_0^{\tau\wedge t} (\eta + A_s m)f'(X_s^A) + \tfrac{1}{2}(\sigma_{\rm S}^2 + \sigma_{\rm I}^2 A_s^2)f''(X_s^A)\,{\rm d}s \\ &\quad + \int_0^{\tau\wedge t} \sigma_{\rm S} f'(X_s^A)\,{\rm d}W_s^{\rm S} + \int_0^{\tau\wedge t} \sigma_{\rm I} A_s f'(X_s^A)\,{\rm d}W_s^{\rm I} \\ &\le f(x) + \int_0^{\tau\wedge t} \sigma_{\rm S} f'(X_s^A)\,{\rm d}W_s^{\rm S} + \int_0^{\tau\wedge t} \sigma_{\rm I} A_s f'(X_s^A)\,{\rm d}W_s^{\rm I} \,. \end{aligned}$$

For $X = X^*$, equality holds. Because $f'(x)$ is bounded by R, the process $\{f(X_t^*)\}$ is a martingale. Thus, $\mathbb{E}[f(X_{\tau\wedge t}^*)] = f(x)$. For an arbitrary strategy we stop the process at $\tau\wedge\tau_n$, where $\tau_n = \inf\{t : |\int_0^{\tau\wedge t} \sigma_{\rm I} A_s f'(X_s^A)\,{\rm d}W_s^{\rm I}| > n\}$. Then $\mathbb{E}[f(X_{\tau\wedge\tau_n\wedge t}^A)] \le f(x)$ because the stochastic integrals are martingales. By bounded convergence, $f(x) \ge \mathbb{E}[f(X_{\tau\wedge t}^A)]$. In particular, $\{f(X_{\tau\wedge t}^A\}$ is a supermartingale. By the martingale convergence theorem, $f(X_{\tau\wedge t}^A)$ converges as $t\to\infty$. We want to show that $\{0,1\}$ are the only possible limits. Let $T = \inf\{s > t : \int_t^s (\sigma_{\rm S}^2 + \sigma_{\rm I}^2 A_v^2)\,{\rm d}v > \sigma_{\rm S}^2\}$. Then $T \le t+1$. The random variable $\int_t^T \sigma_{\rm S}\,{\rm d}W_s^{\rm S} + \int_0^{\tau\wedge t} \sigma_{\rm I} A_s\,{\rm d}W_s^{\rm I}$ is then normally distributed with mean value 0 and variance $\sigma_{\rm S}^2$. Moreover,

$$\Bigl|\int_t^T (\eta + A_v m)\,{\rm d}v\Bigr| \le \int_t^T (\eta + m(1+A_v^2))\,{\rm d}v \le \eta + m + m\sigma_{\rm S}^2/\sigma_{\rm I}^2\;.$$

We conclude that $\mathbb{P}[X_T^A - X_t < 1 \mid \mathcal{F}_t] \ge \delta$ for some $\delta > 0$. It therefore follows analogously as in the proof of Theorem 2.5 that $\{\tau < \infty\}$ or $X_t^A \to \infty$. Letting $t\to\infty$, bounded convergence yields $f(x) \ge \mathbb{E}[f(X_\tau)] = \mathbb{P}[\tau < \infty]$. Thus, $f(x) \ge \delta(x)$. For $X = X^*$ equality holds, and thus $f(x) = \delta(x)$. □

2.2.3 Optimal Investment and Reinsurance

Let us now consider the situation where both investment and reinsurance are possible. The controlled process then fulfils the stochastic differential equation

$${\rm d}X_t^{Ab} = (b_t\theta - (\theta-\eta) + mA_t)\,{\rm d}t + \sigma_{\rm S} b_t\,{\rm d}W_t^{\rm S} + \sigma_{\rm I} A_t\,{\rm d}W_t^{\rm I}\;, \qquad X_0^{Ab} = x\;.$$

The corresponding Hamilton–Jacobi–Bellman equation reads

$$\sup_{A,b}\{\tfrac12(A^2\sigma_{\rm I}^2 + b^2\sigma_{\rm S}^2)f''(x) + (mA + b\theta - (\theta-\eta))f'(x)\} = 0\;. \tag{2.13}$$

We are looking for an increasing, twice continuously differentiable solution with $f(0) = 0$ and $f(\infty) = 1$. The set of admissible strategies are all cadlag adapted processes (A, b) with $b_t \in [0,1]$ such that $\{X_t^{A,b}\}$ is well defined. We still use the filtration $\{\mathcal{F}_t\}$ generated by the process $\{(W_t^{\rm S}, W_t^{\rm I})\}$.

The arguments used above show that a twice continuously differentiable, strictly increasing solution fulfils $f''(x) < 0$. This yields

$$A^*(x) = -\frac{mf'(x)}{\sigma_{\rm I}^2 f''(x)}\;, \qquad b^*(x) = -\frac{\theta f'(x)}{\sigma_{\rm S}^2 f''(x)}\;.$$

The equation to solve becomes

$$-\tfrac12\Bigl(\frac{m^2}{\sigma_{\rm I}^2} + \frac{\theta^2}{\sigma_{\rm S}^2}\Bigr)\frac{f'(x)^2}{f''(x)} - (\theta-\eta)f'(x) = 0\;.$$

Because we are looking for a strictly increasing function, we can divide by $f'(x)$ and obtain the linear differential equation

$$2\sigma_{\rm I}^2\sigma_{\rm S}^2(\theta-\eta)f''(x) + (m^2\sigma_{\rm S}^2 + \theta^2\sigma_{\rm I}^2)f'(x) = 0\;.$$

Theorem 2.8. *Suppose that $\eta < \theta \le \eta + \sqrt{\eta^2 + m^2\sigma_{\rm S}^2/\sigma_{\rm I}^2}$. Then the value function is $\delta(x) = 1 - {\rm e}^{-Rx}$ where*

$$R = \frac{(m^2\sigma_{\rm S}^2 + \theta^2\sigma_{\rm I}^2)}{2\sigma_{\rm I}^2\sigma_{\rm S}^2(\theta - \eta)}\;.$$

The optimal strategy is constant with

$$A^* = \frac{2m\sigma_{\rm S}^2(\theta - \eta)}{(m^2\sigma_{\rm S}^2 + \theta^2\sigma_{\rm I}^2)}\;, \qquad b^* = \frac{2\theta\sigma_{\rm I}^2(\theta - \eta)}{(m^2\sigma_{\rm S}^2 + \theta^2\sigma_{\rm I}^2)}\;.$$

If $\theta \ge \eta + \sqrt{\eta^2 + m^2\sigma_{\rm S}^2/\sigma_{\rm I}^2}$, then $b^* = 1$ and $\delta(x)$ and A^* are given by Theorem 2.7.

Proof. If $\theta \le \eta + \sqrt{\eta^2 + m^2\sigma_{\rm S}^2/\sigma_{\rm I}^2}$, then $b^* \le 1$, and $f(x)$ solves the Hamilton–Jacobi–Bellman equation (2.13). If $\theta > \eta + \sqrt{\eta^2 + m^2\sigma_{\rm S}^2/\sigma_{\rm I}^2}$, then $b^* > 1$. In this case Theorem 2.7 gives a solution. Itô's formula and (2.13) yield

$$f(X_{\tau\wedge t}^{Ab}) \le f(x) + \int_0^{\tau\wedge t} \sigma_{\rm S} b_s f'(X_s^{Ab})\,{\rm d}W_s^{\rm S} + \int_0^{\tau\wedge t} \sigma_{\rm I} A_s f'(X_s^{Ab})\,{\rm d}W_s^{\rm I}\;.$$

The first stochastic integral is a martingale. Using the same arguments as in the proofs of Theorems 2.5 and 2.7 proves the theorem. □

Bibliographical Remarks

Section 2.2.1 is from Schmidli [160]. Section 2.2.2 follows Browne [28], where the Brownian motions $W^{\rm S}$ and $W^{\rm I}$ could be dependent. In the latter paper, the following problems are also solved: the constrained problem where $0 \le A_t \le X_t^A$ has to be fulfilled, minimisation of the expected penalty of ruin $\mathbb{E}[{\rm e}^{-\lambda\tau}]$, and maximisation of an exponential utility in finite time. A similar model was also considered by Markussen and Taksar [129]. Here all capital is invested into the risky asset, i.e., $A_t = X_t^A$.

2.3 Minimising Ruin Probabilities for a Classical Risk Model

Instead of a diffusion approximation we now model the aggregate claim amount as in the classical risk model. The claim arrival process is a Poisson process with rate λ, and the claim sizes $\{Y_i\}$ are iid, positive, and independent of $\{N_t\}$. Therefore, the aggregate claims process $\{\sum_{i=1}^{N_t} Y_i\}$ is a compound Poisson process. We use the notation of Section D.1. The rate of the premia earned by the cedent (that is the first-line insurer) is still denoted by c. But we change the model to allow one to buy reinsurance and to invest the surplus into a risky asset. In order to make the presentation simpler, we assume that the claim size distribution $G(y)$ is continuous. It is possible to relax this assumption. However, the optimal ruin probability will no longer be (twice) continuously differentiable. If investment is allowed, one will have to use viscosity solutions instead of the classical solutions we obtain here.

2.3.1 Optimal Reinsurance

Suppose that the cedent has the possibility to take reinsurance. At any time the cedent can choose a reinsurance from a set $\mathcal{U}$, b_t say. We choose $\mathcal{U}$ as a compact set. For possible reinsurance forms, see Section D.5.

A reinsurance strategy is an adapted process b. We denote the set of possible strategies by $\mathfrak{U}$. If the reinsurance b_t is chosen at time t, a premium has to be paid to the reinsurer. The part of the premium rate left to the cedent is denoted by $c(b_t)$. We assume that $b \mapsto c(b)$ is continuous, that more reinsurance is more expensive, and that full reinsurance (the reinsurer pays all the claim) is more expensive than the first insurance, i.e., the premium rate left to the cedent is strictly negative. Otherwise, one could reinsure the whole portfolio and the ruin probability would become zero. If no reinsurance is chosen, we denote the premium left to the cedent by c, i.e., the rate the policyholders pay to the cedent. We then have $c(b) \le c$ for all b. We assume the net profit condition $c > \lambda\mu$; otherwise, ruin can only be prevented if reinsurance is "too cheap" for some reinsurance possibility. If reinsurance $b \in \mathcal{U}$ is chosen, the part of the claim Y the cedent has to pay is denoted by $s(Y, b)$. The reinsurer pays the amount $Y - s(Y, b)$. We assume that $s(y, b)$ is continuous both in y and in b, and increasing in y. Moreover, we suppose that $0 \le s(y, b) \le y$; hence, it is not possible to reinsure more than the claim, and the reinsurer's payments should be positive. The surplus process then becomes

$$X_t^b = x + \int_0^t c(b_s)\,\mathrm{d}s - \sum_{i=1}^{N_t} s(Y_i, b_{T_i-})\;,$$

where we use the notation of Appendix D. In particular, the Poisson process $\{N_t\}$ models the number of claims until time t and Y_i is the size of the i-th claim. $T_i = \inf\{t : N_t = i\}$ is the occurrence time of the ith claim. The distribution of the claim is $G(y)$ with $G(0) = 0$. Recall that we assume that $G(y)$ is continuous. The filtration $\{\mathcal{F}_t\}$ is the natural filtration of the compound Poisson process $\{\sum_{i=1}^{N_t} Y_i\}$.

For a claim Y_i at time T_i, the reinsurance form $s(\cdot, b_{T_i-})$ chosen prior to the claim applies. Otherwise, one could change the reinsurance form at the claim time to full reinsurance. Then the reinsurer would pay all the claims, while all the premia would go to the cedent.

Examples for reinsurance treaties are

- *Proportional reinsurance:* $\mathcal{U} = [0,1]$ and $s(y,b) = by$. If an expected value principle is used for the reinsurance premium, we have $c(b) = c - (1 - b)(1+\theta)\lambda\mu$ for some $\theta > c/(\lambda\mu) - 1$. If the variance principle is used, then $c(b) = c - \lambda((1-b)\mu + a(1-b)^2(\mu_2 - \mu^2))$ for some $a > (c/\lambda - \mu)/(\mu_2 - \mu^2)$. Here $\mu = \mathbb{E}[Y]$ and $\mu_2 = \mathbb{E}[Y^2]$.
- *Excess of loss reinsurance:* $\mathcal{U} = [0,\infty]$ and $s(y,b) = \min\{y,b\}$. An expected value principle yields $c(b) = c - \lambda(1+\theta)\mathbb{E}[(Y-b)^+]$, and the variance principle yields $c(b) = c - \lambda(\mathbb{E}[(Y-b)^+] + a \operatorname{Var}[\{(Y-b)^+\}^2])$. Here θ and a have to be chosen such that $c(0) < 0$.
- *Proportional reinsurance in a layer:* $\mathcal{U} = \{b \in [0,\infty]^2 \times [0,1] : b_1 < b_2\}$ and $s(y,b) = b_3 \min\{(y-b_1)^+, b_2 - b_1\}$. The premium can be determined similarly as in the previous examples.

The time of ruin is $\tau^b = \inf\{t \ge 0 : X_t^b < 0\}$ and the ruin probability is $\psi^b(x) = \mathbb{P}[\tau^b < \infty]$. As usual in ruin theory, it turns out to be simpler to consider the survival probability $\delta^b(x) = 1 - \psi^b(x)$ because then $\delta^b(x) = 0$ for $x < 0$. The goals are to find $\delta(x) = \sup_{b \in \mathfrak{U}} \delta^b(x)$ and to determine the optimal strategy, if it exists. That is, we minimise the ruin probability.

We start by some general considerations. The next two lemmata are quite technical, and the reader is recommended to skip their proofs upon first reading.

Lemma 2.9. *For any strategy b, with probability 1, either ruin occurs or $X_t^b \to \infty$ as $t \to \infty$.*

Proof. Let b be a strategy. Let $c_0 < 0$ be the premium left to the cedent if full reinsurance is chosen. Let $\mathcal{B} = \{b : c(b) \ge c_0/2\}$. Choose $\varepsilon < -c_0/2$. Let $\kappa = (-c_0 - 2\varepsilon)/(2c - c_0)$. If $\int_t^{t+1} \mathbb{1}_{b_s \in \mathcal{B}} \,\mathrm{d}s \le \kappa$, then

$$X_{t+1}^b \le X_t^b + \kappa c + (1-\kappa)c_0/2 \le X_t^b - \varepsilon\,.$$

Suppose that $\int_t^{t+1} \mathbb{1}_{b_s \in \mathcal{B}} \,\mathrm{d}s > \kappa$. Because both $c(b)$ and $s(y,b)$ are continuous, we can assume that ε is small enough such that

$$\mathbb{P}[\inf_{b \in \mathcal{B}} s(Y,b) > \varepsilon] > 0\,.$$

We also have that

$$\mathbb{P}\Bigl[\int_t^{t+1} \mathbb{1}_{b_s \in \mathcal{B}} \,\mathrm{d}N_s \ge 1 + c/\varepsilon\Bigr] \ge \mathbb{P}[N_\kappa \ge 1 + c/\varepsilon] > 0\,.$$

Thus, $\mathbb{P}[\sum_{i=N_t+1}^{N_{t+1}} s(Y_i, b_{T_i-}) \ge c + \varepsilon] > 0$. Denote a lower bound by $\delta > 0$.

Choose $a > 0$. Let $t_0 = 0$ and $t_{k+1} = \inf\{t \ge t_k + 1 : X_t \le a\}$. As usual, we let $t_{k+1} = \infty$ if $t_k = \infty$ or if $X_t > a$ for all $t \ge t_k + 1$. As we have seen above, we have

$$\mathbb{P}[X_{t_k+1} \le a - \varepsilon \mid \mathcal{F}_{t_k}] \ge \delta\,.$$

Let $V_k = 1\!\!1_{t_k<\infty, X_{t_k+1}<a-\varepsilon}$ and $Z_k = \delta 1\!\!1_{t_k<\infty}$. It follows that the assumptions of Lemma 1.15 are fulfilled. Thus, if $\liminf X_t \le b$, then, for $a = b + \varepsilon/2$, $t_n < \infty$ for all n. Thus, $X_{t_n+1} \le b - \varepsilon/2$ infinitely often. In particular, $\liminf X_t \le b - \varepsilon/2$. We see that $\liminf X_t < \infty$ implies that $\liminf X_t \le -\varepsilon/2$, and therefore ruin occurs. □

Lemma 2.10. *The function $\delta(x)$ is strictly increasing.*

Proof. If $x < z$, it is possible to choose the same strategy for initial capital x and for initial capital z. This proves that $\delta(x)$ is increasing. Suppose that $\delta(x) = \delta(z)$. By the proof of Lemma 2.9 we can conclude that $\delta(x) < 1$. Indeed, if $c(b_t) \le c_0/2$ on the interval $[0, (x + \kappa c)/(-c_0) + \kappa]$ for all t except a set with measure κ, then ruin occurs. Otherwise, the argument of the proof of Lemma 2.9 shows that the aggregate claim until the time $\inf\{t : \int_0^t 1\!\!1_{2c(b_s)\ge c_0}\,\mathrm{d}s > \kappa\}$ is with positive probability strictly larger than $x + \kappa c$. Thus, also in this case ruin occurs with strictly positive probability. The process $\{\delta^b(X^b_{\tau^b\wedge t})\}$ is a martingale. Let us stop the process starting in x at the first time T_z where $X^b_t = z$. Let $\tilde{X}^b_t = X^b_t + z - x$ for $t \le T_z$ and choose an arbitrary strategy $\tilde{b}$ after time T_z. Denote the corresponding characteristics by a tilde sign. Then

$$\tilde{\delta}^b(z) = \mathbb{E}[\tilde{\delta}^b(\tilde{X}_{T_z\wedge\tau})] = \tilde{\delta}^b(2z - x)\mathbb{P}[T_z < \tilde{\tau}^b] \ge \tilde{\delta}^b(2z - x)\mathbb{P}[T_z < \tau^b]\,.$$

Because $\delta^b(x) = \delta^b(z)\mathbb{P}[T_z < \tau^b]$, there is a strategy such that $\mathbb{P}[T_z < \tau^b]$ is arbitrarily close to 1. Because the strategy $\tilde{b}$ is arbitrary, we conclude that $\delta(2z - x) = \delta(z) = \delta(x)$. Thus, $\delta(z)$ would be constant for all $z \ge x$. Because $\delta(z) \to 1$ as $z \to \infty$, this is only possible if $\delta(x) = 1$. But this is a contradiction. Thus, $\delta(x)$ is strictly increasing. □

We start by motivating the Hamilton–Jacobi–Bellman equation. Let $h > 0$ be a (small) number, $u \in \mathcal{U}$, and $\varepsilon > 0$. If $x = 0$, we assume that $c(u) \ge 0$, (otherwise ruin would occur immediately). If $x > 0$ we choose h small enough such that $x + c(u)h > 0$. For each $X_{T_1\wedge h}$ we choose a strategy $\tilde{b}$ such that $\delta^{\tilde{b}}(X_{T_1\wedge h}) > \delta(X_{T_1\wedge h}) - \varepsilon$. We let $b_t = u$ on $(0, T_1 \wedge h]$ and $b_t = \tilde{b}_{t-(T_1\wedge h)}$ for $t > T_1 \wedge h$. This yields by conditioning on $\mathcal{F}_{T_1\wedge h}$, letting $\delta(z) = 0$ for $z < 0$,

$$\begin{aligned}\delta(x) \ge \delta^b(x) &= \mathrm{e}^{-\lambda h}\delta^{\tilde{b}}(x + c(u)h)\\ &\qquad + \int_0^h\int_0^\infty \delta^{\tilde{b}}(x + c(u)t - s(y,u))\,\mathrm{d}G(y)\,\lambda\mathrm{e}^{-\lambda t}\,\mathrm{d}t\\ &\ge \mathrm{e}^{-\lambda h}\delta(x + c(u)h) + \int_0^h\int_0^\infty \delta(x + c(u)t - s(y,u))\,\mathrm{d}G(y)\,\lambda\mathrm{e}^{-\lambda t}\,\mathrm{d}t - \varepsilon\,.\end{aligned}$$

Because this holds for any $\varepsilon > 0$, we can let $\varepsilon = 0$. Reordering the terms yields

$$\frac{\delta(x+c(u)h)-\delta(x)}{h} - \frac{1-\mathrm{e}^{-\lambda h}}{h}\delta(x+c(u)h) + \frac{1}{h}\int_0^h \int_0^\infty \delta(x+c(u)t-s(y,u))\,\mathrm{d}G(y)\lambda\mathrm{e}^{-\lambda t}\,\mathrm{d}t \le 0\,.$$

If we assume that $\delta(x)$ is differentiable, we obtain by letting $h \downarrow 0$

$$c(u)\delta'(x) + \lambda\Bigl[\int_0^\infty \delta(x-s(y,u))\,\mathrm{d}G(y) - \delta(x)\Bigr] \le 0\,.$$

This has to hold for all $u \in \mathcal{U}$. Similarly as in Section 2.4.1 ahead, we could argue that there exists a u such that equality holds in the above equation. We will prove the results differently, and therefore we consider the Hamilton–Jacobi–Bellman equation

$$\sup_{u\in\mathcal{U}}\Bigl\{c(u)f'(x) + \lambda\Bigl[\int_0^\infty f(x-s(y,u))\,\mathrm{d}G(y) - f(x)\Bigr]\Bigr\} = 0\,. \tag{2.14}$$

Here we only consider increasing functions $f(x)$ with $f(x) = 0$ for $x < 0$ and $f(0) > 0$. For the moment we replace the value function δ by a solution f to (2.14) because we do not know whether $\delta(x)$ fulfils (2.14). Because the function for which the supremum is taken is continuous in u and $\mathcal{U}$ is compact, there is for each $x \ge 0$ a value $u(x)$ for which the supremum is attained.

Denote $\bar{\mathcal{U}} = \{b \in \mathcal{U} : c(b) > 0\}$. Because we are interested in strictly increasing solutions to (2.14), it follows from (2.14) that $c(u(x))f'(x) \ge 0$. Thus, $c(u(x)) \ge 0$ follows. If $c(u(x)) \ge 0$, we do not have full insurance and $\mathbb{P}[s(Y,u(x)) > 0] > 0$. Thus, $c(u(x))f'(x) > 0$ follows. In particular, $f'(x) > 0$ and $c(u(x)) > 0$, and (2.14) can be written as

$$\sup_{u\in\bar{\mathcal{U}}}\Bigl\{c(u)f'(x) + \lambda\Bigl[\int_0^\infty f(x-s(y,u))\,\mathrm{d}G(y) - f(x)\Bigr]\Bigr\} = 0\,.$$

We next reformulate (2.14). For $u \in \bar{\mathcal{U}}$, we find that

$$f'(x) \le \frac{\lambda}{c(u)}\Bigl[f(x) - \int_0^\infty f(x-s(y,u))\,\mathrm{d}G(y)\Bigr]\,.$$

For $u = u(x)$, equality holds. Thus,

$$f'(x) = \inf_{u\in\bar{\mathcal{U}}}\Bigl\{\frac{\lambda}{c(u)}\Bigl[f(x) - \int_0^\infty f(x-s(y,u))\,\mathrm{d}G(y)\Bigr]\Bigr\}\,. \tag{2.15}$$

In the same way one can show that (2.14) and (2.15) are equivalent for strictly increasing functions.

Equations (2.14) and (2.15) determine the function f only up to a constant. Instead of the boundary condition $\delta(\infty) = 1$, we can choose $f(0) = 1$ [because we know that $\delta(0) > 0$]. We first show that there is a solution.

Lemma 2.11. *There is a unique solution to (2.15) with $f(0) = 1$. The solution is bounded, strictly increasing, and continuously differentiable.*

Proof. Let
$$\rho(y,u) = \inf\{z : s(z,u) > y\}$$
be the generalised inverse function of $y \mapsto s(y,u)$. This inverse function is well defined because $s(y,u)$ is increasing in y. Using Fubini's theorem, we can reformulate the expression

$$\begin{aligned} f(x) - \int_0^{\rho(x,u)} & f(x - s(y,u)) \, \mathrm{d}G(y) \\ &= 1 - G(\rho(x,u)) + \int_0^x f'(z)[1 - G(\rho(x-z,u))] \, \mathrm{d}z \,, \end{aligned}$$

using the fact that $s(0,u) = 0$. Let $\mathcal{V}$ be the operator acting on positive functions g:

$$\mathcal{V}g(x) = \inf_{u \in \bar{\mathcal{U}}} \Bigl\{ \frac{\lambda}{c(u)} \Bigl[1 - G(\rho(x,u)) + \int_0^x g(z)(1 - G(\rho(x-z,u))) \, \mathrm{d}z \Bigr] \Bigr\} \,.$$

We first show the existence of a solution. Let $\delta_0(x)$ be the survival probability if no reinsurance is taken. The function $\delta_0(x)$ solves

$$\begin{aligned} \delta_0'(x) &= \frac{\lambda}{c} \Bigl[\delta_0(x) - \int_0^\infty \delta_0(x-y) \, \mathrm{d}G(y) \Bigr] \\ &= \frac{\lambda}{c} \Bigl[1 - G(x) + \int_0^x \delta_0'(z)(1 - G(x-z)) \, \mathrm{d}z \Bigr] \,; \end{aligned}$$

see (D.1). Let $g_0(x) = c\delta_0'(x)/(c - \lambda\mu) = \delta_0'(x)/\delta_0(0)$ and define recursively $g_n(x) = \mathcal{V}g_{n-1}(x)$. Note that $g_n(x) > 0$. Then, clearly, $g_1(x) \le g_0(x)$. We show that $g_n(x)$ is decreasing in n. Suppose that $g_{n-1}(x) \le g_n(x)$. Let u_n be a point for which $\mathcal{V}g_{n-1}(x)$ attains the minimum. Such a point exists because the right-hand side of (2.15) [with $f_{n-1}(x) = 1 + \int_0^x g_{n-1}(z) \, \mathrm{d}z$] is continuous in u, the set $\{u : c(u) \ge 0\}$ is compact, and the right-hand side of (2.15) converges to infinity as u approaches a point u_0 where $c(u_0) = 0$. Then

$$\begin{aligned} g_n(x) - g_{n+1}(x) &= \mathcal{V}g_{n-1}(x) - \mathcal{V}g_n(x) \\ &\ge \frac{\lambda}{c(u_n)} \int_0^x (g_{n-1}(z) - g_n(z))[1 - G(\rho(x-z,u_n))] \, \mathrm{d}z \ge 0 \,. \end{aligned}$$

Hence, $g_n(x) \ge g_{n+1}(x)$. Since $g_n(x) > 0$, we have that $g(x) = \lim_{n\to\infty} g_n(x)$ exists pointwise. By bounded convergence,

$$\lim_{n\to\infty} \int_0^x g_n(z)[1 - G(\rho(x-z,u))] \, \mathrm{d}z = \int_0^x g(z)[1 - G(\rho(x-z,u))] \, \mathrm{d}z$$

for each x and u. Let u be a point for which $\mathcal{V}g(x)$ attains its minimum. From

$$\begin{aligned} g_n(x) &= \frac{\lambda}{c(u_n)}\Bigl[1 - G(\rho(x,u_n)) + \int_0^x g_{n-1}(z)[1-G(\rho(x-z,u_n))]\,\mathrm{d}z\Bigr] \\ &\le \frac{\lambda}{c(u)}\Bigl[1 - G(\rho(x,u)) + \int_0^x g_{n-1}(z)[1-G(\rho(x-z,u))]\,\mathrm{d}z\Bigr]\;, \end{aligned}$$

we conclude that $g(x) \le \mathcal{V}g(x)$ by letting $n \to \infty$. Because $g_n(z)$ is decreasing, we find that

$$\begin{aligned} g_n(x) &= \frac{\lambda}{c(u_n)}\Bigl[1 - G(\rho(x,u_n)) + \int_0^x g_{n-1}(z)[1-G(\rho(x-z,u_n))]\,\mathrm{d}z\Bigr] \\ &\ge \frac{\lambda}{c(u_n)}\Bigl[1 - G(\rho(x,u_n)) + \int_0^x g(z)[1-G(\rho(x-z,u_n))]\,\mathrm{d}z\Bigr] \\ &\ge \frac{\lambda}{c(u)}\Bigl[1 - G(\rho(x,u)) + \int_0^x g(z)[1-G(\rho(x-z,u))]\,\mathrm{d}z\Bigr]\;. \end{aligned}$$

This shows that $g(x) \ge \mathcal{V}g(x)$ and therefore $\mathcal{V}g(x) = g(x)$. In particular, $g(x)$ is continuous. Let us define $f(x) = 1 + \int_0^x g(z)\,\mathrm{d}z$. By bounded convergence, $f(x)$ fulfils (2.15). It follows that $f(x)$ is increasing, continuously differentiable, and bounded by $c/(\lambda\mu)$. From (2.15) it follows that $f'(0) > 0$. Let $x_0 = \inf\{z : f'(z) = 0\}$. Because $f(x)$ is strictly increasing in $[0,z]$, we must have $G(x_0) = 1$ and $s(y,u) = 0$ for all points of increase of $G(y)$. But this would be full reinsurance, i.e., $u \notin \bar{\mathcal{U}}$. Thus, $f(x)$ is strictly increasing.

Suppose now that $f_1(x)$ and $f_2(x)$ are solutions to (2.15) with $f_1(0) = f_2(0) = 1$. Denote by $g_i(x) = f_i'(x)$ the derivatives and by $u_i(x)$ a value for which the minimum in (2.15) is attained. Fix $\bar{x} > 0$. Because the right-hand side of (2.15) is continuous in u, and tends to infinity as $c(u)$ approaches zero, we conclude that $c(u_i(x))$ is bounded away from zero on $[0,\bar{x}]$. Let $x_1 = \inf\{\min_i c(u_i(x)) : 0 \le x \le \bar{x}\}/(2\lambda)$ and $x_n = nx_1 \wedge \bar{x}$. It is no loss of generality to assume that $x_1 \le \bar{x}$. Suppose we have proved that $f_1(x) = f_2(x)$ on $[0,x_n]$. This is trivial for $n = 0$. Then for $x \in [x_n, x_{n+1}]$, with $m = \sup_{x_n \le x \le x_{n+1}} |g_1(x) - g_2(x)|$,

$$\begin{aligned} g_1(x) - g_2(x) &= \mathcal{V}g_1(x) - \mathcal{V}g_2(x) \\ &\le \frac{\lambda}{c(u_2(x))}\int_{x_n}^x (g_1(z) - g_2(z))[1 - G(\rho(x-z,u_2(x)))]\,\mathrm{d}z \\ &\le \frac{\lambda m x_1}{c(u_2(x))} \le \frac{m}{2}\;. \end{aligned}$$

Reversing the rôles of $g_1(x)$ and $g_2(x)$, it follows that $|g_1(x) - g_2(x)| \le m/2$. This is only possible for all $x \in [x_n, x_{n+1}]$ if $m = 0$. This shows that $f_1(x) = f_2(x)$ on $[0, x_{n+1}]$. We found that $f_1(x) = f_2(x)$ on $[0,\bar{x}]$. Because $\bar{x}$ is arbitrary, uniqueness follows. □

Denote by $u(x)$ the value of u that minimises (2.15). We first want to show that $u(x)$ is measurable. This is a technical issue and may be skipped upon first reading.

Lemma 2.12. *There is a version of $u(x)$ such that $u(x)$ is measurable.*

Proof. For simplicity the proof is given for the case where $\mathcal{U}$ is one-dimensional. The multidimensional case follows similarly. Let $u(x)$ be the smallest value for which the minimum in (2.15) is taken. The right-hand side of (2.15) is continuous in both u and x. Choose $a \in \mathcal{U}$. Let

$$m_a(x) = \inf_{u \le a} \Bigl\{ \frac{\lambda}{c(u)} \Bigl[f(x) - \int_0^\infty f(x - s(y,u)) \, \mathrm{d}G(y) \Bigr] \Bigr\} .$$

Then $\{u(x) > a\} = \{x : f'(x) < m_a(x)\}$. Thus, $\{u(x) > a\}$ is measurable for all a. This shows that $u(x)$ is a measurable function. □

We can now prove the *verification theorem.*

Theorem 2.13. *Let $f(x)$ be the unique solution to (2.15) with $f(0) = 1$. Then $f(x) = \delta(x)/\delta(0)$. An optimal strategy is given by $b_t^* = u(X_t^*)$, where $u(x)$ is the value that minimises (2.15), and X^* is the surplus process under the optimal strategy.*

Proof. Recall that $u(x)$ also maximises the left-hand side of (2.14). Let b be an arbitrary admissible strategy with a corresponding surplus process X^b. Then

$$f(X^b_{\tau^b \wedge t}) - \int_0^{\tau^b \wedge t} \Bigl(c(b_s) f'(X_s^b) + \lambda \Bigl[\int_0^\infty f(X_s^b - s(y, b_s)) \, \mathrm{d}G(y) - f(X_s^b) \Bigr] \Bigr) \, \mathrm{d}s$$

is a martingale; see the proof of Theorem 11.2.2 of [152]. By (2.14) the process $\{f(X_t^b) 1\!\!1_{\tau^b > t}\}$ is a supermartingale. Thus, $\mathbb{E}[f(X_t^b) 1\!\!1_{\tau^b > t}] = \mathbb{E}[f(X^b_{\tau^b \wedge t})] \le f(x)$. If $b = b^*$, then $\{f(X^*_{\tau^* \wedge t})\}$ is a martingale and $\mathbb{E}[f(X_t^*) 1\!\!1_{\tau^* > t}] = f(x)$. By Lemma 2.9, $X_t^b \to \infty$ on $\{\tau^b = \infty\}$. Because $f(x)$ is bounded, we can let $t \to \infty$. This yields $f(x) \ge f(\infty) \mathbb{P}[\tau^b = \infty] = f(\infty) \delta^b(x)$, or $f(x) = f(\infty) \delta^*(x)$. This shows that $\delta(x) = f(x)/f(\infty)$. For $x = 0$, we obtain $f(\infty) = 1/\delta(0)$. □

Let us now consider proportional reinsurance, $s(y,b) = by$. Equation (2.14) then reads

$$\sup_{u \in \mathcal{U}} \Bigl\{ c(u) f'(x) + \lambda \Bigl[\int_0^{x/u} f(x - uy) \, \mathrm{d}G(y) - f(x) \Bigr] \Bigr\} = 0 .$$

We first obtain the strategy for small x under a reasonable condition.

Lemma 2.14. *Suppose that* $\liminf_{b \uparrow 1} (1-b)^{-1} (c - c(b)) > 0$. *Then there exists $\varepsilon > 0$ such that $u(x) = 1$ for $x \le \varepsilon$.*

Remarks 2.15. i) The condition $\liminf_{b \uparrow 1} (1-b)^{-1} (c - c(b)) > \lambda\mu$ would mean that for b close to 1 the reinsurer charges more than the net premium, i.e., there is a price for the risk. Since it is reasonable to have a price for the risk, the condition of Lemma 2.14 is weak.

ii) Clearly, an economist would reinsure a portfolio if the capital is close to zero. This is in order to have the possibility to pay dividends later. With the ruin probability as optimisation criterion, it is optimal not to reinsure for capital close to zero in order to reach large capital as quickly as possible. Thus, one likes to lower the ruin probability as quickly as possible. ■

Proof. We first note that $f'(0) > 0$ and that $f'(x)$ is continuous. Thus, for x small enough we have $f'(x) > f'(0)/2$. Let $H(x,b) = c(b)f'(x) + \lambda(\mathbb{E}[f(x - bY)] - f(x))$. Then

$$\frac{H(x,1) - H(x,b)}{1-b} = \frac{c - c(b)}{1-b} f'(x) + \lambda \mathbb{E}\Bigl[\frac{f(x-Y) - f(x-bY)}{1-b} \mathbb{I}_{Y \le x/b}\Bigr] .$$

By the mean value theorem, there exists $\zeta(Y) \in (x - Y, x - bY)$ such that

$$f(x - bY) = f(x-Y) + (1-b)Y f'(\zeta(Y)) .$$

By bounded convergence,

$$\lim_{b \uparrow 1} \mathbb{E}\Bigl[\frac{f(x-Y) - f(x-bY)}{1-b} \mathbb{I}_{Y \le x/b}\Bigr] = -\mathbb{E}[Y f'(x-Y) \mathbb{I}_{Y \le x}] .$$

The right-hand side can be made arbitrarily small for x small enough. Because $f'(x) > f'(0)/2$ and $f'(z)$ is uniformly bounded for $z \in [0, x_0]$, the limit is uniform in x. Thus, $H(x,b) < H(x,1)$ for x small enough and b large enough. Now if $u(x) \ne 1$ close to zero, then $\limsup_{x \downarrow 0} u(x) < 1$, i.e., $u(x)$ jumps at zero because $u(0) = 1$. Because $H(x,b)$ is continuous in both x and b, we conclude that $H(0,b) = H(0,1)$ for some $b < 1$. But $H(0,b) = c(b)f'(0) - \lambda f(0)$ is strictly increasing in b. This proves the lemma. □

In the following two examples we choose $s(y,b) = by$ for $b \in [0,1]$, $\lambda = \mu = 1$, and $c(b) = 1.7b - 0.2$. That is, the insurer's safety loading is 0.5, and the reinsurer's safety loading is 0.7. For the numerical calculations, a first solution is obtained by a Euler scheme from (2.15). It turns out that the optimal strategy does not look nice, in particular in the Pareto case. Thus, the operator $\mathcal{V}$ is applied a few times until the strategy obtained stabilises.

Example 2.16. Suppose that the claims are exponentially distributed $G(y) = 1 - \mathrm{e}^{y}$. This is a typical case of small claims. For this claim size distribution the survival function without reinsurance $\delta_0(x) = 1 - 2\mathrm{e}^{-x/3}/3$ is explicitly known. Considering a constant strategy $b_t = b$ with $c(b) > b$, the ruin probability becomes $\psi^b(x) = c^b(x)\mathrm{e}^{-R(b)x}$ for a function $c^b : [0,\infty) \to (0,1)$ with $\lim_{x\to\infty} c^b(x) > 0$. The coefficient $R(b)$ is called the *adjustment coefficient.* From Example 4.3 we find the retention level b^R maximising the adjustment coefficient

$$b^R = \Bigl(1 - \frac{1.5}{1.7}\Bigr)\Bigl(1 + \frac{1}{\sqrt{1.7}}\Bigr) = 0.504847 .$$

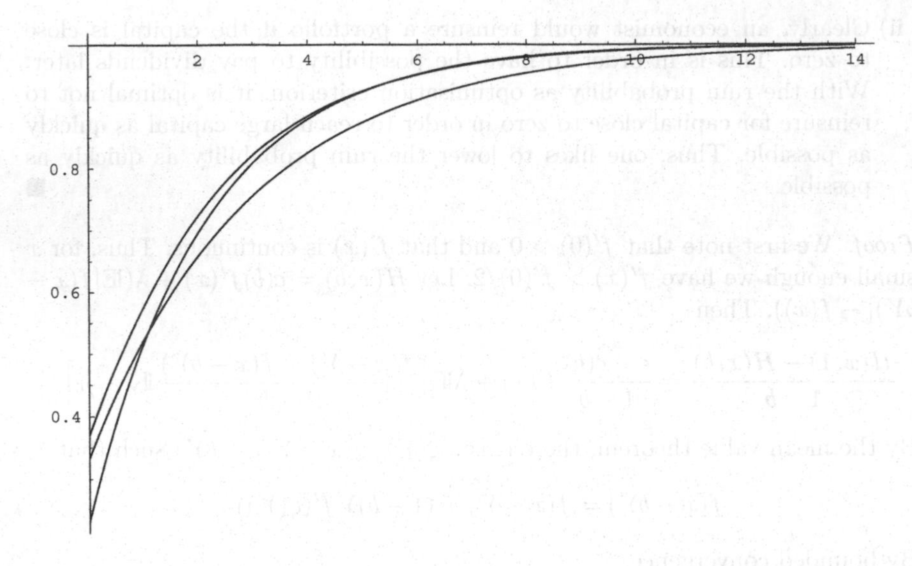

Fig. 2.1. $\delta(x)$ (top graph), $\delta^{\mathrm{R}}(x)$ (middle graph at $x = 4$), and $\delta_0(x)$ (bottom graph at $x = 4$) for exponentially distributed claim sizes.

That is, $R(b^R) = \sup_b R(b)$. Also here $\delta^R(x)$, the survival function for the constant strategy $b_t = b^R$, can be calculated explicitly. If we maximise the adjustment coefficient, we look for the asymptotically optimal constant strategy. We will see in Chapter 4 that $R(b^R)$ is the adjustment coefficient of the optimally controlled risk process. The function $\delta(x)$ has to be obtained numerically. Figure 2.1 shows the $\delta(x)$, $\delta^R(x)$, and $\delta_0(x)$. In Figure 2.2 the optimal reinsurance strategy is calculated. One can see that $u(x)$ converges to b^R; see Corollary 4.13. As known from Lemma 2.14, one does not reinsure close to zero. ■

Example 2.17. Suppose that the claim sizes are Pareto-distributed $G(y) = 1 - (1+y)^{-2}$. Here $\delta_0(x)$ has to be calculated numerically. We now try to find an asymptotically optimal constant strategy. By (D.7)

$$\psi^b(x) \approx \frac{1}{0.7b - 0.2} \frac{b}{1 + x/b} = \frac{1}{(0.7 - 0.2/b)(1 + x/b)}\;.$$

The right-hand side is minimised for $b = 0.4x/(0.7x - 0.2)$. Letting $x \to \infty$ yields the optimal $b^R = 4/7$. Figure 2.3 shows the functions $\delta(x)$, $\delta^R(x)$, and $\delta_0(x)$. As in the example before, $\delta_0(x)$ denotes the survival probability if no reinsurance is taken and $\delta^R(x)$ is the survival probability for the constant strategy $b_t = b^R$. From Theorem 4.37 we know that $\psi^R(x) \sim \psi(x)$ as $x \to \infty$. For moderate values of x, the asymptotic optimal fixed strategy yields the largest ruin probability. This shows that asymptotic optimality does not make

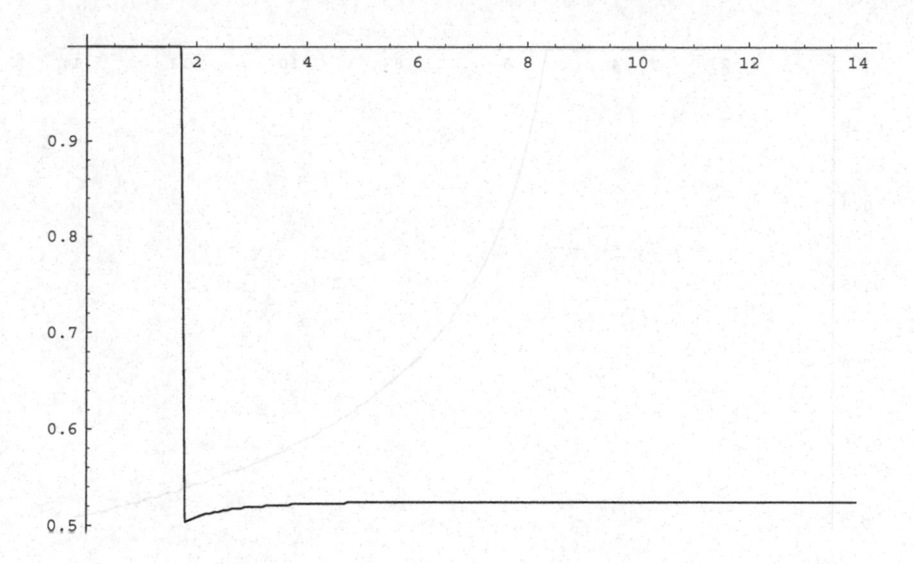

Fig. 2.2. Optimal strategy for exponentially distributed claim sizes.

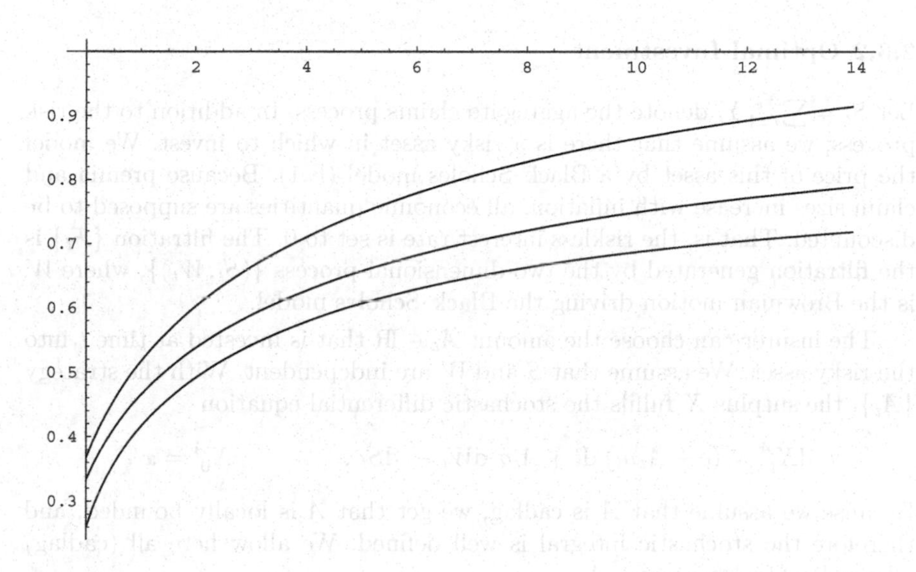

Fig. 2.3. $\delta(x)$ (top graph), $\delta^R(x)$ (bottom graph), and $\delta_0(x)$ (middle graph) for Pareto-distributed claim sizes.

sense for large claims. Clearly, $\delta^R(x) > \delta_0(x)$ for x large enough. The optimal strategy is given by Figure 2.4. In Theorem 4.37 it is shown that $b(x)$ converges to b^R. As we can see in Figure 2.4, the optimal $b(x)$ is at $x = 14$ still quite far from the asymptotically optimal b^R. ∎

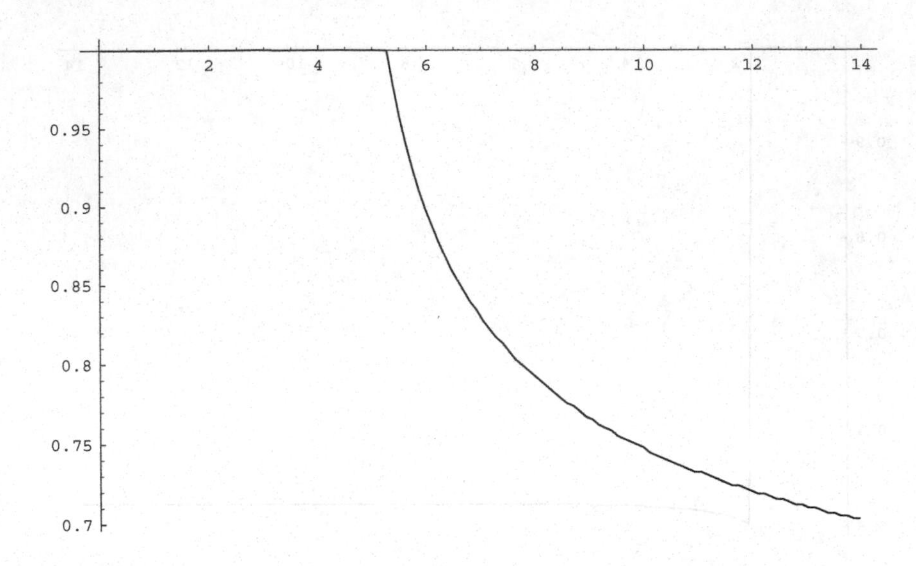

Fig. 2.4. Optimal strategy for Pareto-distributed claim sizes.

2.3.2 Optimal Investment

Let $S_t = \sum_{i=1}^{N_t} Y_i$ denote the aggregate claims process. In addition to the risk process, we assume that there is a risky asset in which to invest. We model the price of this asset by a Black–Scholes model (E.1). Because premia and claim sizes increase with inflation, all economic quantities are supposed to be discounted. That is, the riskless interest rate is set to 0. The filtration $\{\mathcal{F}_t\}$ is the filtration generated by the two-dimensional process $\{(S_t, W_t)\}$, where W is the Brownian motion driving the Black–Scholes model.

The insurer can choose the amount $A_t \in \mathbb{R}$ that is invested at time t into the risky asset. We assume that S and W are independent. With the strategy $\{A_t\}$, the surplus X fulfils the stochastic differential equation

$$\mathrm{d}X_t^A = (c + A_t m)\,\mathrm{d}t + A_t\sigma\,\mathrm{d}W_t - \mathrm{d}S_t\;, \qquad X_0^A = x\;.$$

Because we assume that A is cadlag, we get that A is locally bounded, and therefore the stochastic integral is well defined. We allow here all (cadlag) adapted control processes A.

As before we define the time of ruin by $\tau^A = \inf\{t : X_t^A < 0\}$. The ruin probability is $\psi^A(x)$, and the survival probability is $\delta^A(x) = 1 - \psi^A(x)$. The value function is then $\delta(x) = \sup_{A\in\mathfrak{U}} \delta^A(x)$.

If $c > \lambda\mu$, $\delta^0(0) > 0$. If $c \le \lambda\mu$, choose $A > (\lambda\mu - c)/m$. Use the strategy $A_t = A\mathbb{I}_{t\ge T_1\wedge 1}$. For this strategy we have $\delta^A(0) > 0$. Thus, $\delta(0) > 0$.

We first prove either that $X_t \to \infty$ or that ruin occurs. The proof is quite technical and may be skipped upon first reading.

Lemma 2.18. *Either* $\tau^A < \infty$ *or* $X_t^A \to \infty$.

Proof. Let $t \ge 0$ be some time point. Let $C_s = \sigma^2 \int_t^{t+s} A_v^2 \,\mathrm{d}v$ and $D_s = \inf\{v : C_v \ge s\}$. Then $\int_t^{t+D_s} \sigma A_v \,\mathrm{d}W_v$ is a standard Brownian motion. Consider now

$$X_{t+(D_1\wedge 1)} = X_t + c(D_1 \wedge 1) + m\int_t^{t+(D_1\wedge 1)} A_s \,\mathrm{d}s + \sigma \int_t^{t+(D_1\wedge 1)} A_s \,\mathrm{d}W_s$$
$$- (S_{t+(D_1\wedge 1)} - S_t)\,.$$

Clearly, $c(D_1 \wedge 1) \le c$ and

$$m\int_t^{t+(D_1\wedge 1)} A_s \,\mathrm{d}s \le m\int_t^{t+(D_1\wedge 1)} (1 + A_s^2)\,\mathrm{d}s \le 2m\,.$$

Because $\int_t^{t+D_s} \sigma A_v \,\mathrm{d}W_v$ is a standard Brownian motion, we get that

$$\mathbb{P}\Bigl[\sigma\int_t^{t+(D_1\wedge 1)} A_s \,\mathrm{d}W_s \le 1\Bigr] \ge \mathbb{P}\Bigl[\sup_{0\le s\le 1} W_s \le 1\Bigr] > 0\,.$$

Finally, we have

$$\mathbb{P}\Bigl[\sigma\int_t^{t+D_1} A_s \,\mathrm{d}W_s \le -c - 2m - 1\Bigr] = \frac{1}{\sqrt{2\pi}}\int_{-\infty}^{-c-2m-1} \mathrm{e}^{-y^2/2}\,\mathrm{d}y > 0$$

and

$$\mathbb{P}[S_{t+1} - S_t > c + 2m + 1] > 0\,.$$

We conclude that $\mathbb{P}[X_{t+(D_1\wedge 1)} - X_t \le -1] \ge \delta$ for some $\delta > 0$.

Choose $a > 0$. Now let $t_0 = 0$ and $t_{n+1} = \inf\{t : t \ge t_n + 1 : X_t \le a\}$. Define $s_n = t_n + (D_1^n \wedge 1)$, with D_s^n defined as D_s with $t = t_n$. Let $V_k = \mathbb{1}_{t_n<\infty, X_{s_n}\le a-1}$ and $Z_k = \delta \mathbb{1}_{t_k<\infty}$. It follows now as in the proof of Lemma 2.9 that $\liminf X_t < \infty$ implies that $\tau < \infty$. □

Similarly as in Lemma 2.10, we could prove that $\delta(x)$ is strictly increasing. We next will motivate the Hamilton–Jacobi–Bellman equation.

Let $h > 0$ and choose $A \in \mathbb{R}$ and $\varepsilon > 0$. Let $T = \tau^A \wedge T_1 \wedge h$. At time T a strategy $\tilde{A}$ is chosen such that $\delta^{\tilde{A}}(X_T^A) > \delta(X_T^A) - \varepsilon$. Then

$$\delta(x) \ge \delta^A(x) = \mathbb{E}[\delta^{\tilde{A}}(X_T^A); \tau^A > T] > \mathbb{E}[\delta(X_T^A); \tau^A > T] - \varepsilon\,.$$

The right-hand side does not depend on $\tilde{A}$. Thus, we can let $\varepsilon \downarrow 0$. Note that ruin does not occur if $\tau < T_1$, because there is a strategy for $x = 0$ such that ruin does not immediately occur. It is no loss of generality to suppose that $\tilde{A}$ is such a strategy. We therefore have $\mathbb{E}[\delta(X_T^A); \tau^A > T] = \mathbb{E}[\delta(X_T^A)]$. By Itô's formula,

$$\delta(X_{T-}^A) = \delta(x) + \int_0^T (\tfrac{1}{2}\sigma^2 A^2 \delta''(X_s^A) + (c+mA)\delta'(X_s^A))\,\mathrm{d}s + \int_0^T \sigma A\,\mathrm{d}W_s\,,$$

provided that $\delta(x)$ is twice continuously differentiable. Taking expectations and assuming that the stochastic integral is a martingale, we get

$$\mathbb{E}\Bigl[\int_0^T (\tfrac{1}{2}\sigma^2 A^2 \delta''(X_s^A) + (c+mA)\delta'(X_s^A))\,\mathrm{d}s + (\delta(X_T^A) - \delta(X_{T-}^A))\Bigr] \le 0\,. \tag{2.16}$$

We have that

$$\mathbb{P}[T_1 = T] = 1 - \mathrm{e}^{-\lambda h} - \mathbb{P}[\tau^A < T_1 \le h]\,.$$

If $x = 0$ and $A \neq 0$, then $T = 0$. Let us therefore only consider the case $x > 0$. From (A.3) it follows readily that $h^{-1}\mathbb{P}[\tau^A < T_1]$ tends to zero as $h \to 0$. If we can interchange the limit and integration, we will see that

$$\begin{aligned}\mathbb{E}\Bigl[h^{-1}\int_0^T (\tfrac{1}{2}\sigma^2 A^2 \delta''(X_s^A) + (c+mA)\delta'(X_s^A))\,\mathrm{d}s\Bigr]\\ \longrightarrow \tfrac{1}{2}\sigma^2 A^2 \delta''(x) + (c+mA)\delta'(x)\,.\end{aligned}$$

Moreover,

$$\begin{aligned}&h^{-1}\mathbb{E}[\delta(X_T^A) - \delta(X_{T-}^A)]\\ &\quad= h^{-1}\mathbb{E}\Bigl[\mathbb{1}_{T_1=T}\Bigl\{\int_0^{X_{T-}^A} \delta(X_{T-}^A - y)\,\mathrm{d}G(y) - \delta(X_{T-}^A)\Bigr\}\Bigr]\\ &\quad\longrightarrow \lambda\Bigl[\int_0^x \delta(x-y)\,\mathrm{d}G(y) - \delta(x)\Bigr]\,.\end{aligned}$$

Dividing the inequality (2.16) by h and letting $h \downarrow 0$ yields

$$\tfrac{1}{2}\sigma^2 A^2 \delta''(x) + (c+mA)\delta'(x) + \lambda\Bigl[\int_0^x \delta(x-y)\,\mathrm{d}G(y) - \delta(x)\Bigr] \le 0\,.$$

A is arbitrary. This motivates the Hamilton–Jacobi–Bellman equation

$$\sup_{A\in\mathbb{R}}\Bigl\{\tfrac{1}{2}\sigma^2 A^2 f''(x) + (c+mA)f'(x) + \lambda\Bigl[\int_0^x f(x-y)\,\mathrm{d}G(y) - f(x)\Bigr]\Bigr\} = 0\,. \tag{2.17}$$

We can let $f(0) = 1$. An optimal strategy, if it exists, is expected to be of the form $\{A(X_t)\}$. In order for $\delta(0) > 0$, we need $A(x) \to 0$ as $x \downarrow 0$. This yields $f'(0) = \lambda/c$.

If $f''(x) > 0$ or $f''(x) = 0$ on some interval, then there is no solution to (2.17). Thus, we suppose that $f''(x) < 0$. The supremum is attained in

$$A^*(x) = -\frac{mf'(x)}{\sigma^2 f''(x)}\,.$$

The equation to solve then becomes

$$-\frac{m^2 f'(x)^2}{2\sigma^2 f''(x)} + cf'(x) + \lambda\Big[\int_0^x f(x-y)\,\mathrm{d}G(y) - f(x)\Big] = 0\,. \tag{2.18}$$

Note that because $A^*(0) = 0$, we need $f''(0) = -\infty$.

Before proving the existence of a solution, we prove the *verification theorem.*

Theorem 2.19. *Suppose that $f(x)$ is an increasing, twice continuously differentiable solution to (2.17). Then $f(x)$ is bounded and $f(x) = \delta(x)/\delta(0)$. The strategy of feedback form $A_t^* = A^*(X_t^*)$ is an optimal strategy.*

Proof. Let A be an arbitrary strategy and let X^* denote the surplus process if $A_t = A_t^*$. Choose $n > x$. Define $T = T_n = \inf\{t : X_t \notin [0,n]\}$. Note that $X_{T\wedge t} \in (-\infty, n]$ because the jumps are downwards. The process

$$M_t^1 = \sum_{i=1}^{N_{T\wedge t}} (f(X_{T_i}) - f(X_{T_i-})) - \lambda\int_0^{T\wedge t}\Big(\int_0^{X_s} f(X_s - y)\,\mathrm{d}G(y) - f(X_s)\Big)\,\mathrm{d}s$$

is a martingale; see [26, p. 27]. We write

$$\begin{aligned} f(X_{T\wedge t}) = f(x) + f(X_{T\wedge t}) - f(X_{T_{N_{T\wedge t}}}) + \sum_{i=1}^{N_{T\wedge t}} (f(X_{T_i-}) - f(X_{T_{i-1}})) \\ + M_t^1 + \lambda\int_0^{T\wedge t}\Big(\int_0^{X_s} f(X_s - y)\,\mathrm{d}G(y) - f(X_s)\Big)\,\mathrm{d}s\,. \end{aligned}$$

By Itô's formula,

$$\begin{aligned} f(X_{T_i-}) - f(X_{T_{i-1}}) = \int_{T_{i-1}}^{T_i} \tfrac{1}{2}\sigma^2 A_s^2 f''(X_s) + (c + mA_s)f'(X_s)\,\mathrm{d}s \\ + \int_{T_{i-1}}^{T_i} \sigma A_s f'(X_s)\,\mathrm{d}W_s\,. \end{aligned}$$

The corresponding result holds for $f(X_{T\wedge t}) - f(X_{T_{N_{T\wedge t}}})$. Thus,

$$\begin{aligned} f(X_{T\wedge t}) = f(x) + \int_0^{T\wedge t}\Big(\tfrac{1}{2}\sigma^2 A_s^2 f''(X_s) + (c + mA_s)f'(X_s) \\ + \lambda\Big[\int_0^{X_s} f(X_s - y)\,\mathrm{d}G(y) - f(X_s)\Big]\Big)\,\mathrm{d}s \\ + \int_0^{T\wedge t} \sigma A_s f'(X_s)\,\mathrm{d}W_s + M_t^1\,. \end{aligned}$$

By (2.17) we find that

$$f(X_{T\wedge t}) \le f(x) + \int_0^{T\wedge t} \sigma A_s f'(X_s)\,\mathrm{d}W_s + M_t^1\,,$$

and equality holds for X^*. Let $\{S_m\}$ be a localisation sequence of the stochastic integral, and set $\mathcal{T}_n^m = \mathcal{T}_n \wedge S_m$. Taking expected values yields

$$\mathbb{E}[f(X_{\mathcal{T}_n^m \wedge t})] \le f(x)\,.$$

By bounded convergence, letting $m \to \infty$, and then $t \to \infty$, $\mathbb{E}[f(X_{\mathcal{T}_n})] \le f(x)$. Thus, for $f(0) = 1$,

$$\mathbb{P}[\tau < \mathcal{T}_n, X_\tau = 0] + f(n)\mathbb{P}[\mathcal{T}_n < \tau] = \mathbb{E}[f(X_{\mathcal{T}_n})] \le f(x)\,.$$

Note that $\mathbb{P}[\mathcal{T}_n < \tau] \ge \delta^A(x)$. Because there is a strategy with $\delta^A(x) > 0$, it follows that $f(x)$ is bounded. We can therefore let $n \to \infty$, yielding $\mathbb{E}[f(X_\tau)] \le f(x)$. In particular, we get

$$\delta^A(x) f(\infty) \le \delta^A(x) f(\infty) + \mathbb{P}[\tau < \infty, X_\tau = 0] \le f(x)\,.$$

For X^* we obtain an equality. In particular, $\{f(X_t^*)\}$ is a martingale. It remains to show that $\mathbb{P}[X_\tau^* \ne 0] = 1$. Note first from (2.17) that $G(y)$ must be continuous. Otherwise, the integral in (2.17) is not continuous. Choose $\varepsilon > 0$. Consider the strategy $A(x) = A^*(x)\mathbb{1}_{x \ge \varepsilon}$. Let $T_\varepsilon = \inf\{t : X_t^* < \varepsilon\}$. By the martingale property, $f(x) = f(\infty)\mathbb{P}[T_\varepsilon = \infty] + \mathbb{E}[f(X_{T_\varepsilon}); T_\varepsilon < \tau < \infty]$. The last term is bounded by $f(\varepsilon)\mathbb{P}[T_\varepsilon < \tau < \infty]$. Because $G(y)$ is continuous, it must converge to zero as $\varepsilon \to 0$. Because $\mathbb{P}[T_\varepsilon = \infty] \to \delta^*(x)$, it follows that $f(x) = \delta^*(x) f(\infty)$. Thus, $\delta^*(x) = \delta(x) = f(x)/f(\infty)$. □

We now need to show that there is an increasing, twice continuously differentiable solution to (2.17). Such a solution cannot exist in general. We therefore now assume that $G(x)$ is absolutely continuous and has a bounded density.

Remark 2.20. If $G(x)$ does not have a bounded density, then one has to consider so-called weak solutions. However, Itô's formula no longer applies. One therefore has to use the concept of viscosity solutions; see Section 2.5.3, [61], or [182]. ■

As above, we suppose that $f(0) = 1$ and $f'(0) = \lambda/c$. For the calculations below we let $c = \lambda = 1$. This can always be obtained by rescaling the time and surplus. Because (2.18) depends on m and σ only through m^2/σ^2 it is no loss of generality to let $\sigma = 1$. Let $g(x) = f'(x)$. Using integration by parts, we can rewrite (2.18) as

$$-\frac{m^2 g(x)^2}{2g'(x)} + g(x) - (1 - G(x)) - \int_0^x g(x-z)(1 - G(z))\,\mathrm{d}z = 0\,. \qquad (2.19)$$

We first show that there is a solution close to zero.

Lemma 2.21. *Suppose that $G(x)$ has a bounded density. Then there exists $\kappa > 0$ such that there is a solution to (2.18) on $[0, \kappa)$. Moreover, $f'(x) = \lambda/c - \lambda m\sqrt{x}/(\sigma c^3) + o(\sqrt{x})$ as $x \downarrow 0$.*

Proof. Let $h(x) = g(x^2)$. Then $h(x)$ has to fulfil

$$h'(x) = \frac{m^2 x h(x)^2}{h(x) - (1 - G(x^2)) - 2\int_0^x z h(z)(1 - G(x^2 - z^2))\,dz}\,.$$

Let $Q(v) = \sup_{0<x<\kappa} x^{-1}|v'(x) - v'(0)|$ and $L = \{v \in C^1[0, \kappa] : Q(v) < \infty\}$. Endowed with the norm $\|v\| := \max\{\|v\|_\infty, |v'(0)|, \kappa Q(v)\}$, the space L is a Banach space. $\|\cdot\|_\infty$ denotes the supremum norm. Consider the closed set

$$D_M = \{v \in L : v(0) = 1, v'(0) = -m, \|v - 1\|_\infty \le 1/3, Q(v) \le M\}\,.$$

On D_M we consider the operator

$$\mathcal{V}v(x) := 1 + \int_0^x \frac{m^2 y v(y)^2}{v(y) - (1 - G(y^2)) - 2\int_0^y z v(z)(1 - G(y^2 - z^2))\,dz}\,dy\,.$$

If κ is small enough and M is large enough, then D_M is mapped into itself and the operator $\mathcal{V}$ is a contraction. Thus, there is a fixed point $h(x)$. Because $h(x)$ belongs to D_M, we have $h(x) \ge 2/3$ and the derivative is bounded. The derivative can only become positive if the numerator in the integral becomes zero at some point. But this cannot happen, because the derivative is bounded. Thus, $h'(x) < 0$ on $[0, \kappa]$. The function $g(x) = h(\sqrt{x})$ is then the desired solution and the asymptotics at zero follows. □

The disadvantage of the above approach is that we need both the function $g(x)$ and its derivative. Let us now reformulate (2.18). We can write (with the choice of parameters above)

$$-\frac{g'(x)}{g(x)^2} = \frac{m^2}{2}\,\frac{1}{\int_0^x g(x - z)(1 - G(z))\,dz + 1 - g(x) - G(x)}\,.$$

Integration yields

$$\frac{1}{g(x)} = 1 + \frac{m^2}{2}\int_0^x \frac{dy}{\int_0^y g(y - z)(1 - G(z))\,dz + 1 - g(y) - G(y)}\,.$$

Thus,

$$g(x) = \frac{2}{2 + m^2\int_0^x \{\int_0^y g(y - z)(1 - G(z))\,dz + 1 - g(y) - G(y)\}^{-1}\,dy}\,. \tag{2.20}$$

Note that if there is a decreasing, strictly positive solution to (2.20), we also have an increasing, twice continuously differentiable solution to (2.18) and

(2.17) holds. By Lemma 2.21 there is an interval $[0, \kappa)$ on which (2.20) admits a decreasing strictly positive solution. We will not have problems extending the solution as long as

$$\int_0^x g(x - z)(1 - G(z)) \,\mathrm{d}z + 1 - g(x) - G(x) > 0 \,. \tag{2.21}$$

By the proof of Lemma 2.21, there is an interval $(0, \kappa)$ on which (2.21) is fulfilled.

Lemma 2.22. *Assume that $G(y)$ has a bounded density. Suppose there is $\kappa > 0$ such that there exists a solution to (2.20) on $[0, \kappa)$ such that (2.21) is fulfilled on $(0, \kappa)$. Then the solution can be extended to $[0, \kappa]$, $g(\kappa) > 0$ and (2.21) is also fulfilled at κ.*

Proof. Clearly, $g(x)$ is decreasing and $g(x) > 0$ on $[0, \kappa)$. Therefore, we can extend $g(x)$ to $[0, \kappa]$, interpreting $1/\infty$ as zero. If we can show that (2.21) is fulfilled at κ, we get that $g(\kappa) > 0$. Suppose that

$$\int_0^\kappa g(z)(1 - G(\kappa - z)) \,\mathrm{d}z + 1 - g(\kappa) - G(\kappa) = 0 \,.$$

We obtain $g'(\kappa) = -\infty$. The difference between (2.21) and the above-displayed equation divided by $\kappa - x$ gives

$$\int_0^\kappa \frac{G(\kappa - z) - G(x - z)}{\kappa - x} g(z) \,\mathrm{d}z + \frac{g(\kappa) - g(x)}{\kappa - x} + \frac{G(\kappa) - G(x)}{\kappa - x} > 0 \,.$$

Because $G(y)$ has a bounded density and $g(x)$ is bounded, as $x \uparrow \kappa$, we get $-\infty \ge 0$. This is a contradiction and (2.21) is fulfilled at κ. □

Now we can prove the existence of a solution.

Theorem 2.23. *Assume that $G(y)$ has a bounded density. Then there is a unique increasing, twice continuously differentiable solution to* (2.17).

Proof. Uniqueness follows from the verification theorem 2.19. Let κ be the largest value such that there is a solution to (2.20) on $[0, \kappa)$ and (2.21) is fulfilled on $(0, \kappa)$. From Lemma 2.21 we know that $\kappa > 0$. Suppose that $\kappa < \infty$. By Lemma 2.22, the property holds also at κ. We choose the parameters as above. Let

$$v_0 := 2 + m^2 \int_0^\kappa \frac{\mathrm{d}y}{\int_0^y g(y - z)(1 - G(z)) \,\mathrm{d}z + 1 - g(y) - G(y)}$$

and

$$\xi := \tfrac{1}{2}\Bigl(\int_0^\kappa g(\kappa - z)(1 - G(z)) \,\mathrm{d}z + 1 - g(\kappa) - G(\kappa)\Bigr) > 0 \,.$$

Consider the operator

$$\mathcal{V}h(x) = \frac{2}{v_0 + m^2 \int_\kappa^x \{[\int_0^y h(y-z)(1-G(z))\,\mathrm{d}z + 1 - h(y) - G(y)] \vee \xi\}^{-1}\,\mathrm{d}y}$$

acting on positive, continuous, and decreasing functions $h(u)$ on $[0, \kappa+\zeta)$ with $h(x) = g(x)$ on $[0, \kappa]$ and $h(x) \ge 2(v_0 + m^2(x-\kappa)/\xi)^{-1}$ on $[\kappa, \kappa+\zeta)$. ζ is to be chosen later. Let $h_i(x)$, $i = 1, 2$, be two functions in the space on which $\mathcal{V}$ is defined. To simplify the notation, let

$$I_i(x) = \Bigl[\int_0^x h_i(x-z)(1-G(z))\,\mathrm{d}z + 1 - h_i(x) - G(x)\Bigr] \vee \xi\,.$$

Then for $x > \kappa$,

$$|\mathcal{V}h_1(x) - \mathcal{V}h_2(x)| \le \frac{2m^2 \int_\kappa^x |I_1^{-1}(y) - I_2^{-1}(y)|\,\mathrm{d}y}{(v_0 + m^2 \int_\kappa^x I_1^{-1}(y)\,\mathrm{d}y)(v_0 + m^2 \int_\kappa^x I_2^{-1}(y)\,\mathrm{d}y)}\,.$$

The integral in the numerator can be estimated by

$$\begin{aligned}\int_\kappa^x |I_1^{-1}(y) - I_2^{-1}(y)|\,\mathrm{d}y &\le \xi^{-2} \int_\kappa^x |I_1(y) - I_2(y)|\,\mathrm{d}y \\ &\le \xi^{-2}[(x-\kappa) + \tfrac{1}{2}(x-\kappa)^2] \sup_{\kappa \le y \le \kappa+\zeta} |h_1(y) - h_2(y)|\,.\end{aligned}$$

Therefore, if ζ is small enough, we get that $\mathcal{V}$ is a contraction. Thus, there is a fixed point $h(x)$. By continuity, we can choose ζ small enough such that

$$\int_0^x h(x-z)(1-G(z))\,\mathrm{d}z + 1 - h(x) - G(x) \ge \xi$$

for $x \in [\kappa, \kappa+\zeta]$. This shows that there is a solution to (2.20) on $[0, \kappa+\zeta)$ and (2.21) is fulfilled. This contradicts our assumption that $[0, \kappa)$ is the largest such interval. This proves that $\kappa = \infty$. □

In the following two examples we choose $c = \lambda = \mu = 1$, $m = 0.04$, and $\sigma^2 = 0.01$. Note that the classical net profit condition is not fulfilled. Hence, if investment was not allowed, we would have $\delta^0(x) = 0$, i.e., ruin would occur almost surely. As in Section 2.3.1 the solutions are calculated in the following way. A first version of $g(x)$ is found by an Euler scheme from (2.19). In order to initialise the solution, the function $g(x)$ is chosen as suggested by Lemma 2.21 for x close to zero. Thereafter, (2.20) is iterated until the optimal strategies stabilise.

Example 2.24. Consider exponentially distributed claim sizes $G(x) = 1 - \mathrm{e}^{-x}$. The ruin probability decreases exponentially fast (Figure 2.5). The optimal strategy is given in Figure 2.6. For small x the optimal investment behaves as stated in Lemma 2.21. The investment converges to the asymptotically optimal level; see Corollary 4.27. ■

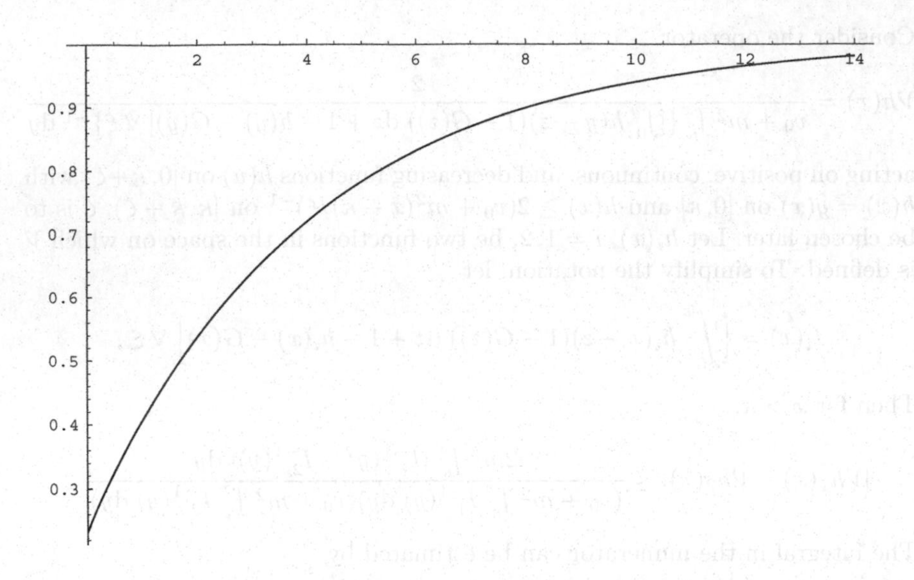

Fig. 2.5. $\delta(x)$ for exponentially distributed claim sizes.

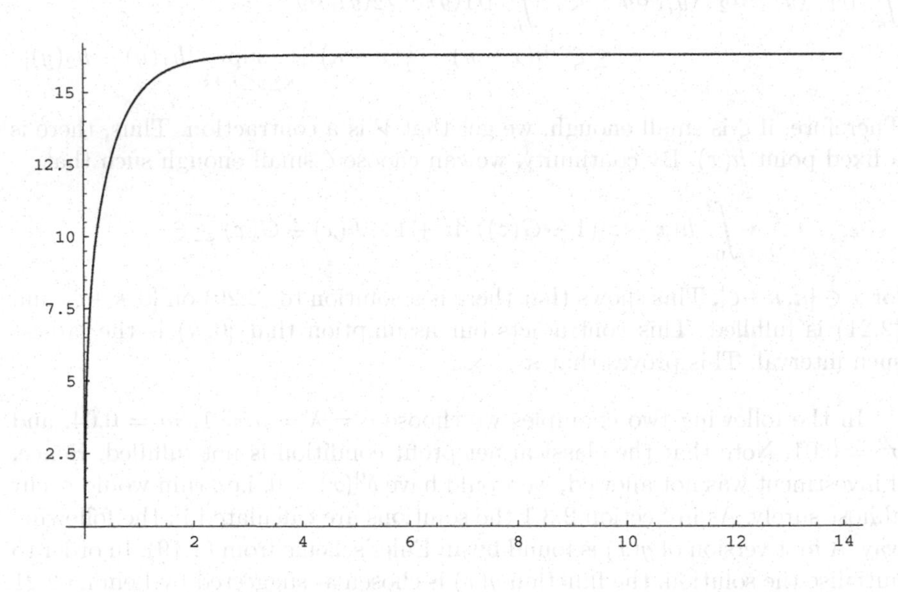

Fig. 2.6. The strategy $A(x)$ for exponentially distributed claim sizes.

Example 2.25. Consider Pareto-distributed claim sizes, $G(x) = 1 - 1/(1+x)^2$. Figure 2.7 indicates that the ruin probability is decreasing slowly. The optimal strategy is given in Figure 2.8. Also, for small x the optimal investment here behaves as stated in Lemma 2.21. For larger x the slope of $A(x)$ is close to a constant. However, $A(14)/14$ is approximately 2.14, where the asymptotic

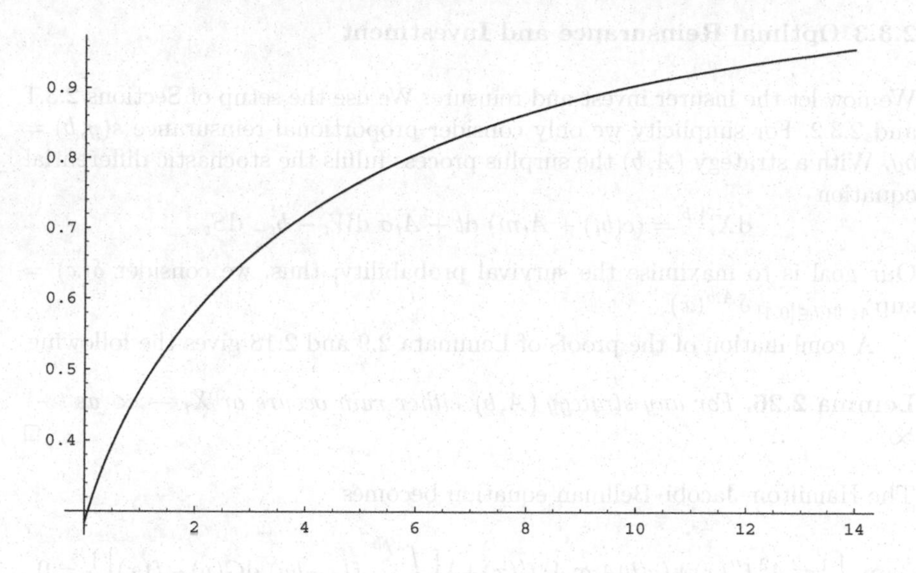

Fig. 2.7. $\delta(x)$ for Pareto-distributed claim sizes.

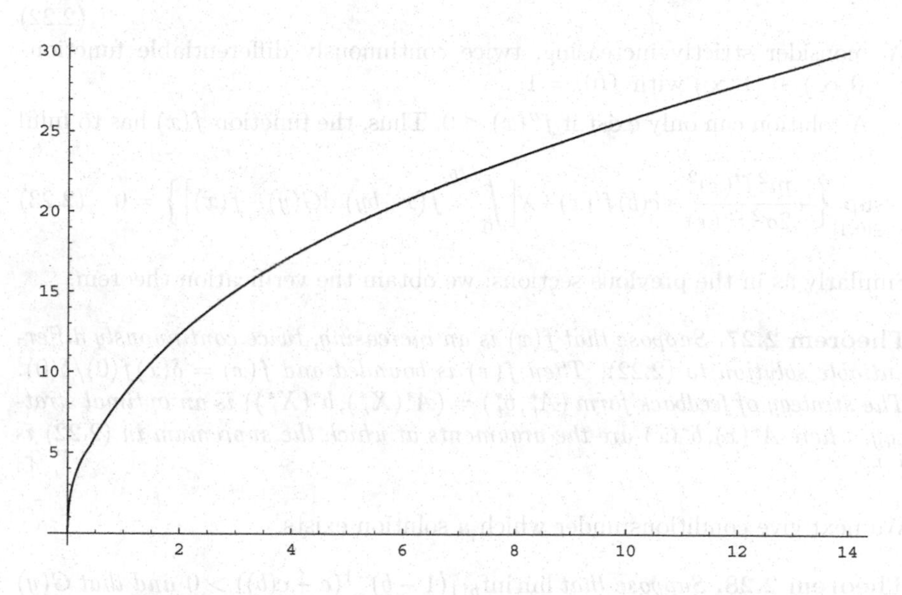

Fig. 2.8. Strategy $A(x)$ for Pareto-distributed claim sizes.

value $\lim_{x\to\infty} A(x)/x$ is approximately 1.33; see Corollary 4.52. We see also here that we are at $x = 14$, still quite far from the asymptotically optimal value. ∎

2.3.3 Optimal Reinsurance and Investment

We now let the insurer invest and reinsure. We use the setup of Sections 2.3.1 and 2.3.2. For simplicity we only consider proportional reinsurance $s(y,b) = by$. With a strategy (A,b) the surplus process fulfils the stochastic differential equation

$$\mathrm{d}X_t^{A,b} = (c(b_t) + A_t m)\,\mathrm{d}t + A_t\sigma\,\mathrm{d}W_t - b_{t-}\,\mathrm{d}S_t\,.$$

Our goal is to maximise the survival probability; thus, we consider $\delta(x) = \sup_{A\in\mathbb{R},b\in[0,1]}\delta^{A,b}(x)$.

A combination of the proofs of Lemmata 2.9 and 2.18 gives the following

Lemma 2.26. *For any strategy (A,b) either ruin occurs or $X_t \to \infty$ as $t \to \infty$.* □

The Hamilton–Jacobi–Bellman equation becomes

$$\sup_{\substack{A\in\mathbb{R}\\ b\in[0,1]}}\Bigl\{\tfrac{1}{2}\sigma^2A^2f''(x)+(c(b)+mA)f'(x)+\lambda\Bigl[\int_0^{x/b} f(x-by)\,\mathrm{d}G(y)-f(x)\Bigr]\Bigr\} = 0\,. \tag{2.22}$$

We consider strictly increasing, twice continuously differentiable functions $f : [0,\infty) \to [1,\infty)$ with $f(0) = 1$.

A solution can only exist if $f''(x) < 0$. Thus, the function $f(x)$ has to fulfil

$$\sup_{b\in[0,1]}\Bigl\{-\frac{m^2f'(x)^2}{2\sigma^2f''(x)}+c(b)f'(x)+\lambda\Bigl[\int_0^{x/b} f(x-by)\,\mathrm{d}G(y)-f(x)\Bigr]\Bigr\} = 0\,. \tag{2.23}$$

Similarly as in the previous sections, we obtain the verification theorem.

Theorem 2.27. *Suppose that $f(x)$ is an increasing, twice continuously differentiable solution to (2.22). Then $f(x)$ is bounded and $f(x) = \delta(x)f(0)/\delta(0)$. The strategy of feedback form $(A_t^*, b_t^*) = (A^*(X_t^*), b^*(X_t^*))$ is an optimal strategy, where $A^*(x), b^*(x)$ are the arguments at which the supremum in (2.22) is taken.* □

We next give conditions under which a solution exists.

Theorem 2.28. *Suppose that $\liminf_{b\uparrow 1}(1-b)^{-1}(c-c(b)) > 0$ and that $G(y)$ has a bounded density. Then there exists a unique increasing, twice continuously differentiable solution to (2.22).*

Proof. From the proof of Lemma 2.14, we know that $b^*(x) = 1$ for x small enough. Note that in (2.14) and (2.23) the terms dependent on b coincide. Therefore, there is a solution to (2.22) on some interval $[0,\kappa)$. As in the proofs of Section 2.3.2, we can assume that $\lambda = \sigma = 1$. Let $g(x) = f'(x)$ and

$$D(x,b) = \int_0^x g(z)(1 - G((x-z)/b))\,\mathrm{d}z + 1 - c(b)g(x) - G(x/b)\,.$$

Then we obtain

$$g(x) = \frac{2}{2 + m^2 \int_0^x \{\inf_{b\in[0,1]} D(y,b)\}^{-1}\,\mathrm{d}y}\,.$$

We show that the solution can be extended to $[0,\kappa]$ and that $\inf_b D(\kappa,b) > 0$. Clearly, $g(\kappa)$ is well defined. Suppose that $\inf_b D(\kappa,b) = 0$. Let x_n be a sequence converging to κ and $b_n = b^*(x_n)$. By choosing a subsequence we can assume that b_n converges to some b_0. Note that by continuity $D(\kappa,b_0) = \inf_b D(\kappa,b) = 0$. Suppose first that $b_0 > 0$. This is only possible if $c(b_0) > 0$ and $g(\kappa) > 0$ because the integral is strictly positive. The same argument as in the proof of Lemma 2.22 yields a contradiction. Thus, $\lim_{x\uparrow\kappa} b^*(x) = 0$ and $g(\kappa) = 0$. Let $\underline{b} = \inf\{b : c(b) \ge 0\}$. There is a x_1 such that $b^*(x) < \underline{b}/2$ for $x > x_1$. Then

$$\frac{1}{g(x)} < \gamma + \frac{m^2}{2}\int_{x_1}^x \frac{1}{-c(\underline{b}/2)g(y)}\,\mathrm{d}y$$

for $x > x_1$ and

$$\gamma = 1 + \frac{m^2}{2}\int_0^{x_1} \{\inf_{b\in[0,1]} D(y,b)\}^{-1}\,\mathrm{d}y\,.$$

By Gronwall's inequality (see [57, p. 498]), we find that

$$\frac{1}{g(x)} < \gamma \mathrm{e}^{\zeta(x-x_1)}$$

for some ζ. This contradicts $g(\kappa) = 0$. The rest of the proof follows now similarly as the proof of Theorem 2.23. □

In the following two examples we choose $\lambda = \mu = 1$, $m = 0.04$, and $\sigma^2 = 0.01$. The premium income is $c(b) = 1.2b - 0.2$. Note that the net profit condition is not fulfilled. Hence, if investment were not possible, we would have $\delta^{0,b}(x) = 0$, i.e., ruin would occur almost surely. The numerical solutions are obtained in the same way as in Section 2.3.2.

Example 2.29. Consider exponentially distributed claim sizes $G(x) = 1 - \mathrm{e}^{-x}$. The ruin probability decreases exponentially fast (Figure 2.9). The optimal strategy is given in Figures 2.10 and 2.11. As we know from the theory, there is no reinsurance for x small. In this area the optimal investment behaves like that stated in Lemma 2.21. Then reinsurance is bought, and the investment is reduced. The retention level and the amount invested converge to the asymptotically optimal levels; see Corollary 4.35. ■

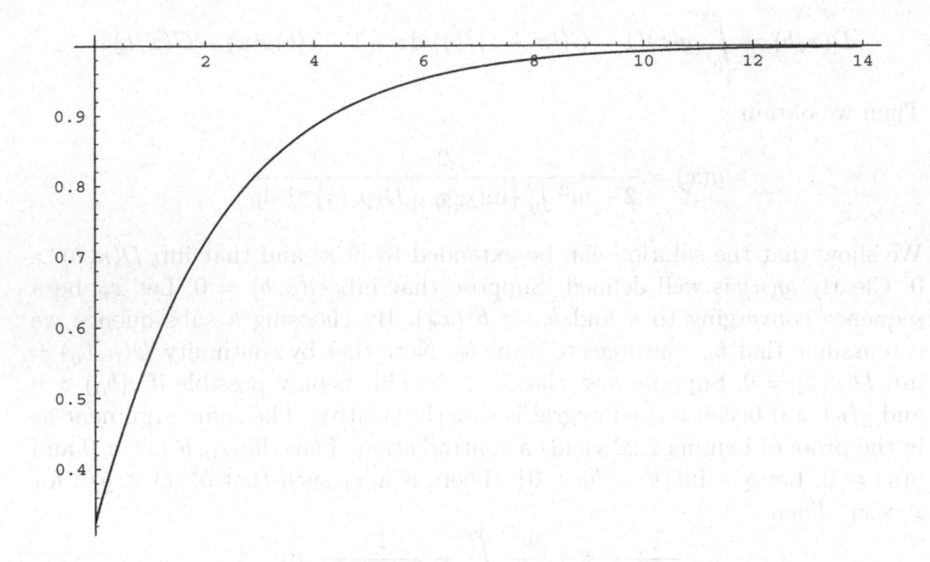

Fig. 2.9. $\delta(x)$ for exponentially distributed claim sizes.

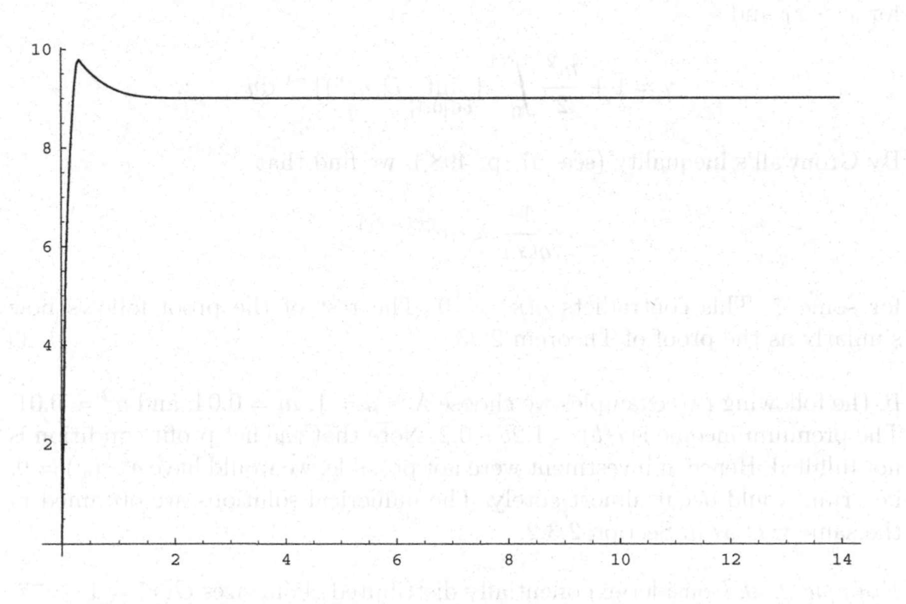

Fig. 2.10. Strategy $A(x)$ for exponentially distributed claim sizes.

Example 2.30. Consider Pareto-distributed claim sizes $G(x) = 1 - 1/(1+x)^2$. Figure 2.12 indicates also that here the ruin probability is decreasing exponentially fast. This is not surprising because choosing a fixed strategy with $b = 0$ such that $Am + c(0) > 0$ gives an exponentially decreasing ruin probability; see (A.2). The optimal strategy is given in Figures 2.13 and 2.14. Also, no

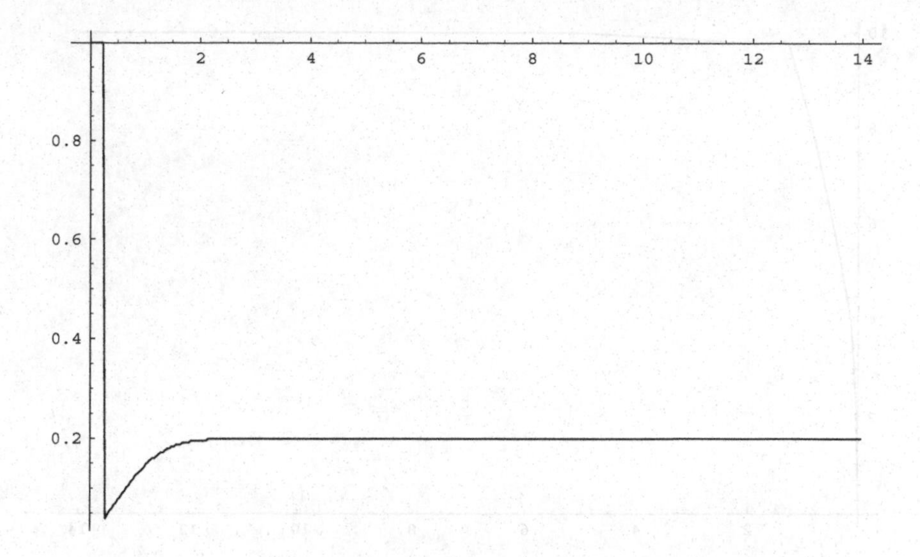

Fig. 2.11. Strategy $b(x)$ for exponentially distributed claim sizes.

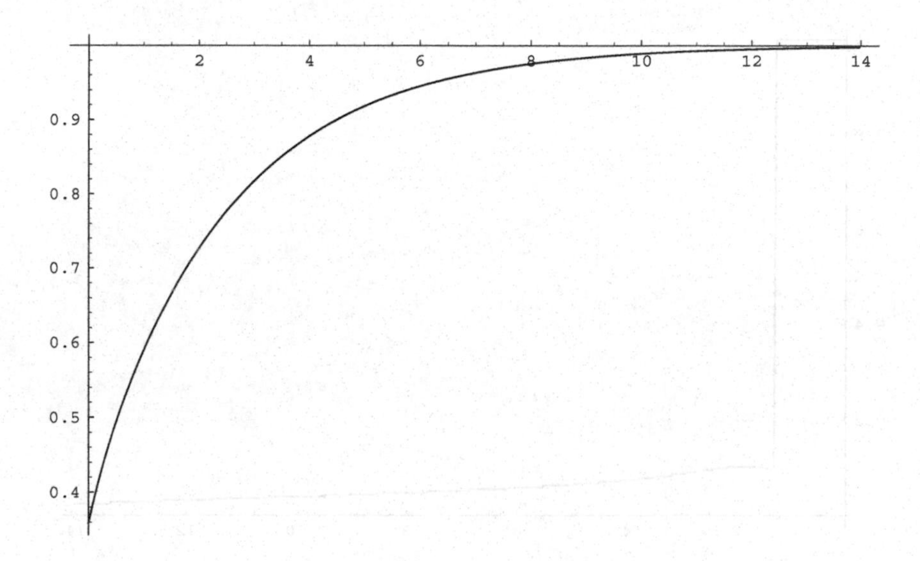

Fig. 2.12. $\delta(x)$ for Pareto-distributed claim sizes.

reinsurance is taken here for x small. The investment strategy for these values of x fulfils the properties stated in Lemma 2.21. Then reinsurance is taken, and from this point the investment strategy no longer changes much. The reinsurance strategy converges to zero (Corollary 4.59). Also, the investment strategy converges to an asymptotically optimal value; see Corollary 4.60. ■

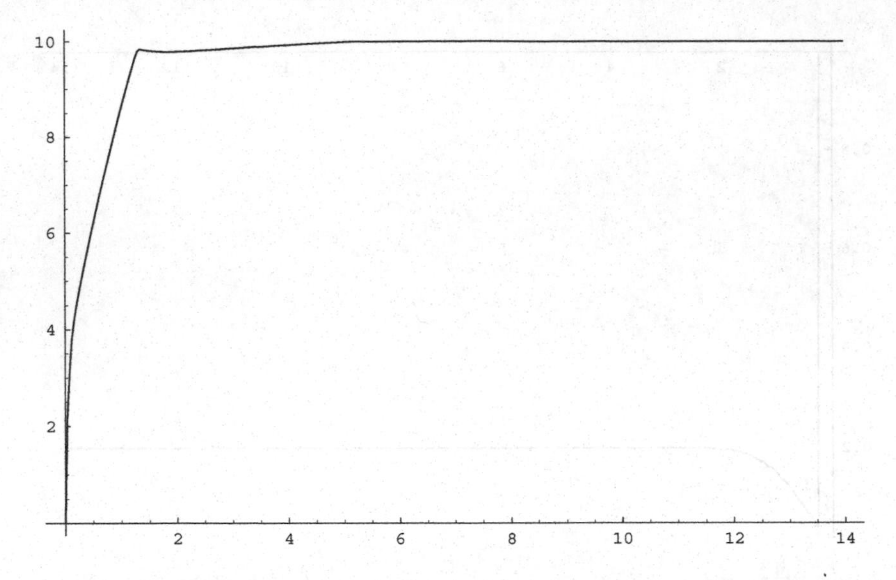

Fig. 2.13. Strategy $A(x)$ for Pareto-distributed claim sizes.

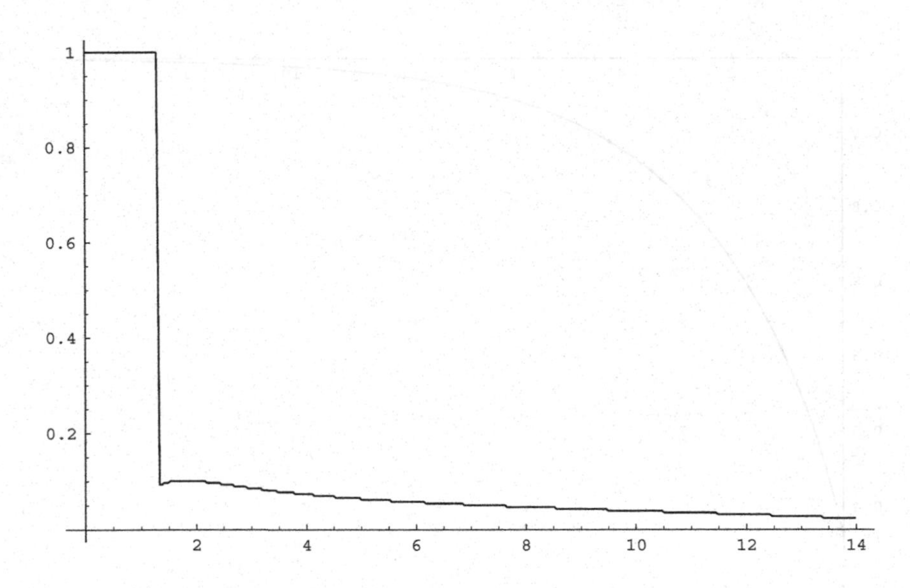

Fig. 2.14. Strategy $b(x)$ for Pareto-distributed claim sizes.

Bibliographical Remarks

Waters [185] considered the problem of minimising the ruin probability in an asymptotic sense and maximised the adjustment coefficient; see also Section 4.1. Section 2.3.1 is taken from [160]. The special case of excess of loss reinsurance is treated in Hipp and Vogt [101]; see also Vogt [184] and Hipp [96]. Section 2.3.2 follows Hipp and Plum [97]. In particular, the proof of Lemma 2.21 is taken from [97]. An alternative proof under slightly stronger assumptions can be found in Hipp and Plum [98]. Their proof is based on a contraction argument and works also in the case of a Markov modulated risk model, where the initial condition is not known. Theorem 2.23 is proved in a similar way. However, the proof of [97] cannot be directly repeated for the case with investment and reinsurance. Section 2.3.3 is from [161]. Hipp and Taksar [100] (see also Hipp [96]) considered minimal ruin probabilities where parts of another portfolio (Cramér–Lundberg model) can be purchased. In addition, they considered the problem where it is only possible to buy part of the other portfolio but not to sell it.

2.4 Optimal Dividends in the Classical Risk Model

In this section we consider the problem of Section 1.2 in a continuous setup. We use the notation of Appendix D. We suppose that ex-dividend the surplus of an insurance portfolio follows a classical Cramér–Lundberg model as in Section D.1. That is, there is a linear premium income. Claims arrive according to a Poisson process with intensity λ, and the claim sizes are iid with distribution function $G(y)$. Claims are positive, i.e., $G(0) = 0$. In order to prevent some minor technical difficulties, we assume that the claim size distribution $G(x)$ is continuous. The filtration $\{\mathcal{F}_t\}$ is the filtration generated by the aggregate claims process $\{\sum_{i=1}^{N_t} Y_i\}$.

When minimising the ruin probability, we were able to show that there is a solution to the Hamilton–Jacobi–Bellman equation. Through the verification theorem it followed then that this solution was the value function and therefore unique. Unfortunately, for the present problem uniqueness does not hold. Hence, our problem is that we cannot be sure whether a solution we may find is the value function. Moreover, we cannot be sure either that the value function solves the Hamilton–Jacobi–Bellman equation. Hence, we need to show directly that the value function solves the Hamilton–Jacobi–Bellman equation. We then have to characterise the value function among all solutions in order that one can be sure that a certain solution to the Hamilton–Jacobi–Bellman equation really is the value function. A direct verification that the value function solves the Hamilton–Jacobi–Bellman equation seems to be difficult. We therefore first restrict the dividend payment processes to absolutely continuous processes with a bounded density. Letting the maximal dividend rate converge to infinity, we will obtain the value function of the general problem and be able to show that it solves the Hamilton–Jacobi–Bellman equation. For a diffusion approximation the optimal dividend payment will be singular (with respect to the Lebesgue measure). The problem is therefore sometimes called a "singular control problem."

Another problem is that the value function is not necessarily differentiable. Indeed, differentiability may not hold for even nice claim size distributions; see the examples at the end of this section. The approach below will give us the property that the density of the value function can only have upward jumps. This will simplify the numerical solution of the problem.

2.4.1 Restricted Dividend Payments

We first assume that the dividends are paid at rate U_t, with $0 \le U_t \le u_0 < \infty$. We allow all dividend rate processes $\{U_t\}$ that are adapted and cadlag. The surplus process of the contract is

$$\Bigl\{X_t^U = x + ct - \sum_{i=1}^{N_t} Y_i - \int_0^t U_s \,\mathrm{d}s\Bigr\}\,.$$

Dividends can only be paid until the time of ruin $\tau^U = \inf\{t : X_t^U < 0\}$. The value of the dividends becomes

$$V^U(x) = \mathbb{E}\Bigl[\int_0^{\tau^U} \mathrm{e}^{-\delta t} U_t \,\mathrm{d}t\Bigr]\,,$$

and the value function is $V(x) = \sup_{U\in\mathfrak{U}} V^U(x)$, where $\mathfrak{U}$ consists of all positive cadlag processes bounded by u_0.

Lemma 2.31. *The function $V(x)$ is bounded by u_0/δ, is increasing, Lipschitz continuous, and therefore absolutely continuous, and converges to u_0/δ as $x \to \infty$.*

Proof. Using the same strategy for two initial capitals shows that $V(x)$ is increasing. Clearly, $V(x) \le \int_0^\infty \mathrm{e}^{-\delta t} u_0 \,\mathrm{d}t = u_0/\delta$. Consider the strategy $U_t = u_0$. As $x \to \infty$, the ruin time τ^U converges to infinity. By bounded convergence, $\mathbb{E}[\mathrm{e}^{-\delta\tau^U}]$ converges to zero. Thus, $\mathbb{E}[\int_0^{\tau^U} \mathrm{e}^{-\delta t} u_0 \,\mathrm{d}t] = (1 - \mathbb{E}[\mathrm{e}^{-\delta\tau^U}])u_0/\delta$ converges to u_0/δ. In particular, $V(x) \ge (1 - \mathbb{E}[\mathrm{e}^{-\delta\tau^U}])u_0/\delta$ converges to u_0/δ. Let h be small. Let $\tilde{U}_t$ be a strategy with initial capital $x + ch$, and define the strategy

$$U_t = \begin{cases} 0, & \text{for } t \le h \text{ or } T_1 \le h,\\ \tilde{U}_{t-h}\,, & \text{for } T_1 \wedge t > h. \end{cases}$$

Taking the supremum over all strategies $\tilde{U}$ yields

$$V(x) \ge \mathrm{e}^{-\lambda h}\mathrm{e}^{-\delta h} V(x + ch) \ge \mathrm{e}^{-(\lambda+\delta)h} V(x)\,. \tag{2.24}$$

Thus, $V(x)$ is right-continuous. Similarly,

$$V(x) \ge V(x - ch) \ge \mathrm{e}^{-(\lambda+\delta)h} V(x)\,,$$

and the left-continuity follows. We get

$$0 \le V(x) - V(x - ch) \le V(x)(1 - e^{-(\lambda+\delta)h}) \le u_0(\lambda + \delta)h/\delta\,,$$

and Lipschitz continuity follows. The argument as given in [49, p. 164] shows that $V(x)$ is absolutely continuous. □

We now prove the Hamilton–Jacobi–Bellman equation. The proof is very technical. Upon first reading, the reader is therefore advised to skip the proof.

Theorem 2.32. *The function $V(x)$ is differentiable from the left and from the right. Its derivatives $V'(x-)$ and $V'(x+)$ from the left and from the right, respectively, fulfil the Hamilton–Jacobi–Bellman equation* (2.26) *below. If $u_0 < c$, then $V(x)$ is continuously differentiable. If $u_0 > c$ and $V(x)$ is not differentiable at x, then $V'(x-) \le 1 < V'(x+)$.*

Proof. Let $h > 0$ and fix $u \in [0, u_0]$. If $x = 0$, we suppose that $u \le c$, and if $x > 0$, we let h be small enough such that $x + (c - u)h \ge 0$, i.e., ruin does not occur because of the dividend payments. Choose $\varepsilon > 0$ and $n \in \mathbb{N}$. Let $x_k = k(x + (c - u)h)/n$ for $0 \le k \le n$. We consider the following strategy:

$$U_t = \begin{cases} u\,, & \text{for } 0 \le t \le T_1 \wedge h\,, \\ U^\varepsilon_{t-T_1\wedge h}\,, & \text{for } t > T_1 \wedge h\,. \end{cases}$$

$\{U^\varepsilon_t\}$ is a strategy for initial capital x_k where $x_k \le X_{T_1\wedge h} < x_{k+1}$ such that $V^\varepsilon(x_k) > V(x_k) - \varepsilon/2$. Thus, $\{U_t\}$ is measurable. By the Lipschitz continuity of $V(x)$, we can choose n large enough such that $V^\varepsilon(x') > V(x') - \varepsilon$ for all $x' \in [0, x + (c - u)h]$. The first claim happens with density $\lambda e^{-\lambda t}$ and T_1 is larger than h with probability $e^{-\lambda h}$. It follows that

$$\begin{aligned} V(x) &\ge e^{-\lambda h}\Bigl[\int_0^h u e^{-\delta t}\,dt + e^{-\delta h} V^\varepsilon(x + (c - u)h)\Bigr] + \int_0^h \Bigl[\int_0^t u e^{-\delta s}\,ds \\ &\qquad + e^{-\delta t} \int_0^{x+(c-u)t} V^\varepsilon(x + (c - u)t - y)\,dG(y)\Bigr] \lambda e^{-\lambda t}\,dt \\ &\ge e^{-\lambda h}\Bigl[\int_0^h u e^{-\delta t}\,dt + e^{-\delta h} V(x + (c - u)h)\Bigr] + \int_0^h \Bigl[\int_0^t u e^{-\delta s}\,ds \\ &\qquad + e^{-\delta t} \int_0^{x+(c-u)t} V(x + (c - u)t - y)\,dG(y)\Bigr] \lambda e^{-\lambda t}\,dt - \varepsilon\,. \end{aligned}$$

The constant ε is arbitrary. We therefore can let it tend to zero. Rearranging the terms and dividing by h yields

$$\begin{aligned} 0 \ge{}& \frac{V(x + (c - u)h) - V(x)}{h} - \frac{1 - e^{-(\delta+\lambda)h}}{h} V(x + (c - u)h) \\ &+ e^{-\lambda h}\frac{1}{h}\int_0^h u e^{-\delta t}\,dt + \frac{1}{h}\int_0^h \Bigl[\int_0^t u e^{-\delta s}\,ds \\ &+ e^{-\delta t} \int_0^{x+(c-u)t} V(x + (c - u)t - y)\,dG(y)\Bigr] \lambda e^{-\lambda t}\,dt\,. \end{aligned} \tag{2.25}$$

Now choose a strategy U such that $V^U(x) \ge V(x) - h^2$. There exists u_h such that $\mathbb{E}[\int_0^{T_1 \wedge h} (U_s - u_h) \mathrm{e}^{-\delta s} \,\mathrm{d}s] = 0$. Let u_t' denote U_t conditioned on $T_1 > s$ and $a(t) = \int_0^t (c - u_s') \,\mathrm{d}s$. In the same way as above, we find that

$$\begin{aligned} 0 \le h + {}& \frac{V(x+a(h)) - V(x)}{h} - \frac{1 - \mathrm{e}^{-(\delta+\lambda)h}}{h} V(x+a(h)) \\ &+ \mathrm{e}^{-\lambda h} \frac{1}{h} \int_0^h u_h \mathrm{e}^{-\delta t} \,\mathrm{d}t + \frac{1}{h} \int_0^h \Bigl[\int_0^t u_h \mathrm{e}^{-\delta s} \,\mathrm{d}s \\ &+ \mathrm{e}^{-\delta t} \int_0^{x+a(t)} V(x + a(t) - y) \,\mathrm{d}G(y) \Bigr] \lambda \mathrm{e}^{-\lambda t} \,\mathrm{d}t \,. \end{aligned}$$

With the exception of the second and fourth terms the terms on the right-hand side of the inequality converge. Choose a sequence h_n tending to zero such that

$$\lim_{n\to\infty} \frac{V(x+a(h_n)) - V(x)}{h_n} = \limsup_{h \downarrow 0} \frac{V(x+a(h)) - V(x)}{h} \,.$$

This limit is finite by the local Lipschitz continuity. Without loss of generality, we can assume that u_{h_n} converges to some value $\tilde{u}$. For this sequence (2.25) holds, and thus the limit of

$$\frac{V(x+a(h_n)) - V(x)}{h_n}$$

is unique. In a slight abuse of notation, denote

$$(c - \tilde{u}) V'(x) = \lim_{n\to\infty} \frac{V(x+a(h_n)) - V(x)}{h_n} \,.$$

Then

$$(c - \tilde{u}) V'(x) + \lambda \Bigl[\int_0^x V(x-y) \,\mathrm{d}G(y) - V(x) \Bigr] - \delta V(x) + \tilde{u} = 0 \,.$$

From (2.25) we conclude that for any u

$$(c - u) V'(x) + \lambda \Bigl[\int_0^x V(x-y) \,\mathrm{d}G(y) - V(x) \Bigr] - \delta V(x) + u \le 0 \,.$$

Thus, we get the Hamilton–Jacobi–Bellman equation

$$\sup_{0 \le u \le u_0} \Bigl\{ (c-u) V'(x) + \lambda \Bigl[\int_0^x V(x-y) \,\mathrm{d}G(y) - V(x) \Bigr] - \delta V(x) + u \Bigr\} = 0 \,. \quad (2.26)$$

We can repeat the argument for any possible limit of

$$\frac{V(x+a(h_n)) - V(x)}{h_n} \,,$$

leading to the same Equation (2.26). In order to distinguish the derivatives, denote the corresponding limit by $(c-\hat{u})\hat{V}'(x)$.

Equation (2.26) is linear in u. Thus, the supremum is taken either in zero or in u_0, unless $V'(x)=1$, in which case any value u will give the same equation. Therefore, there are at most two possible limits. From the definition of u_h we conclude that $u'_s \to \tilde{u}$. The same has to hold for $\hat{u}$; thus, $\hat{u}=\tilde{u}$ if $V'(x)\neq 1$.

If $V'(x)=1$, then

$$c+\lambda\Bigl[\int_0^x V(x-y)\,\mathrm{d}G(y)-V(x)\Bigr]-\delta V(x)=0\,.$$

We conclude that

$$(c-\hat{u})\hat{V}'(x)-(c-\hat{u})=0\,.$$

Suppose that $\hat{V}'(x)\neq 1$. This is only possible if $\hat{u}=c$. Because $\hat{u}$ is either zero or u_0, we conclude that $u_0=c$. As above, we conclude that $\hat{u}=\tilde{u}=c$, because u'_s has to converge to c. We have therefore shown that

$$\limsup_{h\downarrow 0}\frac{V(x+a(h))-V(x)}{h}=(c-\tilde{u})V'(x)\,.$$

Consider now the case $u_0<c$. Then $a(h)>0$ and we have shown differentiability from the right. Suppose that $V'(x)>1$. Consider initial capital $u-ch$. Similarly as above, we conclude that $V(x)$ is differentiable from the left and fulfils (2.26). There are therefore at most two possible derivatives $V'(x)$. Because

$$V'(x)\gtreqless 1\Longleftrightarrow(\lambda+\delta)V(x)-\lambda\int_0^x V(x-y)\,\mathrm{d}G(y)\gtreqless c$$

and the function $V(x)$ is continuous, we see that the derivative from the left coincides with the derivative from the right. That is, $V(x)$ is (continuously) differentiable. Denote this solution by $V_{u_0}(x)$.

It follows by monotone convergence that $\lim_{u_0\uparrow c}V_{u_0}(x)=V_c(x)$. Note that

$$(\lambda+\delta)V_{u_0}(x)-\lambda\int_0^x V_{u_0}(x-y)\,\mathrm{d}G(y)\ge(c-u_0)V'_{u_0}(x)+u_0\,.$$

As $u_0\uparrow c$, the right-hand side tends to c. Thus,

$$(\lambda+\delta)V_c(x)-\lambda\int_0^x V_c(x-y)\,\mathrm{d}G(y)\ge c\,.$$

If equality holds, then

$$\lim_{u_0\uparrow c}\max\{cV'_{u_0}(x),(c-u_0)V'_{u_0}(x)+u_0\}=c\,.$$

Consider a sequence h_n tending to c. By considering a subsequence, we can assume that either $V'_{h_n}(x) > 1$, $V'_{h_n}(x) = 1$, or $V'_{h_n}(x) < 1$ for all n. If $V'_{h_n}(x) > 1$ for all n, then $V'_{h_n}(x)$ tends to 1. Thus, $\limsup V'_{h_n}(x) \le 1$. We see that $V_c(X)$ fulfils the Hamilton–Jacobi–Bellman equation in this case.

If

$$(\lambda+\delta)V_c(x) - \lambda \int_0^x V_c(x-y)\,\mathrm{d}G(y) > c\,,$$

then

$$(\lambda+\delta)V_{u_0}(x) - \lambda \int_0^x V_{u_0}(x-y)\,\mathrm{d}G(y) > c$$

for u_0 large enough. Because $(c-u_0)V'_{u_0}(x) + u_0$ tends to c, we can conclude that $V'_{u_0}(x) > 1$ for u_0 large enough. Thus, it follows that $V'_{u_0}(x)$ converges to some function $f(x)$. By continuity, convergence takes place in a neighbourhood of x, (x_1, x_2), say, and the function $f(z)$ is continuous in (x_1, x_2). Let $x_1 < z \le x_2$. Then

$$\int_{x_1}^z f(v)\,\mathrm{d}v = \lim_{u_0 \uparrow c} \int_{x_1}^z V'_{u_0}(v)\,\mathrm{d}v = \lim_{u_0 \uparrow c} V_{u_0}(z) - V_{u_0}(x_1) = V_c(z) - V_c(x_1)\,.$$

We see that $f(v)$ is a density of $V_c(x)$. Thus, (2.26) also holds for $u_0 = c$.

If $u_0 > c$, then we have shown either differentiability from the right ($\tilde{u} = 0$) or differentiability from the left ($\tilde{u} = u_0$). If $V'(x) = 1$, then we have shown differentiability because we can choose $\tilde{u}$ arbitrarily.

Suppose that $V'(x) < 1$. Starting with initial capital $x + ch$, we can use the same arguments as above showing that one has to choose either $u = 0$ or $u = u_0$. Choosing $u = 0$ shows differentiability from the right. Because $\tilde{u} = 0$ solves the same equation, we conclude that both choices $\tilde{u} = 0$ and $\tilde{u} = u_0$ yield the same equation. That is, both the derivative from the left and the derivative from the right solve (2.26). If $\tilde{u} = 0$ is not a solution, consider the initial capital $x + u_0 h$. Similar arguments as used above show differentiability from the right, that is, $V(x)$ is differentiable at x.

Suppose that $V'(x) > 1$. Consider initial capital $u - ch$. Similar arguments as above show that we have to choose either $u = 0$ or $u = u_0$. If $u = 0$, we see that $V(x)$ is differentiable at x. If $u = u_0$, we would get differentiability from the left, and both choices $\tilde{u} = 0$ and $\tilde{u} = u_0$ solve (2.26). Thus, we conclude that both the derivative from the left and the derivative from the right solve (2.26). Moreover, we see that at points x where $V(x)$ is not differentiable, we have $V'(x-) \le 1 < V'(x+)$. In particular, the derivative from the right and the derivative from the left is are densities. □

As mentioned in the proof above, Equation (2.26) is linear in u. Thus, let

$$u(x) = \begin{cases} 0, & \text{for } V'(x-) > 1, \\ u_0, & \text{for } V'(x-) < 1, \\ \min\{c, u_0\}, & \text{for } V'(x-) = 1. \end{cases} \tag{2.27}$$

Our choice in the case $V'(x-) = 1$ is motivated as follows. If $u_0 < c$, then the choice does not matter, as we will see in the proof below. If $u_0 \ge c$, then $V'(x-) = 1$ implies that $(\lambda+\delta)V(x) - \lambda\int_0^x V(x-y)\,\mathrm{d}G(y) = c$ and

$$V(x) = \frac{c}{\lambda+\delta} + \frac{\lambda}{\lambda+\delta}\int_0^x V(x-y)\,\mathrm{d}G(y)\,.$$

This is the value of the strategy where the incoming premium is paid as dividend until the first claim occurs. After that, the optimal strategy (which we do not know to exist yet) is followed.

The following theorem serves as some sort of a verification theorem.

Theorem 2.33. *Suppose that $f(x)$ is an increasing, bounded, and positive solution to* (2.26). *Then $\lim_{x\to\infty} f(x) = u_0/\delta$. If $u_0 \le c$ or $f'(0) \ge 1$, then $f(x) = V(x)$ and an* optimal dividend strategy *is given by* (2.27).

Proof. The function $f(x)$ must converge to $f(\infty) < \infty$. Then there must be a sequence $\{x_n\}$ tending to infinity such that $f'(x_n)$ tends to zero. Let $u_n = u(x_n)$, where we replace $V'(x-)$ by $f'(x-)$ in the definition of $u(x)$. By (2.27) we can assume that $u(x_n) = u_0$. Letting $n \to \infty$ in (2.26) shows that $f(\infty) = u_0/\delta$. Let U be an arbitrary strategy. From [26, p. 27] we have that the process M' with

$$M'_t = \sum_{i=1}^{N_{\tau^U\wedge t}} (f(X^U_{T_i}) - f(X^U_{T_i-}))\mathrm{e}^{-\delta T_i} - \lambda\int_0^{\tau^U\wedge t} \mathrm{e}^{-\delta s}\Bigl(\int_0^{X^U_s} f(X^U_s - y)\,\mathrm{d}G(y) - f(X^U_s)\Bigr)\,\mathrm{d}s$$

is a martingale. Noting that

$$f(X^U_{T_i-})\mathrm{e}^{-\delta T_i} - f(X^U_{T_{i-1}})\mathrm{e}^{-\delta T_{i-1}} = \int_{T_{i-1}}^{T_i-} [(c-U_s)f'(X^U_s) - \delta f(X^U_s)]\mathrm{e}^{-\delta s}\,\mathrm{d}s$$

and using that T_i- may be replaced by $(T_i \wedge \tau^U \wedge t)-$ and T_{i-1} may be replaced by the last claim time before $T_{i-1}\wedge\tau^U\wedge t$, we get that

$$\Bigl\{f(X^U_{\tau^U\wedge t})\,\mathrm{e}^{-\delta(\tau^U\wedge t)} - \int_0^{\tau^U\wedge t}\Bigl((c-U_s)f'(X^U_s) + \lambda\int_0^{X^U_s} f(X^U_s - y)\,\mathrm{d}G(y) - (\lambda+\delta)f(X^U_s)\Bigr)\mathrm{e}^{-\delta s}\,\mathrm{d}s\Bigr\}$$

is a martingale. If $U = U^*$, the strategy given by (2.27),

$$\Bigl\{f(X^*_{\tau^*\wedge t})\,\mathrm{e}^{-\delta(\tau^*\wedge t)} + \int_0^{\tau^*\wedge t} U^*_s\mathrm{e}^{-\delta s}\,\mathrm{d}s\Bigr\}$$

is a martingale, where we used (2.26). This yields

$$f(x) = \mathbb{E}\Big[f(X^*_{\tau^*\wedge t})\,\mathrm{e}^{-\delta(\tau^*\wedge t)} + \int_0^{\tau^*\wedge t} U^*_s \mathrm{e}^{-\delta s}\,\mathrm{d}s\Big].$$

Letting $t \to \infty$ shows that $f(x) = V^*(x)$. Here we need the additional conditions because we could have $f'(0) < 1$. Then the "optimal" strategy would be to pay dividends at the maximal rate, which immediately leads to ruin. Thus, $f(0)$ would not be the value of the "optimal" strategy. For an arbitrary strategy, (2.26) gives

$$f(x) \ge \mathbb{E}\Big[f(X^U_{\tau^U\wedge t})\,\mathrm{e}^{-\delta(\tau^U\wedge t)} + \int_0^{\tau^U\wedge t} U_s \mathrm{e}^{-\delta s}\,\mathrm{d}s\Big].$$

Letting $t \to \infty$ shows that $f(x) \ge V^U(x)$. Thus, $f(x) = V(x)$. □

As $x \to \infty$, the expression

$$\lambda \int_0^x V(x-y)\,\mathrm{d}G(y) - (\lambda+\delta)V(x)$$

converges to $-u_0$. Thus, (2.26) converges to

$$\lim_{x\to\infty} \max\{cV'(x) - u_0, (c-u_0)V'(x)\} = 0\,.$$

If $u_0 < c$, then this implies that $V'(x)$ converges to zero. In particular, $u(x) = u_0$ for x large enough. If $u_0 > c$, then $V'(x)$ cannot be close to 1 for x large enough. The fact that $V'(x)$ has upward jumps only gives $u(x) = 0$ or $u(x) = u_0$ for x large enough. Because $V(x)$ is bounded, this means that $u(x) = u_0$ for x large enough. In particular, here $V'(x)$ converges to zero. If $u_0 = c$, then we can only conclude that there must be points where $V'(x) < 1$, and the process X^* will be bounded because $u(x) = c$ at a point where $V'(x) < 1$.

If $u_0 < c - \lambda\mu$, then the net profit condition for the process under the strategy $u = u_0$ holds, and $\psi(x) \le \psi^{u_0}(x) < 1$. Thus, ruin does not almost surely occur. If $u_0 \ge c$, the process X^* will be bounded by some value $x_0 = \inf\{z \ge x : V'(z) \le 1\}$. Because $\mathbb{P}[\sum_{i=N_t+1}^{N_{t+1}} Y_i > x_0 + c] > 0$, by the Borel–Cantelli lemma, $\sum_{i=N_t+1}^{N_{t+1}} Y_i > x_0 + c$ will happen for some t. Thus, ruin occurs almost surely. If $c - \lambda\mu \le u_0 < c$, then starting from $x > x_0 = \sup\{z : V'(z) \ge 1\}$, the probability that $\inf_t X^*_t < x_0$ is the ruin probability with initial capital $x - x_0$ and premium rate $c - u_0$. Because the net profit condition is not fulfilled, this probability is 1. In particular, if the process reaches the level $x_0 + c$, it will drop almost surely again below the level x_0. Therefore, $\{X^*_t\}$ returns to $[0, x_0]$ again and again until ruin occurs. By the argument given above, we have that ruin must occur almost surely. Thus, under the optimal strategy ruin occurs almost surely if and only if $u_0 \ge c - \lambda\mu$.

We next look for conditions under which $V(x)$ is differentiable at x in the case $u_0 > c$.

Lemma 2.34. *If* $(\lambda+\delta)V(x) = \lambda\int_0^x V(x-y)\,\mathrm{d}G(y) + c$, *then* $V(x)$ *is differentiable at* x *and* $V'(x) = 1$. *If the derivative from the right or from the left is* 1 *then* $(\lambda+\delta)V(x) = \lambda\int_0^x V(x-y)\,\mathrm{d}G(y) + c$. *If* $(\lambda+\delta)V(x) > \lambda\int_0^x V(x-y)\,\mathrm{d}G(y) + c$ *and there is a sequence* $x_n \uparrow x$ *such that* $u(x_n) = 0$, *then* $V(x)$ *is differentiable at* x *and* $V'(x) > 1$.

Proof. If $(\lambda+\delta)V(x) = \lambda\int_0^x V(x-y)\,\mathrm{d}G(y) + c$, $V'(x-) = 1$, or $V'(x+) = 1$, then we have shown the assertion in the proof of Theorem 2.32. Suppose now that $(\lambda+\delta)V(x) > \lambda\int_0^x V(x-y)\,\mathrm{d}G(y) + c$. By continuity, $(\lambda+\delta)V(z) > \lambda\int_0^z V(z-y)\,\mathrm{d}G(y) + c + \varepsilon$ for $z_0 \le z \le z_1$ for some $\varepsilon > 0$ and $z_0 < x < z_1$. Because $V'(x)$ can only have upward jumps, we conclude that there is $z_2 \in [z_0, z_1]$ such that $u(z) = u_0$ for $z_0 \le z \le z_2$ and $u(z) = 0$ for $z_2 < z \le z_1$. If a sequence x_n as claimed exists, then $x > z_2$ and $V(x)$ is differentiable at x. That $V'(x) > 1$ follows from the definition of $u(x)$. □

If $V(x)$ is not differentiable at x, then $u(z) = u_0$ on some interval $[z_0, x]$, and $u(z) = 0$ on some interval $(x, z_1]$, where $z_0 < x < z_1$.

We remark that if for $u_0 = c$ we have $u_c(x) = 0$ for $x \in [0, x_0)$ for some $x_0 > 0$ and $u_c(x_0) = c$, then also for $u_0 > c$ we have $u_c(x) = 0$ for $x \in [0, x_0)$ and $u_c(x_0) = c$. Indeed, we have $V_c(0) > c/(\lambda+\delta)$. Then also $V_{u_0}(0) \ge V_c(0) > c/(\lambda+\delta)$. That is, $u_c(0) = c$ is not optimal. Therefore, there is $x_1 > 0$ such that $u_c(x) = 0$ for $x \in [0, x_1)$ and $u_c(x_1) = c$. Because the strategy is valid for both $u_0 = c$ and $u_0 > c$, we must have $x_1 = x_0$.

Example 2.35. Let us consider exponentially distributed claim sizes $G(y) = 1 - \mathrm{e}^{-\alpha y}$. Then Equation (2.26) reads

$$\sup_{0\le u\le u_0}\Bigl\{(c-u)V'(x) + \lambda\mathrm{e}^{-\alpha x}\int_0^x V(y)\alpha\mathrm{e}^{\alpha y}\,\mathrm{d}y - (\lambda+\delta)V(x) + u\Bigr\} = 0\,.$$

We need to solve the equation for $u = 0$ and for $u = u_0$. Taking the derivative and plugging in the original equation yields

$$(c-u)V''(x) - (\lambda+\delta-\alpha(c-u))V'(x) - \alpha\delta V(x) + \alpha u = 0\,.$$

With

$$r_{1/2}(u) = \frac{\lambda+\delta-\alpha(c-u) \pm \sqrt{(\lambda+\delta-\alpha(c-u))^2 + 4(c-u)\alpha\delta}}{2(c-u)}\,,$$

the solutions are

$$V_0(x) = A\mathrm{e}^{r_1(0)x} + B\mathrm{e}^{r_2(0)x}\,, \qquad V_1(x) = \frac{u_0}{\delta} + C\mathrm{e}^{r_1(u_0)x} + D\mathrm{e}^{r_2(u_0)x}\,.$$

This is provided that $u_0 \ne c$. For $u_0 = c$ we have to replace the second function by

$$V_1(x) = \frac{u_0}{\delta} + C e^{-\alpha\delta x/(\lambda+\delta)}\,.$$

One easily sees that $-\alpha < r_2(0) < 0 < r_1(0)$. If $u_0 < c$, then also $-\alpha < r_2(u_0) < 0 < r_1(u_0)$. If $u_0 > c$, then $r_1(u_0) < -\alpha < r_2(u_0) < 0$. Let us now consider a point a at which $V'(a) = 1$. At such a point we have $Ar_1(0)e^{r_1(0)a} + Br_2(0)e^{r_2(0)a} = V'(a) = 1$. Thus, $Br_2(0)e^{r_2(0)a} = 1 - Ar_1(0)e^{r_1(0)a}$. In particular, $r_2(0)V(a) = 1 - A(r_1(0) - r_2(0))e^{r_1(0)a}$. The second derivative becomes

$$\begin{aligned}
V''(a) &= Ar_1(0)^2 e^{r_1(0)a} + Br_2(0)^2 e^{r_2(0)a} \\
&= r_2(0) + Ar_1(0)(r_1(0) - r_2(0))e^{r_1(0)a} \\
&= r_1(0) + r_2(0) - r_1(0)r_2(0)V(a) \\
&= \frac{\lambda+\delta}{c} - \alpha + \frac{\alpha\delta}{c}V(a)\,.
\end{aligned}$$

If $x_1 > a$ is another point with $V'(x_1) = 1$, then $V''(x_1) > V''(a)$. If $V''(a) > 0$, then $u(x) = 0$ for $a < x < z_0$ for some $z_0 > a$. Because $V'(x)$ has upward jumps only and $u(z) = u_0$ for z large enough, we get a contradiction. Indeed, there cannot be a point x_1 with $V'(x_1) = 1$ and $V''(x_1) \le 0$. Thus, $V''(a) \le 0$.

We next consider the three cases for u_0.

$u_0 < c$:

Because $V(x)$ is continuously differentiable, there is at most one point a with $V'(a) = 1$. Thus,

$$V(x) = \begin{cases} V_0(x), & \text{for } x < a, \\ V_1(x), & \text{for } x \ge a. \end{cases}$$

Here $a \ge 0$, and we allow $V'(0) < 1$ if $a = 0$. Clearly, $C = 0$ and $D < 0$.

Consider first the case $a = 0$. Plugging $V_1(x)$ into the original equation yields $D = -u_0(\alpha + r_2(u_0))/(\alpha\delta)$. The derivative $V_1'(x) = -u_0 r_2(u_0)(\alpha + r_2(u_0))e^{r_2(u_0)x}/(\alpha\delta)$ is decreasing in x. Therefore, $V_1'(0) \le 1$ in the case $a = 0$. The latter condition can be written as

$$(\lambda+\delta)^2 + \alpha c\Bigl(\frac{c\delta}{u_0} - (\lambda+\delta)\Bigr) \ge 0\,.$$

Thus, for $u_0 \le c\delta/(\lambda+\delta)$, it is always optimal to pay dividends at the maximal rate. If $c\delta/(\lambda+\delta) < u_0 < c$ then if α is small enough, i.e., the mean value of the claim sizes is large, it is also optimal to pay the dividends. The condition is quadratic in δ and increasing in δ. If the net profit condition $\alpha c > \lambda$ is not fulfilled, then it is also always optimal to pay the dividend. If the net profit condition is fulfilled, then it is only optimal always to pay dividends if δ is large enough.

If $a > 0$, we get by plugging in the solution $V(x) = V_0(x)$, $B = -(\alpha + r_2(0))A/(\alpha + r_1(0))$. A is determined from $V_0'(a) = 1$, i.e., $A(r_1(0)e^{r_1(0)a} - r_2(0)(\alpha+r_2(0))e^{r_2(0)a}/(\alpha+r_1(0))) = 1$. The constant D is found from $V_1'(a) = 1$, i.e., $Dr_2(u_0)e^{r_2(u_0)a} = 1$. Finally, $V_1(a) = V_0(a)$ yields an equation for a.

$u_0 = c$:

Because $V(x)$ is continuously differentiable, the value function and the optimal strategy must be of the same form as in the case $u_0 < c$. If $a = 0$, then $V(0) = c/(\lambda + \delta)$, because a dividend at rate c is paid until the first claim. That yields $C = -\lambda c/(\delta(\lambda + \delta))$. From $V'(0) \le 1$, we get the condition

$$\frac{\lambda c}{\delta(\lambda + \delta)} \frac{\alpha \delta}{\lambda + \delta} \le 1 .$$

That is, we get the condition $\lambda c \alpha \le (\lambda + \delta)^2$, which is the limit of the condition in the case $u_0 < c$. Hence, if the claim sizes or the discounting factor are too large, then it is not optimal not to pay dividends. Note that again if the net profit condition is not fulfilled, then dividends will be paid immediately.

If $a > 0$, then we obtain $B = -(\alpha + r_2(0))A/(\alpha + r_1(0))$. The parameter A is obtained from $V_0'(a) = 1$ and the parameter C from $V_1'(a) = 1$, that is, $C = -(\lambda + \delta)\mathrm{e}^{\alpha \delta a/(\lambda + \delta)}/(\alpha \delta)$. Finally, a is determined from $V_0(a) = V_1(a)$. For an explicit expression for $V_0(x)$, see also (2.34).

$u_0 > c$:

One can expect that the solution again is of the same form as in the case $u_0 < c$. If $\lambda c \alpha \le (\lambda + \delta)^2$, then $a = 0$. In this case we obtain C and D because $V_1(x)$ solves (2.26) and because $V_1(0) = c/(\lambda + \delta)$. If $\lambda c \alpha > (\lambda + \delta)^2$, then we know $V_0(x)$ on $[0, a]$. The parameters C and D are found from $V_1(a) = V_0(a)$ and $V_1'(a) = 1$. ∎

2.4.2 Unrestricted Dividend Payments

Introduction

Let $\{D_t\}$ denote the accumulated dividends process. In the situation of the previous section we had $D_t = \int_0^t U_s \,\mathrm{d}s$. The surplus process is then

$$\Bigl\{X_t^D = x + ct - \sum_{i=1}^{N_t} Y_i - D_t\Bigr\} .$$

All increasing adapted cadlag processes D are allowed. The value of the strategy D is

$$V^D(x) = \mathbb{E}\Bigl[\int_{0-}^{\tau^D -} \mathrm{e}^{-\delta t} \,\mathrm{d}D_t\Bigr] ,$$

and $V(x) = \sup_{D \in \mathfrak{U}} V^D(x)$ is the value function. We include the point 0 into the integration area in order to take an immediate dividend $D_0 > 0$ into the value. We do not take a possible dividend at time τ^D into consideration, in order to prevent from ruin caused by a dividend payment.

Remark 2.36. In the literature it is often assumed that D is *caglad*, i.e., it is left-continuous and limits from the right exist. This is in order to prevent from being possible to pay a large amount, leading to ruin. We prevent this possibility here by not taking dividends at the time of ruin into account. In our setup, we can let X^D be the post-dividend process, and therefore it is possible to consider cadlag processes. In particular, to work with right-continuous X^D is simpler than considering processes that are not right-continuous at points where dividends are paid as a lump sum. ■

Let us consider the problem intuitively. If we let $u_0 \to \infty$ in the model of Section 2.4.1, we would expect three types of points: points where no dividend is paid ($V'(x) > 1$), points where the incoming premia are immediately paid as dividends, and points where a dividend is paid at rate $u_0 = \infty$, i.e., a dividend is paid immediately as a lump sum. At points where a dividend is paid immediately, we must have $V'(x) = 1$. We should therefore have $V'(x) \ge 1$. Indeed, if $V'(x) < 1$, it is possible to pay a dividend and to obtain a value strictly larger than $V(x)$,

$$V(x) \ge D_0 + V(x - D_0) = V(x) + D_0 - \int_{x-D_0}^{x} V'(z)\,\mathrm{d}z > V(x)\,.$$

We have the following bounds.

Lemma 2.37. *The function $V(x)$ is locally Lipschitz continuous on $[0,\infty)$, and therefore absolutely continuous. For any x,*

$$x + \frac{c}{\lambda+\delta} \le V(x) \le x + \frac{c}{\delta}$$

and for any $y \le x$, $V(x) - V(y) \ge x - y$.

Proof. The bound for $V(x) - V(y)$ is obtained by paying $x - y$ immediately and then following a strategy for initial capital y. The lower bound of $V(x)$ is obtained by paying x immediately and then paying the premia at rate c until the first claim occurs. Finally, the upper bound comes from the pseudo-strategy of paying x immediately and thereafter paying a dividend at rate c, not stopping at ruin. More specifically, letting $\mathrm{d}D_t = 0$ for $t \ge \tau^D$,

$$\begin{aligned}
V^D(x) &= \mathbb{E}\Big[\int_{0-}^{\infty} \mathrm{e}^{-\delta t}\,\mathrm{d}D_t\Big] = \mathbb{E}\Big[\int_{0-}^{\infty}\int_{t}^{\infty} \delta \mathrm{e}^{-\delta s}\,\mathrm{d}s\,\mathrm{d}D_t\Big] \\
&= \delta\mathbb{E}\Big[\int_{0}^{\infty}\int_{0-}^{s-} \mathrm{e}^{-\delta s}\,\mathrm{d}D_t\,\mathrm{d}s\Big] = \delta\mathbb{E}\Big[\int_{0}^{\infty} D_{s-}\mathrm{e}^{-\delta s}\,\mathrm{d}s\Big] \\
&\le \delta\int_{0}^{\infty} (x + cs)\mathrm{e}^{-\delta s}\,\mathrm{d}s = x + \frac{c}{\delta}\,.
\end{aligned}$$

The locally Lipschitz continuity and absolute continuity follow as in the proof of Lemma 2.31, noting that $V(x)$ is locally bounded. □

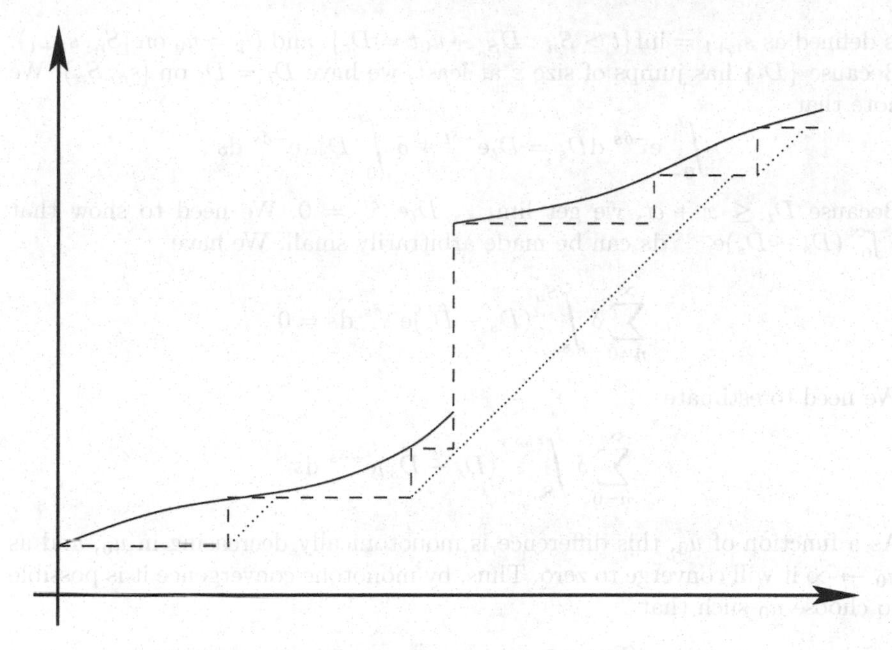

Fig. 2.15. First approximation of a dividend process.

The Value Function

The value function is obtained as a limit of the value function of Section 2.4.1.

Lemma 2.38. *Let $V_u(x)$ be the value function for the restricted dividend strategy in the case $u_0 = u$. Then $\lim_{u\to\infty} V_u(x) = V(x)$.*

Proof. Because $V_u(x)$ is increasing in u, it is converging pointwise. Moreover, the restricted strategy is admissible, and therefore $\lim_{u\to\infty} V_u(x) \le V(x)$. Let $\varepsilon > 0$. Choose a dividend process D such that $V^D(x) \ge V(x) - 2\varepsilon$. We can assume that D is a pure jump process with jumps of at least size ε. That is, we choose first a dividend process $\tilde{D}_t$ such that $V^{\tilde{D}}(x) \ge V(x) - \varepsilon$. The strategy D_t is constructed in the following way, illustrated in Figure 2.15 (dashed line). Let $D_0 = \tilde{D}_0 1\!\!1_{\tilde{D}_0 \ge \varepsilon}$. On $\{\tilde{D}_t < D_{t-} + \varepsilon\}$ we let $\mathrm{d}D_t = 0$, i.e., no dividends are paid. On $\{\tilde{D}_t \ge D_{t-} + \varepsilon\}$ we let $D_t = \tilde{D}_t$, i.e., at time points where the difference between the dividend processes becomes larger than ε, a dividend is paid in order that the accumulated dividend becomes the same. The error made is bounded by ε [because $V^{\tilde{D}}(x) - V^D(x) = \mathbb{E}[\int_0^\infty \delta e^{-\delta s}(\tilde{D}_s - D_s)\,\mathrm{d}s] \le \varepsilon$] i.e., $V^D(x) \ge V(x) - 2\varepsilon$. We need to approximate the value $V^D(x)$; see Figure 2.15 (dotted line). Let $s_0 = 0$. We define a strategy $\hat{D}$, depending on u_0 to be defined later, with $\hat{D}_t = \int_0^t U_s\,\mathrm{d}s$. For $n \in \mathbb{N}$, define $S_n = \inf\{t \ge s_n : D_t \ge \hat{D}_{s_n} + \varepsilon\}$. We let $U_t = 0$ on $[s_n, S_n)$. The variable s_n

is defined as $s_{n+1} = \inf\{t > S_n : \hat{D}_{S_n} + u_0 t = D_t\}$, and $U_t = u_0$ on $[S_n, s_{n+1})$. Because $\{D_t\}$ has jumps of size ε at least, we have $D_t = \hat{D}_t$ on $[s_n, S_n)$. We note that

$$\int_{0-}^{t} e^{-\delta s} \, dD_s = D_t e^{-\delta t} + \delta \int_0^t D_{s-} e^{-\delta s} \, ds \,.$$

Because $D_t \le x + ct$, we get $\lim_{t\to\infty} D_t e^{-\delta t} = 0$. We need to show that $\delta \int_0^\infty (D_s - \hat{D}_s) e^{-\delta s} \, ds$ can be made arbitrarily small. We have

$$\sum_{n=0}^{\infty} \delta \int_{s_n}^{S_n} (D_s - \hat{D}_s) e^{-\delta s} \, ds = 0 \,.$$

We need to estimate

$$\sum_{n=0}^{\infty} \delta \int_{S_n}^{s_{n+1}} (D_s - \hat{D}_s) e^{-\delta s} \, ds \,.$$

As a function of u_0, this difference is monotonically decreasing in u_0, and as $u_0 \to \infty$ it will converge to zero. Thus, by monotone convergence it is possible to choose u_0 such that

$$\mathbb{E}\Bigl[\sum_{n=0}^{\infty} \delta \int_{S_n}^{s_{n+1}} (D_s - \hat{D}_s) e^{-\delta s} \, ds\Bigr] < \varepsilon \,.$$

This shows that

$$V(x) - V^{\hat{D}}(x) < 2\varepsilon + \varepsilon = 3\varepsilon \,.$$

Because ε is arbitrary, this proves the lemma. □

We can now obtain the Hamilton–Jacobi–Bellman equation connected to the problem. The idea of the proof is to let $u_0 \to \infty$ in an equivalent version of (2.26). The proof should be skipped upon first reading.

Theorem 2.39. *The function $V(x)$ is differentiable from the left and from the right on $(0, \infty)$. Both derivatives fulfil the Hamilton–Jacobi–Bellman equation*

$$\max\Bigl\{cV'(x) + \lambda \int_0^x V(x-y) \, dG(y) - (\lambda + \delta) V(x), 1 - V'(x)\Bigr\} = 0 \,. \quad (2.28)$$

Moreover, at points where $V(x)$ is not differentiable, we have $V'(x-) = 1 < V'(x+)$.

Proof. Because we let $u \to \infty$, we restrict ourselves to $u > c$. Equation (2.26) can be written as [see also (2.27)]

$$\max\Bigl\{cV_u'(x) + \lambda \int_0^x V_u(x-y) \, dG(y) - (\lambda + \delta) V_u(x),$$
$$1 + \frac{\lambda \int_0^x V_u(x-y) \, dG(y) - (\lambda + \delta) V_u(x) + c}{u - c} - V_u'(x)\Bigr\} = 0 \,. \quad (2.29)$$

We want to show that $V_u'(x)$ converges pointwise to some function f almost everywhere, where $V_u'(x)$ denotes the derivative either from the left or from the right. The function f constructed below satisfies

$$\max\Bigl\{cf(x) + \lambda \int_0^x V(x-y)\,\mathrm{d}G(y) - (\lambda+\delta)V(x), 1 - f(x)\Bigr\} = 0\,.$$

By bounded convergence,

$$V(x) - V(0) = \lim_{u\to\infty} \int_0^x V_u'(z)\,\mathrm{d}z = \int_0^x \lim_{u\to\infty} V_u'(z)\,\mathrm{d}z = \int_0^x f(z)\,\mathrm{d}z\,.$$

Thus, f is the density of V. This shows that $V(x)$ is differentiable at all points where $f(x)$ is continuous. The proof below will also show that the limits of $f(x)$ from the left and from the right exist.

By bounded convergence,

$$\lim_{u\to\infty} (\lambda+\delta)V_u(x) - \lambda \int_0^x V_u(x-y)\,\mathrm{d}G(y) = (\lambda+\delta)V(x) - \lambda \int_0^x V(x-y)\,\mathrm{d}G(y)\,.$$

Let us first consider the case $(\lambda+\delta)V(x) - \lambda\int_0^x V(x-y)\,\mathrm{d}G(y) = c$. Let $\{u_n\}$ be a sequence converging to infinity such that $\lim_{n\to\infty} V_{u_n}'(x)$ exists. This limit is finite because $cV_u'(x) \le (\lambda+\delta)V_u(x) \le (\lambda+\delta)V(x)$. By (2.29) we find that $\lim_{n\to\infty} V_{u_n}'(x) \le 1$ as well as $\lim_{n\to\infty} V_{u_n}'(x) \ge 1$, depending on whether we take the first or the second term on the left-hand side of (2.29). Because the first term is taken if $V_u'(x) \ge 1$ and the second if $V_u'(x) \le 1$, we find that $\lim_{u\to\infty} V_u'(x) = 1$. Note that by the continuity of $(\lambda+\delta)V_u(x) - \lambda\int_0^x V_u(x-y)\,\mathrm{d}G(y)$, we have that $f(z)$ is continuous at x.

From Equations (2.27) and (2.29) we can conclude that $(\lambda+\delta)V_u(x) - \lambda\int_0^x V_u(x-y)\,\mathrm{d}G(y) \ge c$; therefore, the same holds for $V(x)$. Suppose, therefore, that $(\lambda+\delta)V(x) - \lambda\int_0^x V(x-y)\,\mathrm{d}G(y) > c$. By continuity, there exist $\varepsilon > 0$ and $x_1 < x < x_2$ such that $(\lambda+\delta)V(z) - \lambda\int_0^z V(z-y)\,\mathrm{d}G(y) > c+2\varepsilon$ for all $x_1 \le z \le x_2$. Thus, for u large enough, $(\lambda+\delta)V_u(z) - \lambda\int_0^z V_u(z-y)\,\mathrm{d}G(y) > c+\varepsilon$. Note that $V_u'(z) \ne 1$.

For the optimal strategy there are three possibilities: to pay dividends when $X_t \in [x_1, x_2]$; not to pay dividends when $X_t \in [x_1, x_2]$; or to pay dividends in $[x_1, x_0]$ and not to pay dividends in $[x_0, x_2]$. This is because to change from not paying dividends to paying dividends implies that $V'(z) = 1$ at the change point.

Suppose that there is a sequence u_n converging to ∞ such that the optimal strategy is to pay dividends at rate u_n on $[x_1, x_2]$. Then $\lim_{n\to\infty} V_{u_n}'(z) = 1$. By bounded convergence,

$$V(x_2) - V(x_1) = \lim_{n\to\infty} \int_{x_1}^{x_2} V_{u_n}'(z)\,\mathrm{d}z = x_2 - x_1\,.$$

If there also was a sequence $\{u_n'\}$ converging to ∞ such that $V_{u_n'}'(z) > 1$, then we would have

$$V_{u_n'}'(z) = \frac{(\lambda+\delta)V_{u_n'}(z) - \lambda \int_0^z V_{u_n'}(z-y)\,\mathrm{d}G(y)}{c} \ge \frac{c+\varepsilon}{c} ,$$

and $V(x_2) - V(x_1) \ge (x_2 - x_1)(c+\varepsilon)/c$ would follow. This is not possible. The same argument also shows that a change point is not possible. Thus, $\lim_{u\to\infty} V_u'(x) = 1$. A similar argument shows that

$$\lim_{u\to\infty} V_u'(x) = \frac{(\lambda+\delta)V(x) - \lambda \int_0^x V(x-y)\,\mathrm{d}G(y)}{c}$$

if there is a sequence u_n such that it is optimal not to pay dividends on $[x_1, x_2]$. The last possibility is that $V_u(z)$ is not differentiable on all $[x_1, x_2]$ for u large enough. By Lemma 2.34 there is a unique point $z_u \in [x_1, x_2]$ where $V_u(z)$ is not differentiable. Moreover, $V_u'(z) \le 1$ for $z < z_u$ and $V_u'(z) > 1$ for $z > z_u$. Choose now a sequence u_n such that z_{u_n} converges. If $y = \lim_{n\to\infty} z_{u_n} \ne x$, then the argument given above works on $[x_1, y]$ or $[y, x_2]$ if $x < y$ or $y < x$, respectively. Then $\lim_{u\to\infty} V_u'(x)$ exists. Consider therefore the case $\lim_{n\to\infty} z_{u_n} = x$. In this case the argument above shows that $f(z) = 1$ for $z < x$ and $f(z) > 1$ for $z > 1$. Thus, the left and the right limits of $f(z)$ exist at x. In particular, $f(x-) = 1$. This completes the proof. □

At zero we let $V'(0)$ be the derivative from the right, i.e., $V'(0) = (\lambda + \delta)V(0)/c$. Indeed, by Lemma 2.37 $V(0) \ge c/(\lambda+\delta)$ and therefore $(\lambda+\delta)V(0) \ge c$. By the proof of Theorem 2.39, the first expression on the left-hand side of (2.28) vanishes. Note that if the incoming premium in zero were paid immediately as a dividend, then $V(0) = c/(\lambda+\delta)$, i.e., $V'(x) = 1$.

The Optimal Dividend Strategy

Let

$$\mathfrak{B}_0 = \{x : V'(x-) > 1\} .$$

Here we let $V'(0-) = V'(0+)$. Let $\widetilde{\mathfrak{B}}_c = \{0\}$ if $V(0) = c/(\lambda+\delta)$ and let it be the empty set otherwise. We define the "upper boundary" of $\mathfrak{B}_0$,

$$\mathfrak{B}_c = \widetilde{\mathfrak{B}}_c \cup \{x \notin \mathfrak{B}_0 : \exists \{x_n\} \subset \mathfrak{B}_0,\, x_n \uparrow x\} ,$$

and

$$\mathfrak{B}_\infty = (0,\infty) \setminus (\mathfrak{B}_0 \cup \mathfrak{B}_c) .$$

We can characterise these sets in the following way.

Lemma 2.40. i) *The set $\mathfrak{B}_0$ is open.*

ii) *If $x \in \mathfrak{B}_\infty$, then there exists $\varepsilon > 0$ such that $(x-\varepsilon, x] \subset \mathfrak{B}_\infty$.*

iii) *If $x \notin \mathfrak{B}_\infty$ and there is a sequence $\{x_n\} \subset \mathfrak{B}_\infty$ such that $x_n \to x$, then $x \in \mathfrak{B}_c$.*

iv) *We have $(c\lambda/(\delta(\lambda+\delta)), \infty) \subset \mathfrak{B}_\infty$.*

Proof. i) Let $x \in \mathfrak{B}_0$. Then $V'(x-) > 1$ and we conclude from Theorem 2.39 that $V(x)$ is differentiable at x. But then by continuity it is differentiable in an environment of x, and $V'(z-) > 1$ in an environment.

ii) If $V'(0) = 1$, then $0 \in \mathfrak{B}_c$. Therefore, $x \neq 0$. By the definition of $\mathfrak{B}_c$, there must be $\varepsilon > 0$ such that $(x - \varepsilon, x) \cap \mathfrak{B}_0 = \emptyset$; otherwise, $x \in \mathfrak{B}_c$. This proves the assertion.

iii) We have $V'(x_n-) = 1$. If $V'(x)$, exists the derivative must be 1 and $x \notin \mathfrak{B}_0$. Suppose, therefore, that $V'(x)$ is not differentiable. Then $V'(x-) = 1$ and x is not in $\mathfrak{B}_0$.

iv) Note that because $V(x)$ is strictly increasing, we have $V(x) - \int_0^x V(x - y)\,\mathrm{d}G(y) \geq V(x)(1 - G(x)) \geq 0$. Suppose that there is $x > c\lambda/(\delta(\lambda+\delta))$ such that $V'(x-) > 1$. From Lemma 2.37 we conclude that

$$V(x) \geq x + \frac{c}{\lambda+\delta} > \frac{c\lambda}{\delta(\lambda+\delta)} + \frac{c}{\lambda+\delta} = \frac{c}{\delta}\,.$$

We claim that $V'(z) > 1$ for all $z \geq x$. If this were not the case then $z = \inf\{y > x : V'(y) = 1\} < \infty$. At this point, because $V'(y)$ does not have downward jumps, we also had

$$1 = V'(z) = \frac{(\lambda+\delta)V(z) - \lambda \int_0^z V(z-y)\,\mathrm{d}G(y)}{c} \geq \frac{\delta V(z)}{c} > \frac{\delta V(x)}{c} > 1\,,$$

which is a contradiction. Thus,

$$V'(z) \geq \frac{\delta}{c} V(z)$$

for all $z \geq x$, or $\log(V(z)/V(x)) \geq (z - x)\delta/c$. That means that $V(z)$ is exponentially increasing on $[x, \infty)$. This contradicts Lemma 2.37. Therefore, $V'(x-) = 1$. □

We now define the following strategy D^*. If $x \in \mathfrak{B}_0$, we do not pay dividends. Because $\mathfrak{B}_0$ is open, $t_0 = \inf\{t : x + ct \notin \mathfrak{B}_0\} > 0$. By the definition of $\mathfrak{B}_c$, we have $x + ct_0 \in \mathfrak{B}_c$, that is, a dividend is paid on (t_0, T_1) if $T_1 > t_0$. If $x \in \mathfrak{B}_c$, we pay a dividend $\mathrm{d}D_t^* = c\,\mathrm{d}t$ until the next jump occurs. This means that $X_t^* = x$ for $t < T_1$. If $x \in \mathfrak{B}_\infty$, we pay the sum

$$\Delta D_t^* = x - \sup\{z < x : z \notin \mathfrak{B}_\infty\}\,.$$

Note that by part ii) of Lemma 2.40, $\Delta D_t^* > 0$. By part iii), the jump ends at a point in $\mathfrak{B}_c$, i.e., a continuous dividend will be paid until the next jump. Note that by construction the corresponding dividend process D^* is measurable. Such a strategy is called a *band strategy*.

Theorem 2.41. *The strategy D^* is optimal, i.e., $V^*(x) = V(x)$.*

Proof. We have $X_0^* = x - D_0^*$. Let $J_t = 1\!\!1_{X_t^* \in \mathfrak{B}_c}$. By Lemma 2.40, the process D^* only jumps at the claim times. The process $\{X_t^* e^{-\delta t}\}$ is then a piecewise deterministic Markov process. We do not take the jumps of D^* into the generator. From [26, p. 27], we conclude that the process M' defined as

$$M_t' = \sum_{i=1}^{N_{\tau^*\wedge t}} [V(X_{T_i}^* + \Delta D_{T_i}^*) - V(X_{T_i-}^*)]e^{-\delta T_i} - \int_0^{\tau^*\wedge t} \lambda\Bigl[\int_0^{X_s^*} V(X_s^* - y)\,dG(y) - V(X_s^*)\Bigr]e^{-\delta s}\,ds$$

is a martingale. Moreover, because $V'(x) = 1$ on $\mathfrak{B}_\infty$,

$$V(X_{T_i}^* + \Delta D_{T_i}^*) - V(X_{T_i}^*) = \int_{X_{T_i}^*}^{X_{T_i}^* + \Delta D_{T_i}^*} V'(z)\,dz = \Delta D_{T_i}^*\,.$$

Finally,

$$V(X_{T_i-}^*)e^{-\delta T_i} - V(X_{T_{i-1}}^*)e^{-\delta T_{i-1}} = \int_{T_{i-1}}^{T_i} (cV'(X_s^*) - \delta V(X_s^*))e^{-\delta s} 1\!\!1_{J_s=0}\,ds - \int_{T_{i-1}}^{T_i} \delta V(X_s^*)e^{-\delta s} 1\!\!1_{J_s=1}\,ds\,.$$

This shows that

$$\Bigl\{V(X_{\tau^*\wedge t}^*)e^{-\delta t} + \sum_{i=1}^{N_{\tau^*\wedge t}} \Delta D_{T_i}^* e^{-\delta T_i} - \int_0^{\tau^*\wedge t} \Bigl[cV'(X_s^*) + \lambda\int_0^{X_s^*} V(X_s^* - y)\,dG(y) - (\lambda+\delta)V(X_s^*)\Bigr] 1\!\!1_{J_s=0}\, e^{-\delta s}\,ds - \int_0^{\tau^*\wedge t}\Bigl[\lambda\int_0^{X_s^*} V(X_s^* - y)\,dG(y) - (\lambda+\delta)V(X_s^*)\Bigr] J_s e^{-\delta s}\,ds\Bigr\}$$

is a martingale. On $\mathfrak{B}_c$ we have that both terms on the left-hand side of (2.28) vanish (this also holds for $x = 0$); therefore,

$$\lambda\int_0^{X_s^*} V(X_s^* - y)\,dG(y) - (\lambda+\delta)V(X_s^*) = -c\,.$$

On $\mathfrak{B}_\infty$ the process jumps out immediately. We have $V'(X_s^*) > 1$ on $\{J_s = 0\}$, and the integral over $\{J_s = 0\}$ vanishes. Thus,

$$\Bigl\{V(X_{\tau^*\wedge t}^*)e^{-\delta(\tau^*\wedge t)} + \sum_{i=1}^{N_{\tau^*\wedge t}} \Delta D_{T_i}^* e^{-\delta T_i} + \int_0^{\tau^*\wedge t} cJ_s e^{-\delta s}\,ds\Bigr\}$$

is a martingale. This yields

$$\begin{aligned} V(x) &= V(X_0^*) + D_0 \\ &= \mathbb{E}\Bigl[V(X^*_{\tau^*\wedge t})\mathrm{e}^{-\delta(\tau^*\wedge t)} + \sum_{i=0}^{N_{\tau^*\wedge t}} \Delta D^*_{T_i}\mathrm{e}^{-\delta T_i} + \int_0^{\tau^*\wedge t} cJ_s\mathrm{e}^{-\delta s}\,\mathrm{d}s\Bigr]\,. \end{aligned}$$

Using $V(X_t^*)\mathrm{e}^{-\delta t} \le V(x+ct)\mathrm{e}^{-\delta t} \le (x+ct+c/\delta)\mathrm{e}^{-\delta t}$ is bounded and converges to zero as $t \to \infty$, it follows that

$$\lim_{t\to\infty} \mathbb{E}[V(X^*_{\tau^*\wedge t})\mathrm{e}^{-\delta(\tau^*\wedge t)}] = 0\,.$$

By monotone convergence,

$$\begin{aligned} V(x) &= \lim_{t\to\infty} \mathbb{E}\Bigl[\sum_{i=0}^{N_{\tau^*\wedge t}} \Delta D^*_{T_i}\mathrm{e}^{-\delta T_i} + \int_0^{\tau^*\wedge t} cJ_s\mathrm{e}^{-\delta s}\,\mathrm{d}s\Bigr] \\ &= \mathbb{E}\Bigl[\sum_{i=0}^{N_{\tau^*}} \Delta D^*_{T_i}\mathrm{e}^{-\delta T_i} + \int_0^{\tau^*} cJ_s\mathrm{e}^{-\delta s}\,\mathrm{d}s\Bigr] = V^*(x)\,. \end{aligned}$$

Thus, $V(x) = V^*(x)$. □

Remark 2.42. Let $x_0 = \sup\{x : V'(x) > 1\}$. Then $X_t^* \le x_0$ for all t. Because the distribution of $\sum_{i=0}^{N_1} Y_i$ has unbounded support (note that N_1 is unbounded), we have $\mathbb{P}[\sum_{i=N_{n-1}+1}^{N_n} Y_i > x_0 + c] = \mathbb{P}[\sum_{i=1}^{N_1} Y_i > x_0 + c] > 0$. Because $\{\sum_{i=N_{n-1}+1}^{N_n} Y_i : n \in \mathbb{N}^*\}$ are iid, by the Borel–Cantelli lemma there must be an n such that $\sum_{i=N_{n-1}+1}^{N_n} Y_i > x_0+c$. At time n the process must be strictly negative. Thus, for the optimal strategy ruin occurs almost surely. ■

If $G(y)$ is continuously differentiable, we can find candidate points of $\mathfrak{B}_c$ in the following way.

Proposition 2.43. *Let $v(x) = (\lambda+\delta)^{-1}(c+\lambda\int_0^x V(x-y)\,\mathrm{d}G(y))$. Suppose that $G(y)$ is differentiable at $y = x$. If $x \in \mathfrak{B}_c \setminus \{0\}$, then $v'(x) = 1$.*

Proof. Note that

$$\int_0^x V(x-y)\,\mathrm{d}G(y) = V(0)G(x) + \int_0^x V'(y)G(x-y)\,\mathrm{d}y\,.$$

Because $V'(y)$ is bounded and continuous almost everywhere, the function $v(x)$ is differentiable at all points where $V(x)$ is differentiable. The function $v(x)$ is the value of the following strategy. Until the first claim occurs, the incoming premium is paid as dividend. After that the optimal strategy is followed. If $x \in \mathfrak{B}_c$, then $v(x) = V(x)$. At such a point $x > 0$ one has for $0 < h < x$, $v(x) = V(x) \ge V(x-h) + h \ge v(x-h) + h$. Thus, $v'(x) \ge 1$.

Next consider the following strategy: On the interval $(0, T_1 \wedge h)$ no dividend is paid. On the interval $[T_1 \wedge h, T_1]$ the incoming premium is paid. From T_1 on, the optimal strategy is followed. Let $v(x,h)$ denote the value of this strategy. We obtain

$$v(x,h) = \int_0^h e^{-\delta t} \int_0^{x+ct} V(x+ct-y)\, dG(y) \lambda e^{-\lambda t}\, dt + e^{-\lambda h} e^{-\delta h} v(x+ch)\,.$$

Note that $\lim_{h \downarrow 0} v(x,h) = v(x) = V(x)$. Because $v(x,h) \le V(x)$, we obtain

$$\begin{aligned} 0 &\ge \frac{v(x,h)-v(x)}{h} \\ &= \frac{1}{h}\int_0^h e^{-\delta t} \int_0^{x+ct} V(x+ct-y)\, dG(y)\lambda e^{-\lambda t}\, dt \\ &\quad + \frac{v(x+ch)-v(x)}{h} - \frac{1-e^{-(\lambda+\delta)h}}{h} v(x+ch)\,. \end{aligned}$$

Letting $h \downarrow 0$ yields

$$0 \ge \lambda \int_0^x V(x-y)\, dG(y) + cv'(x) - (\lambda+\delta)V(x) = cv'(x) - c\,,$$

where we used that both terms on the left-hand side of (2.28) are zero for $x \in \mathfrak{B}_c$. This shows that $v'(x) \le 1$. □

Remark 2.44. We only used that $V'(x) = 1$ and $c + \lambda \int_0^x V(x-y)\, dG(y) - (\lambda + \delta)V(x) = 0$. Thus, we have in fact proved that $v'(x) = 1$ for all $x \in \partial\mathfrak{B}_\infty \setminus \{0\}$, where $V(x)$ is differentiable. ■

Solving $v'(x) = 1$ yields candidate points. From these points the values of some candidate strategies can be calculated. However, because $V(x)$ is unknown, how to solve $v'(x) = 1$ is not straightforward. Later we will consider an example where an explicit solution can be obtained.

Characterisation of the Solution

We have not yet been able to determine $V(0)$. Moreover, we have not shown the uniqueness of a solution with a given boundary condition $V(0) = v_0$. It is therefore necessary to characterise the solution.

Theorem 2.45. *$V(x)$ is the minimal solution to (2.28). If $f(x)$ is a solution with only positive jumps of its derivative such that $f(x)e^{-\delta x/c} \to 0$ as $x \to \infty$ and either $f'(0) > 1$ or $f(0) = c/(\lambda+\delta)$, then $f(x) = V(x)$.*

Remarks 2.46. i) It is easy to see that $f(x) = x + c/\delta$ is a solution to (2.28). This solution cannot be $V(x)$. Indeed, by Theorem 2.41 the strategy $D(t) = x + c(T_1 \wedge t)$ would be optimal. But this strategy has the value $V^D(x) = x + c/(\lambda+\delta) < f(x)$.

ii) It is not possible to solve Equation (2.28) with an initial value $f(0) < V(0)$. This means that there must be a point x for which $(\lambda+\delta)f(x)-\lambda\int_0^x f(x-y)\,\mathrm{d}G(y) = c$, and the solution cannot be extended without allowing for $(\lambda+\delta)f(z) - \lambda\int_0^x f(z-y)\,\mathrm{d}G(y) < c$ or $f'(z) < 1$. ■

Proof. Consider the process X^* under the optimal strategy. As in the proof of Theorem 2.41, we find that

$$\Bigl\{f(X^*_{\tau^*\wedge t})\mathrm{e}^{-\delta(\tau^*\wedge t)} + \sum_{i=1}^{N_{\tau^*\wedge t}} (f(X^*_{T_i} + \Delta D^*_{T_i}) - f(X^*_{T_i}))\mathrm{e}^{-\delta T_i}$$
$$- \int_0^{\tau^*\wedge t} \Bigl[cf'(X^*_s) + \lambda\int_0^{X^*_s} f(X^*_s - y)\,\mathrm{d}G(y) - (\lambda+\delta)f(X^*_s)\Bigr]\mathbb{1}_{J_s=0}\mathrm{e}^{-\delta s}\,\mathrm{d}s$$
$$- \int_0^{\tau^*\wedge t} \Bigl[\lambda\int_0^{X^*_s} f(X^*_s - y)\,\mathrm{d}G(y) - (\lambda+\delta)f(X^*_s)\Bigr] J_s\mathrm{e}^{-\delta s}\,\mathrm{d}s\Bigr\}$$

is a martingale. Because $f'(x) \ge 1$, one has $f(x) \ge f(X^*_0) + D^*_0$. Also, at the claim times, $f(X^*_{T_i} + \Delta D^*_{T_i}) \ge f(X^*_{T_i}) + \Delta D^*_{T_i}$. By (2.28),

$$cf'(X^*_s) + \int_0^{X^*_s} f(X^*_s - y)\,\mathrm{d}G(y) - (\lambda+\delta)f(X^*_s) \le 0$$

and

$$\int_0^{X^*_s} f(X^*_s - y)\,\mathrm{d}G(y) - (\lambda+\delta)f(X^*_s) \le -cf'(X^*_s) \le -c\,.$$

This yields

$$f(x) \ge \mathbb{E}\Bigl[f(X^*_t)\mathrm{e}^{-\delta t} + \sum_{i=0}^{N_{\tau^*\wedge t}} \Delta D^*_{T_i}\mathrm{e}^{-\delta T_i} + \int_0^{\tau^*\wedge t} cJ_s\mathrm{e}^{-\delta s}\,\mathrm{d}s\Bigr]$$
$$\ge \mathbb{E}\Bigl[\sum_{i=0}^{N_{\tau^*\wedge t}} \Delta D^*_{T_i}\mathrm{e}^{-\delta T_i} + \int_0^{\tau^*\wedge t} cJ_s\mathrm{e}^{-\delta s}\,\mathrm{d}s\Bigr]\,.$$

By monotone convergence, it follows that $f(x) \ge V^{D^*}(x) = V(x)$.

Suppose now that $f(x)$ is a solution to (2.28) with $f(x)\mathrm{e}^{-\delta x/c} \to 0$. Let $\mathfrak{B}'_0 = \{x : f'(x-) > 1\}$, $\widetilde{\mathfrak{B}}_c = \{0\}$ if $f(0) = c/(\lambda+\delta)$, and $\widetilde{\mathfrak{B}}'_c = \emptyset$ otherwise. Let

$$\mathfrak{B}'_c = \widetilde{\mathfrak{B}}'_c \cup \{x \notin \mathfrak{B}'_0 : \exists\{x_n\} \subset \mathfrak{B}'_0,\ x_n \uparrow x\}\,,$$

and $\mathfrak{B}'_\infty = (0,\infty)\setminus(\mathfrak{B}'_0 \cup \mathfrak{B}'_c)$. The results of Lemma 2.40 parts i)–iv) remain valid. Denote by D' the strategy corresponding to $f(x)$ defined in the same way as D^*, i.e., no dividends if $x \in \mathfrak{B}'_0$, dividends at rate c if $x \in \mathfrak{B}'_c$, and a dividend of size $\Delta D'_t = x - \sup\{z < x : z \notin \mathfrak{B}'_\infty\}$ if $X_t \in \mathfrak{B}'_\infty$. In the same way is in the proof of Theorem 2.41 we find that

$$\Big\{f(X_{\tau\wedge t})\mathrm{e}^{-\delta(\tau\wedge t)} + \sum_{i=1}^{N_{\tau\wedge t}} \Delta D'_{T_i}\mathrm{e}^{-\delta T_i} + \int_0^{\tau\wedge t} cJ'_s\mathrm{e}^{-\delta s}\,\mathrm{d}s\Big\}$$

is a martingale, where $J'_t = \mathrm{1\!I}_{X_t\in\mathfrak{B}'_c}$. Taking expectations and letting $t\to\infty$ yields the assertion provided that $\mathbb{E}[f(X_{\tau\wedge t})\mathrm{e}^{-\delta(\tau\wedge t)}]\to 0$. We have that

$$f(X_{\tau\wedge t})\mathrm{e}^{-\delta(\tau\wedge t)} \le f(x+ct)\mathrm{e}^{-\delta t} = f(x+ct)\mathrm{e}^{-\delta(x+ct)/c}\mathrm{e}^{\delta x/c}$$

is bounded by our assumption and tends to zero as $t\to\infty$. This proves the theorem. □

Remark 2.47. A solution of the form $f(x) = x + \kappa$ indicates that a dividend at rate c should be paid if $X_t = 0$. Because $-(\lambda+\delta)f(0) = -\kappa(\lambda+\delta)$, the process

$$\Big\{\kappa\mathrm{e}^{-\delta t}\mathrm{1\!I}_{t<T_1} + \int_0^{T_1\wedge t} \kappa(\lambda+\delta)\mathrm{e}^{-\delta s}\,\mathrm{d}s\Big\}$$

is a martingale. The integral then only corresponds to the value of the dividends if $\kappa(\lambda+\delta) = c$. If $f'(0) > 1$, then the strategy proposed above does not pay a dividend at zero. This explains the condition on the solution in 0. Moreover, the above proof implies that a solution $f(x)$ to (2.28) can only be bounded by a linear function, if the initial value is correct, or if $f(x)$ is linear for x close to zero. ■

For the initial capital zero there are two possibilities. Either a dividend may be paid in zero, i.e., $V(0) = c/(\lambda+\delta)$, or no dividend is paid, i.e., there is a value $x_0 = \inf\{x : V'(x) = 1\} > 0$, and the barrier strategy paying capital exceeding x_0 is optimal for $x\in[0,x_0]$. The value $V(0)$ can therefore be found by comparing the barrier strategies with a barrier at x_0 for all x_0. On $[0,x_0]$ one therefore has to solve the equation

$$cf'(x) + \lambda\int_0^x f(x-y)\,\mathrm{d}G(y) - (\lambda+\delta)f(x) = 0\,. \tag{2.30}$$

Let us therefore first consider Equation (2.30).

Lemma 2.48. *There is a unique solution to* (2.30) *with* $f(0) = 1$.

Proof. Let $x_1 = \inf\{x : f'(x)\le 0\}$. Then on $[0,x_1)$ we get

$$f'(x) = \frac{1}{c}\Big[(\lambda+\delta)f(x) - \lambda\int_0^x f(x-y)\,\mathrm{d}G(y)\Big] > \frac{\delta}{c}f(x)\,.$$

Thus, $x_1 = \infty$ and $f(x) > \mathrm{e}^{\delta x/c}$ is exponentially increasing. On the other hand, $f'(x)\le(\lambda+\delta)f(x)/c$, and so there is also an exponential upper bound $f(x)\le\mathrm{e}^{(\lambda+\delta)x/c}$. We first construct a solution on the interval $[0,x_0]$ for any $x_0 > 0$. Because there is an exponential upper bound, it will follow from uniqueness that the solution exists on $[0, \infty)$. We let $f(x_0) = 1$. The solution with initial value 1 is then $f(x)/f(0)$. Let $\tau_0 = \inf\{t > 0 : X_t^0 > x_0\}$ where X^0 is the process without paying dividends. Define the function $f(x) = \mathbb{E}[\mathrm{e}^{-\delta\tau_0} \mathbb{I}_{\tau_0 < \tau}]$. Clearly, $f(x_0) = 1$. Suppose that $x < x_0$ and let $h < (x_0 - x)/c$. We consider the process on $[0, T_1 \wedge h]$. Then by the Markov property,

$$f(x) = \mathrm{e}^{-\lambda h}\mathrm{e}^{-\delta h} f(x + ch) + \int_0^h \int_0^{x+ct} \mathrm{e}^{-\delta t} f(x + ct - y) \,\mathrm{d}G(y) \lambda \mathrm{e}^{-\lambda t} \,\mathrm{d}t \;.$$

Letting $h \downarrow 0$ shows that $f(x)$ is right-continuous. Reordering of the terms and dividing by h yields

$$\frac{f(x + ch) - f(x)}{h} - \frac{1 - \mathrm{e}^{-(\lambda+\delta)h}}{h} f(x + ch) + \lambda \frac{1}{h} \int_0^h \mathrm{e}^{-(\lambda+\delta)t} \int_0^{x+ct} f(x + ct - y) \,\mathrm{d}G(y) \,\mathrm{d}t = 0 \;.$$

Letting $h \downarrow 0$ shows that $f(x)$ is differentiable from the right, and the derivative from the right solves (2.30). Replacing x by $x - ch$ shows in the same way that $f(x)$ is left-continuous and differentiable from the left. The two derivatives coincide. Thus, $f(x)$ is a differentiable solution to (2.30).

Suppose now that $f(x)$ solves (2.30) with $f(0) = 1$. As in the proof of Theorem 2.41, it follows that

$$\Bigl\{ f(X_{\tau_0 \wedge \tau \wedge t}) \mathrm{e}^{-\delta(\tau_0 \wedge t)} - \int_0^{\tau_0 \wedge \tau \wedge t} \Bigl[c f'(X_s) + \int_0^{X_s} f(X_s - y) \,\mathrm{d}G(y) - (\lambda + \delta) f(X_s) \Bigr] \mathrm{e}^{-\delta s} \,\mathrm{d}s \Bigr\}$$

is a martingale, i.e., $\{f(X_{\tau_0 \wedge \tau \wedge t}) \mathrm{e}^{-\delta(\tau_0 \wedge t)}\}$ is a martingale. This yields by bounded convergence

$$f(x) = \mathbb{E}[f(X_{\tau_0 \wedge \tau}) \mathrm{e}^{-\delta\tau_0}] = f(x_0) \mathbb{E}[\mathrm{e}^{-\delta\tau_0} \mathbb{I}_{\tau_0 < \tau}]$$

because $f(X_\tau) = 0$ and $\tau_0 \wedge \tau < \infty$. Note that $f(x_0)$ is determined by the boundary condition $f(0) = 1$. Because x_0 is arbitrary, the solution is unique. □

We call a strategy a *barrier strategy* with a barrier at x_0 if $D_0 = (x - x_0)^+$ and $\Delta D_t = c \mathbb{I}_{X_t = x_0}$, that is, all surplus exceeding the level x_0 is paid as a dividend. We next find the value of the barrier strategy with a barrier at x_0.

Lemma 2.49. *Let $f(x)$ denote the solution to (2.30) with $f(0) = 1$. The value of the barrier strategy with a barrier at x_0 is $V_{x_0}(x) = f(x)/f'(x_0)$ if $x \le x_0$ and $V_{x_0}(x) = V_{x_0}(x_0) + x - x_0$ if $x > x_0$.*

Proof. Because any capital larger than x_0 will be paid as a dividend immediately, $V_{x_0}(x)$ is the correct value for $x > x_0$ provided that $V_{x_0}(x_0)$ is the correct value for the initial value x_0. Note that for $x \le x_0$, $V_{x_0}(x)$ solves (2.30) and $V'_{x_0}(x_0) = 1$. Let $J_s = \mathbb{1}_{X_s = x_0}$. As in the proof of Theorem 2.41, we get that

$$\Bigl\{V_{x_0}(X_{\tau\wedge t})\mathrm{e}^{-\delta t} + \int_0^{\tau\wedge t} cJ_s \mathrm{e}^{-\delta s}\,\mathrm{d}s\Bigr\}$$

is a martingale. Thus,

$$V_{x_0}(x) = \mathbb{E}\Bigl[V_{x_0}(X_{\tau\wedge t})\mathrm{e}^{-\delta t} + \int_0^{\tau\wedge t} cJ_s \mathrm{e}^{-\delta s}\,\mathrm{d}s\Bigr]\,,$$

and the result follows by letting $t \to \infty$. □

We now can determine $V(0)$.

Proposition 2.50. *Let $f(x)$ denote the solution to* (2.30) *with $f(0) = 1$. The value $V(0)$ is given by*

$$V(0) = \sup_{x\ge 0} (f'(x))^{-1}\,.$$

Proof. We know from Theorem 2.41 that the optimal strategy is a barrier strategy, with the possibility of a barrier at zero. By Lemma 2.49, the value at zero of a barrier strategy with a barrier at x_0 is $1/f'(x_0)$. Note that as a solution to (2.30) the result also holds for $x_0 = 0$ because $V_0(0) = c/(\lambda+\delta)$. Because any barrier strategy is admissible, we have to maximise $1/f'(x_0)$. □

The solution can be constructed in the following way. Let $f_0(x)$ be the solution to (2.30) and $x_0 = \sup\{x : f_0'(x) = \inf_y f_0'(y)\}$. Then we let

$$V_0(x) = \begin{cases} f_0(x)/f_0'(x_0)\,, & \text{for } x \le x_0,\\ f_0(x_0)/f_0'(x_0) + x - x_0\,, & \text{for } x > x_0. \end{cases}$$

If $V_0(x)$ solves (2.28), then $V(x) = V_0(x)$. Suppose that we have constructed $V_n(x)$ and x_n. For each $a > x_n$, let $f_{n+1}(x;a)$ be the function with $f_{n+1}(x;a) = V_n(x)$ for $x \le a$, and $f_{n+1}(x;a)$ solves (2.30) for $x > a$. Then we have to choose

$$a = \inf\Bigl\{x > x_0 : \inf_{y>x} f'_{n+1}(y;x) = 1\Bigr\}\,,$$

where the derivative is taken with respect to the first argument. Namely, if a is chosen to be smaller, then the derivative will not reach 1 again. If the derivative is chosen to be larger, then the derivative will reach a value smaller than 1. These conclusions can be derived directly from (2.30). Let $x_{n+1} = \sup\{x : f'(x;a) = 1\}$. The function V_{n+1} is defined as

$$V_{n+1}(x) = \begin{cases} f_{n+1}(x;a)\,, & \text{for } x \le x_{n+1},\\ f_{n+1}(x_{n+1};a) + x - x_{n+1}\,, & \text{for } x > x_{n+1}. \end{cases}$$

If $V_{n+1}(x)$ solves (2.28), then $V(x) = V_{n+1}(x)$. If not, the construction can proceed in the same way.

Exponentially Distributed Claim Sizes

Suppose now that $G(y) = 1 - \mathrm{e}^{-\alpha y}$. Even though we have already calculated the strategy on the interval $[0, a]$ in Example 2.35, here we do the calculations again with the tools we developed for the unrestricted case. We start by discussing for which cases the surplus is immediately paid as dividends, i.e., $V(x) = x + c/(\lambda + \delta)$. The first expression of the left-hand side of (2.28) is

$$\lambda\Bigl(\frac{c}{\lambda+\delta} - \frac{1}{\alpha}\Bigr)(1 - \mathrm{e}^{-\alpha x}) - \delta x\,. \tag{2.31}$$

If $\alpha c \le \lambda + \delta$, then this is negative and $x + c/(\lambda + \delta)$ solves (2.28), i.e., it is equal to $V(x)$. If $\alpha c > \lambda + \delta$, then the function (2.31) is strictly concave. It is therefore negative if and only if its derivative in zero is negative, i.e., $\lambda(c\alpha/(\lambda+\delta)-1) \le \delta$, or equivalently, $\lambda c\alpha \le (\lambda+\delta)^2$. Thus, $V(x) = x+c/(\lambda+\delta)$ if and only if $\lambda c\alpha \le (\lambda + \delta)^2$.

Let us now assume that $\lambda c\alpha > (\lambda+\delta)^2$. The function $G(y)$ is continuously differentiable. Therefore, all points in $\mathfrak{B}_c \setminus \{0\}$ can be found from $v'(x) = 1$.

$$\begin{aligned}
v(x) &= \frac{c}{\lambda+\delta} + \frac{\lambda}{\lambda+\delta}\int_0^x V(x-y)\alpha\mathrm{e}^{-\alpha y}\,\mathrm{d}y \\
&= \frac{c}{\lambda+\delta} + \frac{\lambda\alpha}{\lambda+\delta}\mathrm{e}^{-\alpha x}\int_0^x V(y)\mathrm{e}^{\alpha y}\,\mathrm{d}y\,,
\end{aligned}$$

and by replacing the integral by the definition of $v(x)$, using $v(x) = V(x)$,

$$1 = v'(x) = -\frac{\lambda\alpha^2}{\lambda+\delta}\mathrm{e}^{-\alpha x}\int_0^x V(y)\mathrm{e}^{\alpha y}\,\mathrm{d}y + \frac{\lambda\alpha}{\lambda+\delta}V(x) = \frac{c\alpha}{\lambda+\delta} - \frac{\alpha\delta}{\lambda+\delta}V(x)\,.$$

This gives

$$v(x) = V(x) = \frac{c\alpha - \lambda - \delta}{\alpha\delta}\,. \tag{2.32}$$

Because $V(x)$ is a strictly increasing function, this is only possible for at most one point, i.e., $\mathfrak{B}_c \setminus \{0\}$ consists of at most one point, a say. Because $\mathfrak{B}_c \setminus \{0\} \ne \emptyset$, a point $a > 0$ exists. If dividends were paid for x close to zero, then $V(x) = x + c/(\lambda + \delta)$ close to zero. The considerations above show that this is not possible if $c\alpha > (\lambda+\delta)^2/\lambda$. Thus, no dividend is paid for $0 < x < a$, and $V(a + x) = V(a) + x$ for $x \ge 0$.

On $(0, a)$ the function satisfies

$$cV'(x) = (\lambda + \delta)V(x) - \lambda\mathrm{e}^{-\alpha x}\int_0^x V(y)\alpha\mathrm{e}^{\alpha y}\,\mathrm{d}y\,. \tag{2.33}$$

The right-hand side is differentiable. Therefore,

$$cV''(x) = (\lambda + \delta)V'(x) + \lambda\alpha\mathrm{e}^{-\alpha x}\int_0^x V(y)\alpha\mathrm{e}^{\alpha y}\,\mathrm{d}y - \alpha\lambda V(x)\,.$$

Plugging in (2.33) yields

$$cV''(x) = (\lambda + \delta - \alpha c)V'(x) + \alpha\delta V(x)\,.$$

The solution to this differential equation is

$$V(x) = C_1 e^{\theta_1 x} + C_2 e^{-\theta_2 x}\,,$$

where θ_1 and $-\theta_2$ solve

$$c\theta^2 - (\lambda + \delta - \alpha c)\theta - \alpha\delta = 0\,,$$

that is,

$$\theta_1 = \frac{\sqrt{(\alpha c - \lambda - \delta)^2 + 4\alpha\delta c} - (\alpha c - \lambda - \delta)}{2c}\,,$$
$$\theta_2 = \frac{\sqrt{(\alpha c - \lambda - \delta)^2 + 4\alpha\delta c} + (\alpha c - \lambda - \delta)}{2c}\,.$$

Note that

$$\theta_2 = \frac{\sqrt{(\alpha c + \lambda + \delta)^2 - 4\alpha\lambda c} + (\alpha c - \lambda - \delta)}{2c} < \alpha\,.$$

Plugging $V(x)$ into (2.33) gives

$$C_2 = -\frac{\alpha - \theta_2}{\alpha + \theta_1} C_1\,.$$

From $V'(a) = 1$ we find that

$$C_1 = \frac{\alpha + \theta_1}{\theta_1(\alpha + \theta_1)e^{\theta_1 a} + \theta_2(\alpha - \theta_2)e^{-\theta_2 a}}\,.$$

This yields

$$V(a) = \frac{(\alpha + \theta_1)e^{\theta_1 a} - (\alpha - \theta_2)e^{-\theta_2 a}}{\theta_1(\alpha + \theta_1)e^{\theta_1 a} + \theta_2(\alpha - \theta_2)e^{-\theta_2 a}} = \frac{\alpha c - \lambda - \delta}{\alpha\delta}\,,$$

where we used (2.32) in the second equality. We find that

$$e^{(\theta_1+\theta_2)a} = \frac{-(\alpha c - \lambda - \delta)\theta_2^2 + \alpha(\alpha c - \lambda - 2\delta)\theta_2 + \alpha^2\delta}{-(\alpha c - \lambda - \delta)\theta_1^2 - \alpha(\alpha c - \lambda - 2\delta)\theta_1 + \alpha^2\delta} = \frac{(\lambda + \delta)\theta_2^2 - \alpha\delta\theta_2}{(\lambda + \delta)\theta_1^2 + \alpha\delta\theta_1}\,.$$

Thus, we have

$$a = \frac{1}{\theta_1 + \theta_2} \log \frac{\theta_2((\lambda + \delta)\theta_2 - \alpha\delta)}{\theta_1((\lambda + \delta)\theta_1 + \alpha\delta)}$$

and

$$V(x) = \begin{cases} \frac{(\alpha+\theta_1)e^{\theta_1 x} - (\alpha-\theta_2)e^{-\theta_2 x}}{\theta_1(\alpha+\theta_1)e^{\theta_1 a} + \theta_2(\alpha-\theta_2)e^{-\theta_2 a}}, & \text{for } \alpha\lambda c > (\lambda + \delta)^2 \text{ and } x < a, \\ \frac{\alpha c - \lambda - \delta}{\alpha\delta} + x - a, & \text{for } \alpha\lambda c > (\lambda + \delta)^2 \text{ and } x \ge a, \\ x + \frac{c}{\lambda+\delta}, & \text{for } \alpha\lambda c \le (\lambda + \delta)^2. \end{cases} \tag{2.34}$$

Figure 2.16 shows the value function if $\alpha = \lambda = 1$, $\delta = 0.1$, and $c = 2$. The barrier is at $a = 4.21407$.

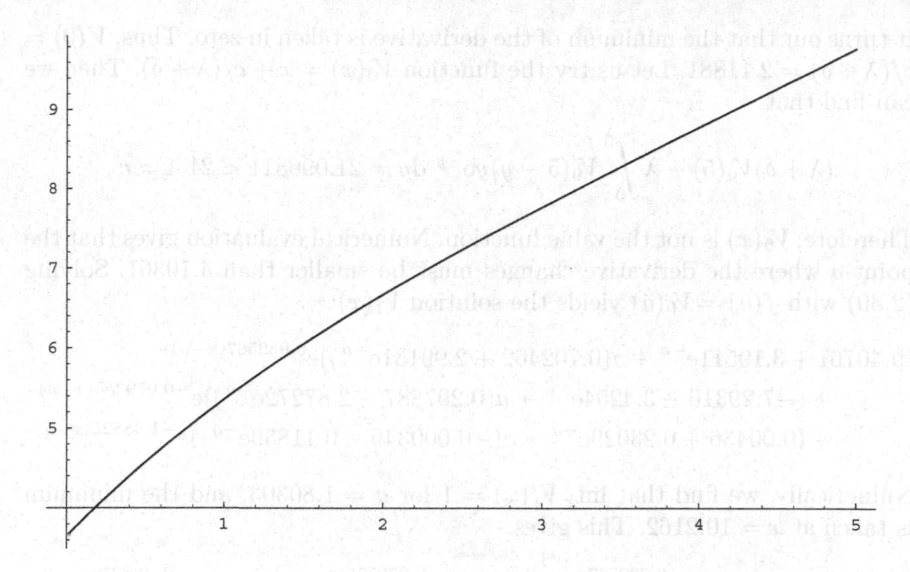

Fig. 2.16. Value function for exponentially distributed claim sizes.

Some Further Examples

Mixed exponentially distributed claim sizes

Let us consider an example with $\lambda = c = 1$, $\delta = 0.1$, and $G(y) = 1 - \frac{1}{2}(e^{-y} + e^{-2y})$. The solution to (2.30) with $f(0) = 1$ is (obtained, for example, via the Laplace transform)

$$f(x) = 2.039864e^{0.261423x} - 0.958743e^{-0.445973x} - 0.081121e^{-1.71545x} .$$

The minimum of the derivative is taken at $x = 0.911789$, and $f'(0.911789) = 0.990644$. This yields

$$V_0(x) = 2.05913e^{.261423x} - 0.967798e^{-0.445973x} - 0.081887e^{-1.71545x} .$$

Direct verification shows that

$$V(x) = \begin{cases} V_0(x) , & \text{for } x \le 0.911789 , \\ x + 1.04001, & \text{for } x > 0.911789. \end{cases}$$

Gamma-distributed claim sizes

The following example is taken from [12]. Let $\lambda = 10$, $c = 21.4$, $\delta = 0.1$, and $G(x) = 1 - (y+1)e^{-y}$. The solution to (2.30) with $f(0) = 1$ is

$$f(x) = 5.947983e^{0.039567x} - 5.058731e^{-0.079355x} + 0.110748e^{-1.48825x} .$$

It turns out that the minimum of the derivative is taken in zero. Thus, $V(0) = c/(\lambda+\delta) = 2.11881$. Let us try the function $V_0(x) = x + c/(\lambda+\delta)$. Then we can find that

$$(\lambda+\delta)V_0(5) - \lambda\int_0^5 V_0(5-y)y\mathrm{e}^{-y}\,\mathrm{d}y = 21.096811 < 21.4 = c\,.$$

Therefore, $V_0(x)$ is not the value function. Numerical evaluation gives that the point a where the derivative changes must be smaller than 4.10361. Solving (2.30) with $f(a) = V_0(a)$ yields the solution $V_1(x)$:

$$\begin{aligned}(9.40761 &+ 3.19511\mathrm{e}^{-a} + a(0.702462 + 2.99131\mathrm{e}^{-a}))\mathrm{e}^{0.039567(x-a)}\\ &+ (-7.29316 - 3.4254\mathrm{e}^{-a} + a(0.297887 - 2.87272\mathrm{e}^{-a}))\mathrm{e}^{-0.079355(x-a)}\\ &+ (0.00436 + 0.23029\mathrm{e}^{-a} + a(-0.000349 - 0.11859\mathrm{e}^{-a}))\mathrm{e}^{-1.48825(x-a)}\,.\end{aligned}$$

Numerically, we find that $\inf_x V_1'(x) = 1$ for $a = 1.80303$, and the minimum is taken at $x = 10.2162$. This gives

$$V_1(x) = 11.2571\mathrm{e}^{0.039567x} - 9.43151\mathrm{e}^{-0.079355x} + 0.094314\mathrm{e}^{-1.48825x}\,.$$

The value function is therefore

$$V(x) = \begin{cases} x + 2.11881\,, & \text{for } x \le 1.80303,\\ V_1(x)\,, & \text{for } 1.80303 < x \le 10.2162,\\ x + 2.45582\,, & \text{for } x > 10.2162.\end{cases}$$

Note that $V_0'(1.80303) = 1 < V_1'(1.80303) = 1.11742$, that is, $V(x)$ is not differentiable at a.

Bibliographical Remarks

The problem was first considered by Gerber [72]. He showed by discretisation that the optimal strategy exists and is a band strategy. He found the bounds of Lemma 2.37 and proved the continuity of the value function via (2.24). Proposition 2.43 and the example with exponentially distributed claim sizes are also taken from there. Optimal barrier strategies were also considered in Bühlmann [29], Dassios and Embrechts [38], and Tobler [181]. Dickson and Waters [47] considered the mth moment of the dividend payoff for a barrier strategy. Azcue and Muler [12] considered optimal dividend strategies when, in addition, the insurer has the possibility to reinsure the insurance portfolio. Thonhauser and Albrecher [180] extended the value function by a penalty $-\ell\mathrm{e}^{-\delta\tau}$ for early ruin with $\ell > 0$.

Modifications of the model also were considered. Gerber [75, 77] considered a barrier strategy with a linear dividend barrier with slope in $(0, c)$. He obtained an upper bound and an explicit solution in the case of exponentially distributed claim sizes. Siegl and Tichy [169, 170] obtained the value of the dividends and the ruin probability in the case of Erlang-distributed claim sizes. Albrecher and Tichy [3] proved that the algorithm introduced in [170] converges. Dickson and Waters [47] considered barrier strategies where the shareholders have to cover the deficit at ruin in both cases where business stops at ruin and business can be continued after ruin. The first model is also discussed by Gerber, Shiu, and Smith [80], which considered the optimal dividend barriers with and without costs at ruin. Højgaard [103] maximised dividend payouts through a control of the premia. A dividend barrier is fixed, but lowering premia increases the number of customers.

For a barrier strategy, Irbäck [109] studied the distribution of the ruin time and found the asymptotic distribution if the barrier is at a high level.

2.5 Optimal Dividends for a Diffusion Approximation

In this section we replace the Cramér–Lundberg model with a diffusion model, i.e., under a strategy with accumulated dividends D, the surplus is

$$X_t^D = x + mt + \sigma W_t - D_t ,$$

where W is a standard Brownian motion. That is, we approximate the risk model by a diffusion approximation; see Section D.3. The filtration $\{\mathcal{F}_t\}$ is the filtration generated by the Brownian motion. Even though we could solve the problem directly, we proceed as in Section 2.4, i.e., we first only allow $D_t = \int_0^t U_s \,\mathrm{d}s$ with $U_s \in [0, u_0]$. This helps us to understand the Hamilton–Jacobi–Bellman equation (2.40).

2.5.1 Restricted Dividend Payments

Consider the process

$$X_t^U = x + mt + \sigma W_t - \int_0^t U_s \,\mathrm{d}s .$$

We only allow adapted processes U with $0 \le U_t \le u_0$. The process is stopped at the time $\tau^U = \inf\{t \ge 0 : X_t^U < 0\}$. The value of the process is $V^U(x) = \mathbb{E}[\int_0^{\tau^U} U_s \mathrm{e}^{-\delta s} \,\mathrm{d}s]$, and $V(x) = \sup_U V^U(x)$. Note that $V(0) = 0$ because ruin will happen immediately by the fluctuations of the Brownian motion. As in Section 2.4.1, we find that $V(x) \le u_0/\delta$, and V is a bounded function. Let $y > x$, U be a strategy for the starting value x, and $\hat{U}$ be a strategy for initial capital $y - x$, where the process is determined by the Brownian motion $\{W_{\tau_x+t} - W_{\tau_x}\}$. Then the strategy with $\tilde{U}_t = U_t$ for $t \le \tau_x$, $\tilde{U}_t = \hat{U}_{t-\tau_x}$ for $\tau_x < t \le \tau_y$ is a strategy for initial capital y. This shows that $V(y) \ge V(x) + \mathbb{E}[\mathrm{e}^{-\delta\tau_x}]V(y - x)$. In particular, $V(x)$ is strictly increasing. It is easy to see that $\lim_{x\to\infty} V(x) = u_0/\delta$.

We start by motivating the Hamilton–Jacobi–Bellman equation. Let $\varepsilon > 0$. Then for each $x > 0$ there is a dividend strategy U^x such that $V^{U^x}(x) \ge V(x) - \varepsilon$. Fix $0 \le u \le u_0$ and $h > 0$ and consider the dividend strategy

$$U_t = \begin{cases} u, & \text{if } 0 \le t \le \tau \wedge h, \\ U^{X_h}_{t-h}, & \text{if } t > h \text{ and } \tau > h. \end{cases}$$

We do not worry here about measurability. Then we find that

$$\begin{aligned} V(x) \ge V^U(x) &= \mathbb{E}\Bigl[\int_0^{\tau\wedge h} u e^{-\delta s}\,ds\Bigr] + \mathbb{E}\Bigl[\mathbb{1}_{\tau > h}\int_h^\tau U_s e^{-\delta(s+h)}\,ds\Bigr] \\ &= u\mathbb{E}\Bigl[\frac{1-e^{-\delta(\tau\wedge h)}}{\delta}\Bigr] + e^{-\delta h}\mathbb{E}[\mathbb{1}_{\tau>h} V^{U^{X_h}}(X_h)] \\ &\ge u\frac{1-\mathbb{E}[e^{-\delta(\tau\wedge h)}]}{\delta} + e^{-\delta h}\mathbb{E}[\mathbb{1}_{\tau>h}(V(X_h)-\varepsilon)] \\ &\ge u\frac{1-\mathbb{E}[e^{-\delta(\tau\wedge h)}]}{\delta} + e^{-\delta h}\mathbb{E}[V(X_{\tau\wedge h}) - \varepsilon]\;. \end{aligned}$$

Here we used that $V(X_\tau) = V(0) = 0$. Because ε is arbitrary, we have

$$V(x) \ge u\frac{1-\mathbb{E}[e^{-\delta(\tau\wedge h)}]}{\delta} + e^{-\delta h}\mathbb{E}[V(X_{\tau\wedge h})]\;.$$

Assume now that $V(x)$ is twice continuously differentiable. Then by Itô's formula,

$$\begin{aligned} V(X_{\tau\wedge h}) = V(x) &+ \int_0^{\tau\wedge h} \sigma V'(X_s)\,dW_s \\ &+ \int_0^{\tau\wedge h}\Bigl((m-u)V'(X_s) + \frac{\sigma^2}{2}V''(X_s)\Bigr)\,ds\;. \end{aligned}$$

We assume that $\{\int_0^t V'(X_s)\,dW_s\}$ is a martingale. Then the expectation of the stochastic integral is zero. This yields

$$\begin{aligned} &u\mathbb{E}\Bigl[\frac{1-e^{-\delta(\tau\wedge h)}}{\delta h}\Bigr] - \frac{1-e^{-\delta h}}{h}V(x) \\ &\qquad + e^{-\delta h}\mathbb{E}\Bigl[\frac{1}{h}\int_0^{h\wedge\tau}\Bigl((m-u)V'(X_s) + \frac{\sigma^2}{2}V''(X_s)\Bigr)\,ds\Bigr] \le 0\;. \end{aligned}$$

Assume that it is allowed to interchange the limit and expectation. Then, letting $h \to 0$ gives

$$\tfrac{1}{2}\sigma^2 V''(x) + (m-u)V'(x) - \delta V(x) + u \le 0\;.$$

This inequality must be true for all $0 \le u \le u_0$; thus,

$$\sup_{0\le u\le u_0}\;\{\tfrac{1}{2}\sigma^2 V''(x) + (m-u)V'(x) - \delta V(x) + u\} \le 0\;.$$

We will not argue that the reversed inequality also holds. That will follow automatically after solving the equation and verifying that its solution is the solution to our problem. We therefore now consider the *Hamilton–Jacobi–Bellman equation*

$$\sup_{0\le u\le u_0} \{\tfrac{1}{2}\sigma^2 f''(x) + (m-u)f'(x) - \delta f(x) + u\} = 0\,. \tag{2.35}$$

Thus,

$$u(x) = \begin{cases} 0, & \text{if } f'(x) > 1, \\ \in [0, u_0], & \text{if } f'(x) = 1, \\ u_0, & \text{if } f'(x) < 1, \end{cases}$$

is the argument u at which the supremum is taken. If $f'(x) = 1$, the equation does not depend on u.

We next want to find a twice continuously differentiable solution to (2.35). There will be intervals on which $u(x) = 0$ and intervals on which $u(x) = u_0$. These solutions can be put together at points where $f'(x) = 1$. We therefore solve the two equations separately. On $\{f'(x) > 1\}$ the solution is of the form

$$f_1(x) = A\mathrm{e}^{\theta_1(0)x} + B\mathrm{e}^{-\theta_2(0)x}\,,$$

and on $\{f'(x) < 1\}$ of the form

$$f_2(x) = \frac{u_0}{\delta} + C\mathrm{e}^{\theta_1(u_0)x} + D\mathrm{e}^{-\theta_2(u_0)x}\,,$$

where

$$\theta_1(u) = \frac{\sqrt{(m-u)^2 + 2\delta\sigma^2} - (m-u)}{\sigma^2} > 0 \tag{2.36a}$$

and

$$\theta_2(u) = \frac{\sqrt{(m-u)^2 + 2\delta\sigma^2} + (m-u)}{\sigma^2} > 0\,. \tag{2.36b}$$

Let x_0 be a point where $f_1'(x_0) = 1$. This yields

$$A\theta_1(0)\mathrm{e}^{\theta_1(0)x_0} - B\theta_2(0)\mathrm{e}^{-\theta_2(0)x_0} = 1\,,$$

giving an expression for B. Plugging this expression into the solution yields

$$f_1(x_0) = A\frac{\theta_1(0) + \theta_2(0)}{\theta_2(0)}\mathrm{e}^{\theta_1(0)x_0} - \frac{1}{\theta_2(0)}\,.$$

The second derivative becomes

$$\begin{aligned} f_1''(x_0) &= A\theta_1(0)(\theta_1(0) + \theta_2(0))\mathrm{e}^{\theta_1(0)x_0} - \theta_2(0) \\ &= \theta_1(0)\theta_2(0)f(x_0) - (\theta_2(0) - \theta_1(0)) \\ &= \frac{2}{\sigma^2}(\delta f(x_0) - m)\,. \end{aligned}$$

Suppose for the moment that $f_1''(x_0) > 0$. Because $f'(x)$ is increasing, one must have $f(x) = f_1(x)$ on some interval $[x_0, x_0 + \varepsilon]$. Then at any point $x > x_0$ where $f'(x) = 1$ we would have $f''(x) = f_1''(x) = 2(\delta f(x) - m)/\sigma^2 > 2(\delta f(x_0) - m)/\sigma^2 > 0$. Thus, $f(x) = f_1(x)$ on $[x_0, \infty)$. $f(x)$ is bounded, and thus $A = 0$. It is strictly increasing, and thus $B < 0$. Because $f(x) > 0$ for $x > 0$, we have $B > 0$. This is a contradiction. This shows that either $f(x) = f_1(x)$, $f(x) = f_2(x)$, or there is a point x_0 such that $f(x) = f_1(x)$ for $x \le x_0$ and $f(x) = f_2(x)$ for $x \ge x_0$. By the argument used above, we have that $f(x) = f_1(x)$ is not possible.

Because $0 \le f(x) < u_0/\delta$, we find that $C = 0$. Because $f(x)$ is increasing, we have $D < 0$. If $f(x) = f_2(x)$ for all x, the solution is $f(x) = u_0(1 - \mathrm{e}^{-\theta_2(u_0)x})/\delta$ because $f(0) = 0$. Then it must hold that $f'(x) = u_0\theta_2(u_0)\mathrm{e}^{-\theta_2(u_0)x}/\delta \le 1$ for all x. Therefore, $\delta \ge u_0\theta_2(u_0)$. Suppose now that an $x_0 > 0$ exists. $f'(x_0) = 1$ yields $D = -\mathrm{e}^{\theta_2(u_0)x_0}/\theta_2(u_0)$, or $0 < f(x_0) = u_0/\delta - 1/\theta_2(u_0)$. Thus, $\delta < u_0\theta_2(u_0)$. $0 = f(0) = f_1(0)$ yields $B = -A$. $f_1'(x_0) = 1$ determines $A = (\theta_1(0)\mathrm{e}^{\theta_1(0)x_0} + \theta_2(0)\mathrm{e}^{-\theta_2(0)x_0})^{-1}$. x_0 is now determined from the equation

$$\begin{aligned} f_1(x_0) &= \frac{\mathrm{e}^{\theta_1(0)x_0} - \mathrm{e}^{-\theta_2(0)x_0}}{\theta_1(0)\mathrm{e}^{\theta_1(0)x_0} + \theta_2(0)\mathrm{e}^{-\theta_2(0)x_0}} = \frac{\mathrm{e}^{(\theta_1(0)+\theta_2(0))x_0} - 1}{\theta_1(0)\mathrm{e}^{(\theta_1(0)+\theta_2(0))x_0} + \theta_2(0)} \\ &= f_2(x_0) = \frac{u_0}{\delta} - \frac{1}{\theta_2(u_0)}\,. \end{aligned}$$

The latter is equivalent to

$$\mathrm{e}^{(\theta_1(0)+\theta_2(0))x_0} = \frac{1 + (\frac{u_0}{\delta} - \frac{1}{\theta_2(u_0)})\theta_2(0)}{1 - (\frac{u_0}{\delta} - \frac{1}{\theta_2(u_0)})\theta_1(0)}\,. \tag{2.37}$$

Thus, we can find a solution if and only if the numerator is strictly positive. Note that

$$\theta_1(0) = \frac{\sqrt{m^2 + 2\delta\sigma^2} - m}{\sigma^2} < \frac{(m + \delta\sigma^2/m) - m}{\sigma^2} = \frac{\delta}{m}$$

and

$$\frac{u_0}{\delta} - \frac{\sigma^2}{\sqrt{(u_0 - m)^2 + 2\delta\sigma^2} + m - u_0} = \frac{u_0 + m - \sqrt{(u_0 - m)^2 + 2\delta\sigma^2}}{2\delta}\,. \tag{2.38}$$

From $u_0 - m < \sqrt{(u_0 - m)^2 + 2\delta\sigma^2}$, it now follows that x_0 is well defined. Simple calculus gives that

$$\begin{aligned} x_0 &= \frac{\log(1 + (\frac{u_0}{\delta} - \frac{1}{\theta_2(u_0)})\theta_2(0)) - \log(1 - (\frac{u_0}{\delta} - \frac{1}{\theta_2(u_0)})\theta_1(0))}{\theta_1(0) + \theta_2(0)} \\ &= \frac{\log\theta_2(0) + \log(\theta_2(0) - \theta_2(u_0)) - \log\theta_1(0) - \log(\theta_1(0) + \theta_2(u_0))}{\theta_1(0) + \theta_2(0)} \end{aligned}$$

is well defined. The solution we found is

$$f(x) = \begin{cases} u_0\delta^{-1}(1-\mathrm{e}^{-\theta_2(u_0)x_0}), & \text{if } \delta \ge u_0\theta_2(u_0), \\ \frac{\mathrm{e}^{\theta_1(0)x}-\mathrm{e}^{-\theta_2(0)x}}{\theta_1(0)\mathrm{e}^{\theta_1(0)x_0}+\theta_2(0)\mathrm{e}^{-\theta_2(0)x_0}}, & \text{if } \delta < u_0\theta_2(u_0) \text{ and } x \le x_0, \\ \frac{u_0}{\delta} - \theta_2(u_0)^{-1}\mathrm{e}^{-\theta_2(u_0)(x-x_0)}, & \text{if } \delta < u_0\theta_2(u_0) \text{ and } x > x_0. \end{cases} \tag{2.39}$$

Note that because $f_1(x_0) = f_2(x_0)$ and $f_1'(x_0) = f_2'(x_0) = 1$, it follows from the differential equations that $f_1''(x_0) = f_2''(x_0)$. The function $f(x)$ is hence indeed twice continuously differentiable.

We note that the condition $\delta < u_0\theta_2(u_0)$ is equivalent to $\delta\sigma^2 < 2u_0m$. Hence, the optimal dividend strategy is $U_t = u_0$ if δ or σ is too large, or if u_0 or m is too small.

The following verification theorem shows that $f(x) = V(x)$.

Theorem 2.51. *The function $f(x)$ is identical with $V(x)$, and an optimal dividend strategy U^* is given by $U_t^* = u_0 \mathbb{1}_{X_t^* > x_0}$.*

Proof. Let U be some arbitrary strategy. By Itô's formula,

$$\begin{aligned} &\mathrm{e}^{-\delta(\tau\wedge t)} f(X_{\tau\wedge t}) \\ &\quad = f(x) + \int_0^{\tau\wedge t} \mathrm{e}^{-\delta s}\sigma f'(X_s)\,\mathrm{d}W_s \\ &\qquad + \int_0^{\tau\wedge t} \mathrm{e}^{-\delta s}[-\delta f(X_s) + (m-U_s)f'(X_s) + \tfrac{1}{2}\sigma^2 f''(X_s)]\,\mathrm{d}s \\ &\quad \le f(x) - \int_0^{\tau\wedge t} \mathrm{e}^{-\delta s}U_s\,\mathrm{d}s + \int_0^{\tau\wedge t} \sigma f'(X_s)\,\mathrm{d}W_s \end{aligned}$$

where we used that f fulfils (2.35). In the case $U = U^*$, equality holds. The derivative $f'(x)$ is bounded. Thus (see Proposition A.5), $\{\int_0^{\tau\wedge t} \sigma f'(X_s)\,\mathrm{d}W_s\}$ is a martingale. Taking expected values yields

$$f(x) \ge \mathbb{E}\Bigl[\mathrm{e}^{-\delta(\tau\wedge t)} f(X_{\tau\wedge t}) + \int_0^{\tau\wedge t} \mathrm{e}^{-\delta s}U_s\,\mathrm{d}s\Bigr],$$

with equality if $U = U^*$. If $\tau \le t$, then $f(X_\tau) = 0$. Because $f(x)$ is bounded, $\lim_{t\to\infty} \mathbb{E}[\mathrm{e}^{-\delta(\tau\wedge t)} f(X_{\tau\wedge t})] = 0$. This yields

$$f(x) \ge \mathbb{E}\Bigl[\int_0^\tau \mathrm{e}^{-\delta s}U_s\,\mathrm{d}s\Bigr] = V^U(x),$$

with equality if $U = U^*$. Thus, $V^U(x) \le f(x)$, i.e., $V(x) \le f(x)$. Because $f(x) = V^{U^*}(x)$, it follows also that $V(x) \ge f(x)$. This proves the theorem. □

If $u_0 < m$, consider the strategy $U_t = u_0$. Clearly, $X_t^{u_0} \le X_t^*$ for all t. Because $\psi^{u_0}(x) = \exp\{-2(m-u_0)x/\sigma^2\} < 1$ for all $x > 0$, it follows that $\psi^*(x) \le \psi^{u_0}(x) < 1$ and it is possible that ruin does not occur. Suppose that $m \le u_0$. If we start at x_0+1, the level x_0 is reached almost surely. Thus, $\liminf_{t\to\infty} X_t \le x_0$. Let $T_0 = \inf\{t : X_t \le x_0 + 1\}$ and $T_n = \inf\{t \ge T_{n-1} + 1 : X_t \le x_0 + 1\}$. Let $A_n = \{\inf\{mt + \sigma(W_{T_n+t} - W_{T_n}) : 0 \le t \le 1\} < -x_0 - 1\}$. Then $\mathbb{P}[A_n] = \delta > 0$, and the $\{A_n\}$ are independent. Thus, A_n occurs infinitely often by the Borel–Cantelli lemma. But if A_n occurs, then ruin occurs. Thus, ruin occurs almost surely under the optimal strategy if and only if $m \le u_0$.

2.5.2 Unrestricted Dividend Payments

Let us now allow all increasing adapted cadlag processes as aggregate dividend processes. We will proceed as in Section 2.4.2. Let $V_u(x)$ be the value function in the case $u_0 = u$ is the maximal dividend rate. For an increasing cadlag process $\{D_t\}$, we define the value $V^D(x) = \mathbb{E}[\int_{0-}^{\tau^D-} e^{-\delta t} \, dD_t]$. The value function becomes $V(x) = \sup_D V^D(x)$, where the supremum is taken over all increasing cadlag processes. Clearly, $V(x) \ge V_u(x)$. In the same way as in Lemma 2.38, we can prove the following

Lemma 2.52. *We have* $\lim_{u\to\infty} V_u(x) = V(x)$. □

We therefore now let $u \to \infty$. $\theta_i(0)$ does not depend on u. The parameter $\theta_2(u)$ can be written as

$$\theta_2(u) = \frac{2\delta}{\sqrt{(u-m)^2 + 2\delta\sigma^2} + u - m}\;,$$

and $\theta_2(u) \to 0$ follows. Let us now consider the value $V_u(x_0)$. This value is strictly negative if $\delta > u\theta_2(u)$. Using (2.38) yields

$$\begin{aligned}\frac{u}{\delta} - \frac{1}{\theta_2(u)} &= \frac{u + m - \sqrt{(u-m)^2 + 2\delta\sigma^2}}{2\delta} \\ &= \frac{(u+m)^2 - (u-m)^2 - 2\delta\sigma^2}{2\delta(u + m + \sqrt{(u-m)^2 + 2\delta\sigma^2})} \\ &= \frac{2um - \delta\sigma^2}{\delta(u + m + \sqrt{(u-m)^2 + 2\delta\sigma^2})}\;,\end{aligned}$$

and we see that $V_u(x_0) \to m/\delta$. Thus, $u\theta_2(u) > \delta$ for u large enough. x_0 is determined by (2.37), which has the limit

$$e^{(\theta_1(0)+\theta_2(0))x_0} = \frac{1 + m\theta_2(0)/\delta}{1 - m\theta_1(0)/\delta} = \frac{\delta + m\theta_2(0)}{\delta - m\theta_1(0)}\;.$$

The limit of x_0 is found from this equation. For $x \ge x_0$, we need the limit of

$$f_{2,u}(x) = \frac{u}{\delta} - \frac{1}{\theta_2(u)} + \frac{1}{\theta_2(u)}(1 - \mathrm{e}^{-\theta_2(u)(x - x_0(u))}) ,$$

where we use the notation $x_0(u)$ to show the dependence on u. This yields the limit for $x \ge x_0 = x_0(\infty)$:

$$V(x) = \frac{m}{\delta} + x - x_0 .$$

Together with Lemma 2.52, we have proved the following

Theorem 2.53. *The value function is*

$$V(x) = \begin{cases} \frac{\mathrm{e}^{\theta_1(0)x} - \mathrm{e}^{-\theta_2(0)x}}{\theta_1(0)\mathrm{e}^{\theta_1(0)x_0} + \theta_2(0)\mathrm{e}^{-\theta_2(0)x_0}}, & \text{if } x \le x_0, \\ \frac{m}{\delta} + (x - x_0), & \text{if } x > x_0, \end{cases}$$

where

$$x_0 = \frac{1}{\theta_1(0) + \theta_2(0)}(\log(\delta + m\theta_2(0)) - \log(\delta - m\theta_1(0))) .$$

□

We already found the value in x_0, $V(x_0) = m/\delta$. As the derivative from the right is 1, from the left it also becomes 1. Thus, $V(x)$ is differentiable. The second derivative from the right is zero, and from the left it is

$$\begin{aligned} \frac{\theta_1(0)^2\mathrm{e}^{\theta_1(0)x_0} - \theta_2(0)^2\mathrm{e}^{-\theta_2(0)x_0}}{\theta_1(0)\mathrm{e}^{\theta_1(0)x_0} + \theta_2(0)\mathrm{e}^{-\theta_2(0)x_0}} &= \frac{\theta_1(0)^2\frac{\delta + m\theta_2(0)}{\delta - m\theta_1(0)} - \theta_2(0)^2}{\theta_1(0)\frac{\delta + m\theta_2(0)}{\delta - m\theta_1(0)} + \theta_2(0)} \\ &= \frac{m\theta_1(0)\theta_2(0) - \delta(\theta_2(0) - \theta_1(0))}{\delta} \\ &= \delta^{-1}\Bigl(m\frac{2\delta\sigma^2}{\sigma^4} - \delta\frac{2m}{\sigma^2}\Bigr) = 0 . \end{aligned}$$

This shows that $V(x)$ is twice continuously differentiable. Intuitively, if the derivatives in the restricted case converge to the derivatives in the unrestricted case, we can write Equation (2.35) as

$$\begin{aligned} \max\{\tfrac{1}{2}\sigma^2 f''(x) + mf'(x) - \delta f(x), \\ u^{-1}(\tfrac{1}{2}\sigma^2 f''(x) + (m-u)f'(x) - \delta f(x) + u)\} = 0 . \end{aligned}$$

Letting $u \to \infty$ gives us the following

Proposition 2.54. *The function $V(x)$ fulfils the Hamilton–Jacobi–Bellman equation*

$$\max\{\tfrac{1}{2}\sigma^2 V''(x) + mV'(x) - \delta V(x), 1 - V'(x)\} = 0 . \tag{2.40}$$

Proof. This follows by a direct verification using that $V(x) > V(x_0) = m/\delta$ whenever $x > x_0$. □

We now want to find the optimal strategy.

Theorem 2.55. *The optimal dividend strategy D^* is to pay any capital larger than x_0, i.e., $D_t^* = \max\{\sup_{0\le s\le t\wedge\tau} Y_s - x_0, 0\}$, where $Y_t = x + mt + \sigma W_t$.*

Proof. If $x > x_0$, then $D_0^* = x - x_0$. Because $V(x) = V(x_0) + x - x_0$, it is enough to prove the result for $x \le x_0$. Then $X_t^* \le x_0$. The process D^* is continuous and increasing and therefore of bounded variation. Thus, by Itô's formula,

$$\begin{aligned}
\mathrm{e}^{-\delta(\tau\wedge t)}V(X_{\tau\wedge t}^*) &= V(x) + \int_0^{\tau\wedge t} [mV'(X_s^*) + \tfrac{1}{2}\sigma^2 V''(X_s^*) - \delta V(X_s^*)]\mathrm{e}^{-\delta s}\,\mathrm{d}s \\
&\quad - \int_0^{\tau\wedge t} \mathrm{e}^{-\delta s}V'(X_s^*)\,\mathrm{d}D_s^* + \int_0^{\tau\wedge t} V'(X_s^*)\mathrm{e}^{-\delta s}\sigma\,\mathrm{d}W_s \\
&= V(x) + \int_0^{\tau\wedge t} V'(X_s^*)\mathrm{e}^{-\delta s}\sigma\,\mathrm{d}W_s - \int_0^{\tau\wedge t} \mathrm{e}^{-\delta s}\,\mathrm{d}D_s^*\,,
\end{aligned}$$

where we used Equation (2.40) and that D^* only increases at points where $X_s^* = x_0$, i.e., $V'(X_s^*) = 1$. The integral with respect to the Brownian motion is a martingale because $V'(x)$ is bounded. Taking expected values yields

$$V(x) = \mathbb{E}\Big[\mathrm{e}^{-\delta(\tau\wedge t)}V(X_{\tau\wedge t}^*) + \int_0^{\tau\wedge t} \mathrm{e}^{-\delta s}\,\mathrm{d}D_s^*\Big]\,.$$

We next let $t \to \infty$. Because $V(X_{\tau\wedge t}^*)$ is bounded, we can interchange the limit and integration. Thus, $V(x) = V^{U^*}(x)$ because $V(X_\tau) = 0$. □

It should be noted that for unrestricted dividends, ruin also occurs almost surely. This can be seen because the process X^* is bounded. Alternatively, one could consider the process X whose dividend process is $D_t = u_0 \mathbb{1}_{X_t > x_0}$, where x_0 is chosen as in the present section. Then $X_t \ge X_t^*$ for all t because $D_t \le D_t^*$. Therefore, also $\tau^* \le \tau^U < \infty$.

2.5.3 A Note on Viscosity Solutions

This section should be skipped upon first reading. An important tool in stochastic optimisation is viscosity solutions. We do not use the technique here for simplicity. Even though the method seems much simpler, there are some problems connected with it; see the remarks at the end of this section. But we want to illustrate the method here with a simple example. This example is not a control problem. But it is possible to see how the theory of viscosity solutions readily yields the solution to the considered problem if the solution is not twice continuously differentiable or if differentiability is not known.

Suppose that we consider the process

$$X_t = x + mt + \sigma W_t - u \int_0^t \mathbb{1}_{X_s > x_0} \,\mathrm{d}s\,,$$

where $x_0 > 0$ and $u > m$ is a constant. In contrast to Section 2.5.1, x_0 does not need to be the optimal dividend barrier. Our goal is to calculate the value function

$$V(x) = u\mathbb{E}\Bigl[\int_0^\tau \mathbb{1}_{X_s > x_0} \mathrm{e}^{-\delta s} \,\mathrm{d}s\Bigr]\,.$$

Let τ_x and τ_y be the ruin times for initial capital x and y, respectively. If y tends to x, then τ_y tends to τ_x. Thus, $V(x)$ is continuous.

From Section 2.5.1 we know that the problem is connected to the equation

$$\tfrac{1}{2}\sigma^2 V''(x) + mV'(x) - \delta V(x) + u(1 - V'(x))\mathbb{1}_{x > x_0} = 0\,. \tag{2.41}$$

In order for $V(x)$ to be twice continuously differentiable, it is necessary that $V'(x_0) = 1$. From the definition of $V(x)$, we also obtain that $V(0) = 0$ and $V(\infty) = u/\delta$.

Solving Equation (2.41) on $[0, x_0]$ yields

$$V(x) = A\mathrm{e}^{\theta_1(0)x} + B\mathrm{e}^{-\theta_2(0)x}\,,$$

where $\theta_i(u)$ is defined in (2.36). From $V(0) = 0$, we conclude that $B = -A$, and from $V'(x_0) = 1$, we conclude that

$$V(x) = \frac{\mathrm{e}^{\theta_1(0)x} - \mathrm{e}^{-\theta_2(0)x}}{\theta_1(0)\mathrm{e}^{\theta_1(0)x_0} + \theta_2(0)\mathrm{e}^{-\theta_2(0)x_0}}\,.$$

On $[x_0, \infty)$ we find that

$$V(x) = \frac{u}{\delta} + C\mathrm{e}^{\theta_1(u)x} + D\mathrm{e}^{-\theta_2(u)x}\,.$$

From $V(\infty) = u/\delta$, we conclude that $C = 0$. $V'(x_0) = 1$ implies that

$$V(x) = \frac{u}{\delta} - \frac{\mathrm{e}^{-\theta_2(u)x}}{\theta_2(u)\mathrm{e}^{-\theta_2(u)x_0}}\,.$$

The two solutions coincide at $x = x_0$ exactly if the second derivatives coincide:

$$\frac{\theta_1^2(0)\mathrm{e}^{\theta_1(0)x_0} - \theta_2^2(0)\mathrm{e}^{-\theta_2(0)x_0}}{\theta_1(0)\mathrm{e}^{\theta_1(0)x_0} + \theta_2(0)\mathrm{e}^{-\theta_2(0)x_0}} = -\frac{\theta_2^2(u)\mathrm{e}^{-\theta_2(u)x_0}}{\theta_2(u)\mathrm{e}^{-\theta_2(u)x_0}} = -\theta_2(u)\,.$$

Solving for x_0 yields

$$\mathrm{e}^{(\theta_1(0)+\theta_2(0))x_0} = \frac{\theta_2(0)}{\theta_1(0)}\frac{\theta_2(0) - \theta_2(u)}{\theta_1(0) + \theta_2(u)}\,.$$

We see that $V(x)$ can only be twice continuously differentiable at x_0 if x_0 is the optimal barrier.

Let us now assume that we know $V(x_0)$.

Lemma 2.56. i) *If* $x < x_0$*, then*

$$V(x) = \frac{\mathrm{e}^{\theta_1(0)x} - \mathrm{e}^{-\theta_2(0)x}}{\mathrm{e}^{\theta_1(0)x_0} - \mathrm{e}^{-\theta_2(0)x_0}} V(x_0) \; .$$

ii) *If* $x > x_0$*, then*

$$V(x) = \frac{u}{\delta} - \frac{\mathrm{e}^{-\theta_2(u)x}}{\mathrm{e}^{-\theta_2(u)x_0}} \Bigl(\frac{u}{\delta} - V(x_0)\Bigr) \; .$$

Proof. i) Let $T = \inf\{t : X_t \notin [0, x_0]\}$ and

$$f(x) = \frac{\mathrm{e}^{\theta_1(0)x} - \mathrm{e}^{-\theta_2(0)x}}{\mathrm{e}^{\theta_1(0)x_0} - \mathrm{e}^{-\theta_2(0)x_0}} V(x_0) \; .$$

Note that $T < \infty$. The function $f(x)$ is twice continuously differentiable and solves (2.41) on $[0, x_0]$. From Itô's formula we find that

$$\begin{aligned} f(X_T)\mathrm{e}^{-\delta T} &= f(x) + \int_0^T \sigma f'(X_t)\mathrm{e}^{-\delta t} \, \mathrm{d}W_t \\ &\quad + \int_0^T (\tfrac{1}{2}\sigma^2 f''(X_t) + m f'(X_t) - \delta f(X_t))\mathrm{e}^{-\delta t} \, \mathrm{d}t \\ &= f(x) + \int_0^T \sigma f'(X_t)\mathrm{e}^{-\delta t} \, \mathrm{d}W_t \; . \end{aligned}$$

The left-hand side is bounded. Thus, the integral on the right-hand side is also bounded. It therefore has mean value 0. We find that

$$f(x) = \mathbb{E}[f(X_T)\mathrm{e}^{-\delta T}] = \mathbb{E}[V(x_0)\mathrm{e}^{-\delta T}; X_T = x_0] = V(x) \; .$$

ii) Let $T = \inf\{t : X_t < x_0\}$ and

$$f(x) = \frac{u}{\delta} - \frac{\mathrm{e}^{-\theta_2(u)x}}{\mathrm{e}^{-\theta_2(u)x_0}} \Bigl(\frac{u}{\delta} - V(x_0)\Bigr) \; .$$

Note that $T < \infty$ and $f(x)$ is a bounded, twice continuously differentiable solution to (2.41) on $[x_0, \infty)$. From Itô's formula, we find that

$$\begin{aligned} f(X_T)\mathrm{e}^{-\delta T} &= \int_0^T \sigma f'(X_t)\mathrm{e}^{-\delta t} \, \mathrm{d}W_t \\ &\quad + \int_0^T (\tfrac{1}{2}\sigma^2 f''(X_t) + (m-u) f'(X_t) - \delta f(X_t))\mathrm{e}^{-\delta t} \, \mathrm{d}t \\ &= f(x) + \int_0^T \sigma f'(X_t)\mathrm{e}^{-\delta t} \, \mathrm{d}W_t - \int_0^T u\mathrm{e}^{-\delta t} \, \mathrm{d}t \; . \end{aligned}$$

Also, here the stochastic integral is bounded and therefore a martingale. Thus,

$$\begin{aligned} f(x) &= \mathbb{E}\Big[f(X_T)\mathrm{e}^{-\delta T} + \int_0^T u\mathrm{e}^{-\delta t}\,\mathrm{d}t\Big] \\ &= \mathbb{E}\Big[V(x_0)\mathrm{e}^{-\delta T} + \int_0^T u\mathrm{e}^{-\delta t}\,\mathrm{d}t\Big] = V(x)\,. \end{aligned}$$

□

We conclude now that $V(x)$ is twice continuously differentiable for $x \neq x_0$. We note that all derivatives from the left and from the right exist. Let x^{o} denote the optimal barrier. $V(x)$ is twice differentiable at x_0 if and only if $x_0 = x^{\mathrm{o}}$. In the latter case we have $V'(x_0) = 1$.

For the first derivative from the left we get

$$V'(x_0-) = \frac{\theta_1(0)\mathrm{e}^{\theta_1(0)x_0} + \theta_2(0)\mathrm{e}^{-\theta_2(0)x_0}}{\mathrm{e}^{\theta_1(0)x_0} - \mathrm{e}^{-\theta_2(0)x_0}} V(x_0)\,.$$

The derivative from the right is

$$V'(x_0+) = \frac{\theta_2(u)\mathrm{e}^{-\theta_2(u)x_0}}{\mathrm{e}^{-\theta_2(u)x_0}}\Big(\frac{u}{\delta} - V(x_0)\Big) = \theta_2(u)\Big(\frac{u}{\delta} - V(x_0)\Big)\,.$$

One can readily see that $V(x_0)$ is increasing in x_0. Indeed, starting at the barrier x_0, it takes more time until ruin for a larger x_0. If $x_0' > x_0$, then the process $\{x_0' + mt + \sigma W_t - D_t\}$ is still alive when ruin for $\{X_t\}$ occurs, and there is a strictly positive probability that the process with a dividend barrier at x_0' will reach x_0' again. Moreover, $V'(x_0+)$ is decreasing in x_0. Thus, if $x_0 < x^{\mathrm{o}}$, then $V'(x_0+) > 1$. If $x_0 > x^{\mathrm{o}}$, then $V'(x_0+) < 1$.

Let $\varphi(x)$ be a twice continuously differentiable function with $\varphi(x_0) = V(x_0)$ and $\varphi(x) \ge V(x)$. Such a function exists if $V'(x_0-) \ge V'(x_0+)$. We call $\varphi(x)$ a *test function*. We consider the process starting at x_0.

Let $\varepsilon, h > 0$ and $T = \inf\{t : |X_t - x_0| > \varepsilon\} \wedge h$. From Itô's formula, we conclude that

$$\begin{aligned} V(X_T)\mathrm{e}^{-\delta T} &\le \varphi(X_T)\mathrm{e}^{-\delta T} = V(x_0) + \int_0^T \sigma\varphi'(X_t)\mathrm{e}^{-\delta t}\,\mathrm{d}W_t \\ &\quad + \int_0^T (\tfrac{1}{2}\sigma^2\varphi''(X_t) + (m - u\mathbb{I}_{X_t > x_0})\varphi'(X_t) - \delta\varphi(X_t))\mathrm{e}^{-\delta t}\,\mathrm{d}t\,. \end{aligned}$$

The integrand of the stochastic integral is bounded. Thus, the stochastic integral is a martingale and has mean value 0. From

$$V(x_0) = \mathbb{E}\Big[V(X_T)\mathrm{e}^{-\delta T} + \int_0^T u\mathrm{e}^{-\delta t}\mathbb{I}_{X_t > x_0}\,\mathrm{d}t\Big]\,,$$

we find that

$$\mathbb{E}\Big[\int_0^T (\tfrac{1}{2}\sigma^2\varphi''(X_t)+m\varphi'(X_t)-\delta\varphi(X_t)+u(1-\varphi'(X_T))\mathbb{1}_{X_t>x_0})e^{-\delta t}\, dt\Big] \geq 0\,.$$

The integrand is bounded. If we divide by h and let $h \downarrow 0$, we find that

$$\tfrac{1}{2}\sigma^2\varphi''(x_0) + m\varphi'(x_0) - \delta\varphi(x_0) + u(1-\varphi'(x_0)) \lim_{h\downarrow 0} \frac{1}{h}\mathbb{E}\Big[\int_0^h \mathbb{1}_{X_t>x_0}\, dt\Big] \geq 0\,.$$

Suppose that $V'(x_0-) > V'(x_0+)$. Then we can choose $\varphi(x)$ such that $\varphi(x) = V(x)$ on $[0, x_0]$. This implies that

$$u(1-\varphi'(x_0)) \lim_{h\downarrow 0} \frac{1}{h}\mathbb{E}\Big[\int_0^h \mathbb{1}_{X_t>x_0}\, dt\Big] \geq 0\,.$$

We should note that

$$\frac{1}{h}\mathbb{E}\Big[\int_0^h \mathbb{1}_{X_t>x_0}\, dt\Big] = \frac{1}{h}\int_0^h \mathbb{P}[X_t > x_0]\, dt = \frac{1}{h}\int_0^h \mathbb{P}[X_t \geq x_0]\, dt\,.$$

Because $\mathbb{P}[X_t > x_0]$ is continuous in t, the limit exists. The properties of the Brownian motion imply that the limit is in $(0,1)$. Thus, $V'(x_0-) = \varphi'(x_0) \leq 1$. We can also choose $\varphi(x)$ such that $\varphi(x) = V(x)$ on $[x_0, \infty)$. Then

$$-u(1-\varphi'(x_0))\Big\{1 - \lim_{h\downarrow 0} \frac{1}{h}\mathbb{E}\Big[\int_0^h \mathbb{1}_{X_t\leq x_0}\, dt\Big]\Big\} \geq 0\,.$$

This implies that $V'(x_0+) = \varphi'(x_0) \geq 1$. This is a contradiction, and we conclude that $V'(x_0-) \leq V'(x_0+)$.

Let $\varphi(x)$ be a twice continuously differentiable function with $\varphi(x_0) = V(x_0)$ and $\varphi(x) \leq V(x)$. Such a test function exists because $V'(x_0-) \leq V'(x_0+)$. In the same way as above, we conclude that

$$\tfrac{1}{2}\sigma^2\varphi''(x_0) + m\varphi'(x_0) - \delta\varphi(x_0) + u(1-\varphi'(x_0)) \lim_{h\downarrow 0} \frac{1}{h}\mathbb{E}\Big[\int_0^h \mathbb{1}_{X_t>x_0}\, dt\Big] \leq 0\,.$$

If $V'(x_0-) < V'(x_0+)$, then we can choose $\varphi(x) = V(x)$ on $[0, x_0]$. We conclude then that $V'(x_0-) \geq 1$. Choosing $\varphi(x) = V(x)$ on $[x_0, \infty)$, we conclude that $V'(x_0+) \leq 1$. This is again a contradiction. Thus, $V'(x_0-) = V'(x_0+)$, and $V(x)$ is continuously differentiable. We thus find that

$$\begin{aligned} V(x_0) &= \frac{\theta_2(u)u/\delta}{\frac{\theta_1(0)e^{\theta_1(0)x_0}+\theta_2(0)e^{-\theta_2(0)x_0}}{e^{\theta_1(0)x_0}-e^{-\theta_2(0)x_0}} + \theta_2(u)} \\ &= \frac{\theta_2(u)(e^{\theta_1(0)x_0} - e^{-\theta_2(0)x_0})u/\delta}{\theta_1(0)e^{\theta_1(0)x_0} + \theta_2(0)e^{-\theta_2(0)x_0} + \theta_2(u)(e^{\theta_1(0)x_0} - e^{-\theta_2(0)x_0})}\,. \end{aligned}$$

We now conclude that

$$x_0 \gtreqless x^o \Longleftrightarrow V'(x_0) \lesseqgtr 1 \Longleftrightarrow V''(x_0-) \lesseqgtr V''(x_0+)\;,$$

where we used (2.41).

Let $\varphi(x)$ be a twice continuously differentiable function such that $\varphi(x_0) = V(x_0)$ and $\varphi(x) \ge V(x)$. Then the function $\varphi(x)$ must fulfil $\varphi'(x_0) = V'(x_0)$ and $\varphi''(x) \ge \max\{V''(x_0-), V''(x_0+)\}$. Thus,

$$\tfrac{1}{2}\sigma^2\varphi''(x_0) + m\varphi'(x_0) - \delta\varphi(x_0) \ge \tfrac{1}{2}\sigma^2 V''(x_0-) + mV'(x_0) - \delta V(x_0) = 0\,.$$

We call a function $V(x)$ with the property that for each $x_1 \ge 0$ such that for each twice continuously differentiable test function $\varphi(x)$ with $\varphi(x_1) = V(x_1)$ and $\varphi(x) \ge V(x)$ we have

$$\tfrac{1}{2}\sigma^2\varphi''(x_1) + m\varphi'(x_1) - \delta\varphi(x_1) + u(1-\varphi'(x_1))\mathbb{I}_{x_1 > x_0} \ge 0$$

a *viscosity subsolution* to (2.41). Note that at any point where $V(x)$ is twice continuously differentiable it is automatically a viscosity subsolution.

Let $\varphi(x)$ be a twice continuously differentiable function such that $\varphi(x_0) = V(x_0)$ and $\varphi(x) \le V(x)$. Then the function $\varphi(x)$ must fulfil $\varphi'(x_0) = V(x_0)$ and $\varphi''(x) \le \min\{V''(x_0-), V''(x_0+)\}$. Thus,

$$\tfrac{1}{2}\sigma^2\varphi''(x_0) + m\varphi'(x_0) - \delta\varphi(x_0) \le \tfrac{1}{2}\sigma^2 V''(x_0-) + mV'(x_0) - \delta V(x_0) = 0\,.$$

Analogously, we call a function $V(x)$ a *viscosity supersolution* to (2.41) if for each x_1 and any function $\varphi(x)$ such that $\varphi(x_1) = V(x_1)$ and $\varphi(x) \le V(x)$ for all x we have that

$$\tfrac{1}{2}\sigma^2\varphi''(x_1) + m\varphi'(x_1) - \delta\varphi(x_1) + u(1-\varphi'(x_1))\mathbb{I}_{x_1 > x_0} \le 0\,.$$

This condition is automatically fulfilled if $V(x)$ is twice continuously differentiable in x_1. Because $V(x)$ is both a super- and a subsolution, we call it a *viscosity solution* to (2.41).

In this section we did not consider a control problem. This is because many problems may arise with control problems where the solutions are not *classical* solutions. We have been lucky with the problems considered in this book. When we wanted to apply Itô's formula, we always had twice continuously differentiable value functions, which means classical solutions. If this had not been the case, the method of viscosity solutions had been a possibility to circumvent this problem. One then needs to show that the viscosity solution is the value function. That a viscosity solution is an upper bound of the value function can usually be shown similarly as in the verification theorems of this book. To show that it is a lower bound of the value function often is more tricky. If an optimal strategy exists, application of the Itô–Tanaka formula (see [149, p. 223]) may give both the lower bound and the existence of an optimal strategy. However, it may often be the case that no optimal strategy exists. This means that the natural optimal strategy (the optimiser in the

Hamilton–Jacobi–Bellman equation) is not optimal. In this case one has to construct ε-optimal strategies and to show that their values deviate from the viscosity solution by at most by ε. And of course, one should not forget to prove that the value function is a viscosity solution to the Hamilton–Jacobi–Bellman equation.

Bibliographical Remarks

Shreve, Lehoczky, and Gaver [168] considered a general diffusion process

$$X_t^D = x + \int_0^t m(X_s^D)\,\mathrm{d}s + \int_0^t \sigma(X_s^D)\,\mathrm{d}W_s - D_t\;,$$

where $m(x)$ and $\sigma(x)$ are linearly bounded, differentiable functions with Lipschitz continuous derivatives. They maximised the expression

$$V^D(x) = \mathbb{E}\Bigl[\int_{0-}^{\tau-} \mathrm{e}^{-\delta t}\,\mathrm{d}D_t - \mathrm{e}^{-\delta\tau} P\Bigr]$$

for some penalty $P \in \mathbb{R}$. They showed that if there is a concave function $f(x)$ solving $m(x)f'(x) + \frac{1}{2}\sigma^2(x)f''(x) - \delta f(x) = 0$ with $f'(U^*) = 1$ for some U^* and $f(0) = -P$, then $f(x)$ is the value function and the optimal strategy is a barrier strategy with a barrier at U^*. As an example, the model considered here was solved. Later, our problem was also considered by Asmussen and Taksar [11]. Cai et al. [30] considered a diffusion approximation where interest is earned on the surplus. A barrier strategy is applied and the barrier is chosen in an optimal way. Moreover, an alternative model was considered where it is possible to borrow money at a higher rate when the surplus becomes negative. Thonhauser and Albrecher [180] considered the problem where in addition early ruin is punished by a term $-\ell\mathrm{e}^{-\delta\tau}$ for some $\ell > 0$. A similar problem was considered by Dayananda [41, 42]. The surplus is modelled by a diffusion approximation and reinsurance is possible. If the surplus becomes negative the shareholders have to invest new capital at a larger cost. The average of the dividend payments minus investment is the value function, which is then maximised. Asmussen, Højgaard, and Taksar [9] considered the model where in addition (cheap) reinsurance can be bought and found the optimal reinsurance and dividend policy. Shreve, Lehoczky, and Gaver [168] also considered the problem where in addition money can be added to avoid ruin, i.e.,

$$X_t^{DL} = x + \int_0^t m(X_s^{DL})\,\mathrm{d}s + \int_0^t \sigma(X_s^{DL})\,\mathrm{d}W_s - D_t + L_t\;,$$

where $\{L_t\}$ is an increasing process. The process has to fulfil $X_t^{DL} \ge 0$. They maximised

$$V^{DL} = \mathbb{E}\Bigl[\int_{0-}^{\infty} e^{-\delta t}\, dD_t - \vartheta \int_{0-}^{\infty} e^{-\delta t}\, dL_t\Bigr]\;,$$

where $\vartheta \ge 1$. This results in an additional barrier at zero where the process is reflected. The boundary conditions become $f'(0) = \vartheta$, $f'(U^*) = 1$. If $\vartheta = 1$, then the optimal process is $X_t^* \equiv 0$. The special case of constant $m(x) = m$ and $\sigma(x) = \sigma$ has already been treated by Harrison and Taylor [94]. Paulsen [145, 146] solved the problem, where there are in addition fixed costs for paying dividends and/or for reinvestment of money in order not to get ruined.

A variation of the model was considered by Hubalek and Schachermayer [107]. They optimised the utility of the dividends similarly to Merton's problem of Section 3.1.1 for a HARA utility. The limits of $\alpha \to 1$ yield the problem considered here.

Højgaard and Taksar [104, 105] considered a model where (cheap) reinsurance can be bought. They optimised the quantity $\int_0^\tau e^{-\delta t} X_t^b\, dt$, a discounted average of the future surplus.

3

Problems in Life Insurance

In this chapter we consider processes in continuous time, i.e., $I = \mathbb{R}_+$.

Life insurance is a field that is very close to mathematical finance. This is mainly the case because the performance of a contract is measured by the return on investment. Mortality risk is not a big problem because the strong law of large numbers applies. Mortality risk can therefore be "hedged away" through a large portfolio. But the investment risk works on all of the contracts in the same way. Financial risk therefore has to be hedged in the same way as in mathematical finance.

Usually, the guaranteed return for the policyholder is quite low. The benefit of the policy lies then in the actual obtained return, which usually is put on the policyholder's account via a bonus scheme. The fund manager thus has the duty to invest the surplus in such a way that the return is high but the risk is still small. This is exactly the duty of a fund manager in an investment fund, too. However, there is a difference. Because of the interest-rate guarantee, there must be enough reserves to cover the liability. Otherwise, the supervisor would close the company.

Thus, life insurance is an excellent field in which to apply stochastic control theory. The investor will have to choose a strategy such that the policyholders as well as the equity holders of the company are satisfied. This means that both investment and bonus payments should be optimal. In practise, risk capital also has to be allocated in an optimal way, dividends have to be paid, etc, in such a way that there is enough money for investment in order for a high return to be achieved. On the other hand, the surplus should not be too high, because of taxation reasons.

We start by considering the classical Merton problem from mathematical finance. This problem is used as a skeleton for the problems from life insurance. We then will consider a single life insurance contract and the bonus problem in life insurance. As a last model we consider a diffusion approximation to the reserves of a pension scheme. There we will define a loss function if the surplus or the premia deviate from some predefined target values. We then want to minimise the expected loss.

For most of the problems we will not be able to solve the Hamilton–Jacobi–Bellman equation. Our verification theorems are thus only of the form "If a solution exists then ...." In specific situations it may be possible to find a solution. But the reader should keep in mind that it is important to show the existence of a solution with the required properties. Otherwise, we cannot be sure whether the value function really solves the Hamilton–Jacobi–Bellman equation under consideration.

We will work, as for diffusion processes before, with the filtration of the Brownian motion, possibly augmented with the information from the life insurance contract. It should be remarked that the natural filtration to consider is the filtration of the observed processes. This filtration might be smaller. The solutions to the Hamilton–Jacobi–Bellman equation are always adapted to the filtration generated by the observed processes. Thus, for our applications the (possibly for some strategies) different filtrations are not a problem. A discussion on this point can be found in [113, pp. 187–189].

3.1 Merton's Problem for Life Insurers

An agent has an initial capital x. On the market there is a riskless bond given by (E.3) and a risky asset given by (E.1). We assume that $m > \delta$. The present wealth at time t of the agent is denoted by X_t. The agent can choose how much is invested into the risky asset. By θ_t we denote the fraction of wealth that is invested into the risky asset, i.e., $\theta_t X_t$ is the amount invested into the risky asset, $(1-\theta_t)X_t$ is the amount invested into the riskfree bond. The agent can then also choose how much of the wealth is consumed. The goal will be to maximise the expected utility of the consumption. If consumption after time T is allowed, we model this by measuring the future consumption by a terminal utility.

3.1.1 The Classical Merton Problem

We suppose here that the agent has no liabilities. That is, there is money to invest, but there are no liabilities from an insurance contract.

Optimal Consumption

If the agent consumes the wealth at rate $\{c_t\}$, the wealth process fulfils

$$\mathrm{d}X_t^{\theta,c} = [(1-\theta_t)\delta + m\theta_t]X_t^{\theta,c}\,\mathrm{d}t + \sigma\theta_t X_t^{\theta,c}\,\mathrm{d}W_t - c_t\,\mathrm{d}t\,.$$

We allow all adapted cadlag controls $\{(\theta_t, c_t)\}$ such that $c_t \ge 0$ and there is a unique solution $\{X_t^{\theta,c}\}$ to the above stochastic differential equation. We further restrict to strategies such that $X_t^{\theta,c} \ge 0$, i.e., the agent is not allowed to have debts. This means that if there is no money left the agent can no longer invest. We will further have to assume that our probability space $(\Omega, \mathcal{F}, \mathbb{P})$ is chosen in such a way that for the optimal strategy found below a unique solution $\{X_t^*\}$ exists. The filtration $\{\mathcal{F}_t\}$ is the Brownian filtration, that is, the smallest right-continuous filtration such that $\{W_t\}$ is adapted. Because we are not going to change the measure for the present model, we can complete the filtration, i.e., $\mathcal{F}_0$ contains all sets $A \in \mathcal{F}$ such that $\mathbb{P}[A] = 0$. Recall that $\mathcal{F}$ is complete, and $B \subset A \in \mathcal{F}$ with $\mathbb{P}[A] = 0$ therefore implies that $B \in \mathcal{F}$.

Let us define $\tau = \inf\{t : X_t^{\theta,c} = 0\}$. Consumption can only be positive, $c_t \ge 0$. The agent has a utility function $\phi(c,t)$ that measures how glad he is about the consumption. Because the agent prefers higher consumption, the utility function $\phi(c,t)$ is assumed to be strictly increasing. Because the agent is risk-averse, the utility $\phi(c,t)$ is assumed to be strictly concave. In other words, a monetary unit means less to the agent if c is large than if c is small. Finally, because the agent's preferences do not change rapidly, we assume that $\phi(c,t)$ is continuous in t. Note that as a concave function $\phi(c,t)$ is continuous in c. For simplicity of the notation we norm the utility functions such that $\phi(0,t) = 0$. To avoid some technical problems we suppose that $\phi(c,t)$ is continuously differentiable with respect to c and that $\lim_{c\to\infty} \phi_c(c,t) = 0$, where $\phi_c(c,t)$ denotes the derivative with respect to c.

The value of the strategy $\{(\theta_t, c_t)\}$ is

$$V^{\theta,c}(x) = \mathbb{E}\Bigl[\int_0^{\tau\wedge T} \phi(c_t, t)\,dt\Bigr],$$

and the value function becomes $V(x) = \sup_{\theta,c} V^{\theta,c}(x)$. For simplicity of notation, we will omit the stopping time τ. We just make the convention that $c_t = 0$ for $t \ge \tau$.

We start with proving some basic properties of $V(x)$.

Lemma 3.1. *The function $V(x)$ is strictly increasing and concave with boundary value $V(0) = 0$.*

Proof. If $x = 0$, any strategy with $\theta_t \ne 0$ would immediately lead to ruin. Hence, no wealth can be consumed. Thus, $V(0) = \int_0^T \phi(0,t)\,dt = 0$. Let (θ, c) be a strategy for initial capital x and let $y > x$. We denote the corresponding wealth process starting in x by X. Let $c'_t = c_t + \mathbb{I}_{t<y-x}$. Use the strategy (θ, c') for initial capital y. Because $c'_t > c_t$ for $t < y - x$, we get $V(y) > V(x)$.

Let $z = (1-\alpha)x + \alpha y$ for $\alpha \in (0,1)$. Let (θ, c) be a strategy for initial capital x and (φ, b) be a strategy for initial capital y. Consider the strategy $\pi_t = (1-\alpha)\theta_t + \alpha\varphi_t$ and $e_t = (1-\alpha)c_t + \alpha b_t$. Then $X_t^{\pi,e} = (1-\alpha)X_t^{\theta,c} + \alpha X_t^{\varphi,b}$. The value of the strategy now becomes

$$\begin{aligned}V(z) \ge V^{\pi,e}(z) &= \mathbb{E}\Bigl[\int_0^T \phi((1-\alpha)c_t + \alpha b_t, t)\,\mathrm{d}t\Bigr] \\ &\ge \mathbb{E}\Bigl[\int_0^T (1-\alpha)\phi(c_t,t) + \alpha\phi(b_t,t)\,\mathrm{d}t\Bigr] = (1-\alpha)V^{\theta,c}(x) + \alpha V^{\varphi,b}(y)\,.\end{aligned}$$

Taking the supremum on the right-hand side shows concavity: $V(z) \ge (1-\alpha)V(x) + \alpha V(y)$. □

In order to treat the problem, we introduce

$$V^{\theta,c}(t,x) = \mathbb{E}\Bigl[\int_t^T \phi(c_s,s)\,\mathrm{d}s \Bigm| X_t = x\Bigr]\,.$$

The set of admissible strategies are all cadlag processes $\{(\theta_s, c_s) : s \ge t\}$ adapted to the filtration $\{\mathcal{F}_s : s \ge t\}$. The problem is then to maximise dividends over (t,T). This is basically the same problem as maximising over $(0,T)$. The value function becomes $V(t,x) = \sup_{\theta,c} V^{\theta,c}(t,x)$. The function we are looking for is $V(x) = V(0,x)$.

We next want to find the Hamilton–Jacobi–Bellman equation. For notational simplicity we let $t = 0$. The equation at $t > 0$ can then just be obtained by small changes in the derivation below.

Let $h > 0$ and θ and c be some constants. We consider the strategy

$$\theta_t = \begin{cases} \theta, & \text{if } 0 \le t < \tau \wedge h, \\ \theta^\varepsilon_{t-h}, & \text{if } \tau \wedge t \ge h, \\ 0, & \text{if } t \ge \tau,\ \tau < h, \end{cases} \qquad c_t = \begin{cases} c, & \text{if } 0 \le t < \tau \wedge h, \\ c^\varepsilon_{t-h}, & \text{if } \tau \wedge t \ge h, \\ 0, & \text{if } t \ge \tau,\ \tau < h. \end{cases}$$

Here $(\theta^\varepsilon, c^\varepsilon)$ is a strategy for initial capital X_h such that $V(h, X_h) < V^\varepsilon(h, X_h) + \varepsilon$. The value of the dividends is then

$$\begin{aligned}V(0,x) \ge V^{\theta,c}(0,x) &= \mathbb{E}\Bigl[\int_0^{\tau\wedge h} \phi(c,t)\,\mathrm{d}t + V^{\theta,c}(\tau\wedge h, X_{\tau\wedge h})\Bigr] \\ &> \mathbb{E}\Bigl[\int_0^{\tau\wedge h} \phi(c,t)\,\mathrm{d}t + V(\tau\wedge h, X_{\tau\wedge h})\Bigr] - \varepsilon\,.\end{aligned}$$

We can let ε tend to zero. From Itô's formula we find, provided that $V(t,x)$ is twice continuously differentiable in x and continuously differentiable in t,

$$\begin{aligned}V(\tau\wedge h, X_{\tau\wedge h}) = V(0,x) &+ \int_0^{\tau\wedge h} (\{[(1-\theta)\delta + m\theta]X_t - c\}V_x(t,X_t) \\ &+ \tfrac{1}{2}\sigma^2\theta^2 X_t^2 V_{xx}(t,X_t) + V_t(t,X_t))\,\mathrm{d}t \\ &+ \int_0^{\tau\wedge h} \sigma\theta X_t V_x(t,X_t)\,\mathrm{d}W_t\,.\end{aligned}$$

The derivatives are denoted by subscripts. If the integral with respect to Brownian motion is a martingale, we find that

$$\mathbb{E}\Big[\int_0^{\tau\wedge h}\phi(c,t)\,\mathrm{d}t+\int_0^{\tau\wedge h}\{[(1-\theta)\delta+m\theta]X_t-c\}V_x(t,X_t)$$
$$+\tfrac{1}{2}\sigma^2\theta^2X_t^2V_{xx}(t,X_t)+V_t(t,X_t)\,\mathrm{d}t\Big]\le 0\,.$$

Dividing by h and letting $h\downarrow 0$, provided that the limit and integration can be interchanged, yields

$$\phi(c,0)+\{[(1-\theta)\delta+m\theta]x-c\}V_x(0,x)+\tfrac{1}{2}\sigma^2\theta^2x^2V_{xx}(0,x)+V_t(0,x)\le 0\,.$$

This inequality has to hold for all θ, c. This motivates the *Hamilton–Jacobi–Bellman equation*

$$\sup_{\theta,c}\{\phi(c,t)+\{[(1-\theta)\delta+m\theta]x-c\}V_x(t,x)+\tfrac{1}{2}\sigma^2\theta^2x^2V_{xx}(t,x)+V_t(t,x)\}=0\,. \tag{3.1}$$

The function $V:[0,T]\times\mathbb{R}_+\to\mathbb{R}_+$ should be twice continuously differentiable, strictly increasing and concave in x, and continuously differentiable in t with the boundary conditions $V(t,0)=V(x,T)=0$. We already know that $V_{xx}(t,x)\le 0$. If $V_{xx}(t,x)=0$, then a solution exists only if $V_x(t,x)=0$. Because we look for a solution that is strictly increasing in x, the set on which $V_{xx}(t,x)=0$ should have measure zero. Let us therefore assume that $V_{xx}(t,x)<0$. Because the left-hand side of (3.1) is quadratic in θ, we find that

$$\theta^*(t,x)=-\frac{(m-\delta)V_x(t,x)}{\sigma^2xV_{xx}(t,x)}\,.$$

$\theta^*(t,x)$ will be our candidate for the optimal strategy. We see that $\theta^*(t,x)>0$, provided that our assumptions hold.

The left-hand side of (3.1) is strictly concave in c. Thus, there is a unique point $c^*(t,x)$ where the supremum is taken, $\phi_c(c^*(t,x),t)=V_x(t,x)$, or $c^*(t,x)=0$ if $\phi_c(0,t)<V_x(t,x)$. Note that by our assumptions $c^*(t,x)<\infty$, because $V_x(t,x)>0$ if a solution to (3.1) exists and $\phi_c(c,t)\to 0$ as $c\to\infty$.

Intuitively there should be an upper bound on the fraction of money invested. If, for example, double the wealth is invested into the risky asset, one can easily lose double the wealth. We will therefore assume in the next result that θ_t^* is bounded.

Theorem 3.2. *Suppose that there is a solution $f(t,x)$ to (3.1) that is twice continuously differentiable and increasing in x, and continuously differentiable in t with boundary conditions $f(T,x)=0$. Then $V(t,x)\le f(t,x)$. If*

$$\theta^*(t,x)=-\frac{(m-\delta)f_x(t,x)}{\sigma^2xf_{xx}(t,x)} \tag{3.2}$$

is bounded and $f(t,0)=0$ for all t, then $V(t,x)=f(t,x)$ and an optimal strategy is given by $\{(\theta^*(t,X_t^*),c^*(t,X_t^*))\}$.

Proof. Choosing $\theta = c = 0$ in (3.1) shows that $f_t(t,x) \le 0$. Thus, $f(t,x)$ is decreasing in t. By the boundary condition we conclude that $f(t,x) \ge 0$. Let $\{(\theta_t, c_t)\}$ be an arbitrary strategy. Consider the process $\{f(s, X_s)\}$ conditioned on $X_t = x$. By Itô's formula,

$$f(s,X_s) = f(t,x) + \int_t^s (f_t(v,X_v) + \{[(1-\theta_v)\delta + m\theta_v]X_v - c_v\} f_x(v,X_v)$$
$$+ \tfrac{1}{2}\sigma^2\theta_v^2 X_v^2 f_{xx}(v,X_v))\,\mathrm{d}v + \int_t^s \sigma\theta_v X_v f_x(v,X_v)\,\mathrm{d}W_v\,.$$

By (3.1) we get

$$f(s,X_s) + \int_t^s \phi(c_v,v)\,\mathrm{d}v \le f(t,x) + \int_t^s \sigma\theta_v X_v f_x(v,X_v)\,\mathrm{d}W_v\,.$$

Let $\{\tau_n\}$ be a localisation sequence of the stochastic integral. Then

$$\mathbb{E}\Bigl[f(\tau_n \wedge T, X_{\tau_n\wedge T}) + \int_t^{\tau_n\wedge T} \phi(c_v,v)\,\mathrm{d}v\Bigr] \le f(t,x)\,.$$

Because $f(\tau_n \wedge T, X_{\tau_n\wedge T}) \ge 0$, we have

$$\mathbb{E}\Bigl[\int_t^{\tau_n\wedge T} \phi(c_v,v)\,\mathrm{d}v\Bigr] \le f(t,x)\,.$$

Monotone convergence yields

$$V^{\theta,c}(t,x) = \mathbb{E}\Bigl[\int_t^{T} \phi(c_v,v)\,\mathrm{d}v\Bigr] \le f(t,x)\,.$$

Taking the supremum over all strategies implies that $V(t,x) \le f(t,x)$.

Suppose now that $\theta_t^*(t,X_t^*)$ is bounded by a value $\bar\theta$, say. Note that by (3.1) $\theta_t^*(t,X_t^*) > 0$ whenever $X_t^* > 0$. Choose $\varepsilon < x$ and define $\tau_\varepsilon = \inf\{s > t : X_t^* < \varepsilon\}$. Then $f_x(t,x) \le f_x(t,\varepsilon)$ by concavity. That is, $f_x(t,x)$ is bounded on $[0,T] \times [\varepsilon,\infty)$. From (3.1) we find by Itô's formula that

$$f(s,X_s^*) = f(t,x) - \int_t^s \phi(c_v^*,v)\,\mathrm{d}v + \int_t^s \sigma\theta_v^* X_v^* f_x(v,X_v^*)\,\mathrm{d}W_v\,. \tag{3.3}$$

Also by Itô's formula, we have that

$$X_t^2 = \int_0^t (2\{[(1-\theta_s)\delta + m\theta_s]X_s - c_s\}X_s + \sigma^2\theta_s^2X_s^2)\,\mathrm{d}s + \int_0^t 2\sigma\theta_s X_s^2\,\mathrm{d}W_s\,.$$

If the process is stopped at $\tau^n = \inf\{t : X_t \notin [\varepsilon, n]\}$, the stochastic integral becomes a martingale for bounded $\{\theta_t\}$. We thus have

$$\mathbb{E}[X_{\tau^n\wedge t}^2] = \mathbb{E}\Bigl[\int_0^{\tau_n\wedge t} (2\{[(1-\theta_s)\delta + m\theta_s]X_s - c_s\}X_s + \sigma^2\theta_s^2X_s^2)\,\mathrm{d}s\Bigr]\,.$$

Therefore, there is an exponential upper bound of $\mathbb{E}[X^2_{\tau^n\wedge t}]$ that is independent of n. Thus, this bound holds for $\mathbb{E}[X^2_t]$ and $\int_0^T \mathbb{E}[X^2_t]\,dt < \infty$. Proposition A.5 thus implies that the stochastic integral in (3.3) is a martingale. That is,

$$\mathbb{E}\Big[f(\tau_\varepsilon \wedge T, X^*_{\tau_\varepsilon\wedge T}) + \int_t^{\tau_\varepsilon\wedge T} \phi(c_v, v)\,\mathrm{d}v\Big] = f(t,x)\,.$$

The second term converges monotonically as $\varepsilon \to 0$. The first term can be written as

$$\mathbb{E}[f(\tau_\varepsilon, \varepsilon); \tau_\varepsilon \le T] + \mathbb{E}[f(T, X^*_T); \tau_\varepsilon > T]\,.$$

The first term is bounded by $f(0,\varepsilon)\mathbb{P}[\tau_\varepsilon \le T]$, which is uniformly bounded. It thus converges to $\mathbb{E}[f(\tau,0);\tau \le T] = 0$. The second term is monotone and converges to $\mathbb{E}[f(T, X^*_T); \tau \ge T] = 0$. The latter equality follows from the boundary condition. □

Example 3.3. Let us consider the utility functions

$$\phi(c,t) = c^\alpha \mathrm{e}^{-\kappa(1-\alpha)t}\,,$$

where $\alpha \in (0,1)$ and $\kappa \in \mathbb{R}$. We try a solution of the form $V(t,x) = g(t)^{1-\alpha}x^\alpha$ for a continuously differentiable function $g : [0,T] \to \mathbb{R}_+$ with $G(T) = 0$. Writing $\mathrm{e}^{-\kappa(1-\alpha)t}$ and $g(t)^{1-\alpha}$ is for convenience only. Then

$$\theta^*(t,x) = \frac{m-\delta}{(1-\alpha)\sigma^2}$$

becomes independent of t and x. For the dividend rate we obtain

$$c^*(t,x) = \frac{\mathrm{e}^{-\kappa t}}{g(t)}x\,.$$

The Hamilton–Jacobi–Bellman equation (3.1) reads

$$\frac{\mathrm{e}^{-\alpha\kappa t}}{g(t)^\alpha}x^\alpha \mathrm{e}^{-(1-\alpha)\kappa t} + \Big(\delta + \frac{(m-\delta)^2}{2(1-\alpha)\sigma^2} - \frac{\mathrm{e}^{-\kappa t}}{g(t)}\Big)\alpha x^\alpha g(t)^{1-\alpha} + (1-\alpha)g(t)^{-\alpha}g'(t)x^\alpha = 0\,.$$

The function $g(t)$ is thus the solution to

$$(1-\alpha)g'(t) + \alpha\Big(\delta + \frac{(m-\delta)^2}{2(1-\alpha)\sigma^2}\Big)g(t) + (1-\alpha)\mathrm{e}^{-\kappa t} = 0\,,$$

with the boundary condition $g(T) = 0$. We thus find that

$$g(t) = \frac{1-\alpha}{r^*-\kappa}\mathrm{e}^{-\kappa t}\Big[\mathrm{e}^{(r^*-\kappa)(T-t)} - 1\Big]$$

with

$$r^* = \frac{\alpha}{1-\alpha}\Bigl(\delta + \frac{(m-\delta)^2}{2(1-\alpha)\sigma^2}\Bigr)\,,$$

provided that $r^* \neq \kappa$. If $r^* = \kappa$, then $g(t) = (1-\alpha)(T-t)e^{-r^* t}$.

The rate $c^*(t,x)/x$ tends to infinity as $t \to T$. This has to be the case because it cannot be optimal to have any capital left at time T. ■

Optimal Final Utility

Suppose now that we want to maximise

$$V^\theta(x) = \mathbb{E}[\phi(X_T^\theta)]$$

for some utility function $\phi(x)$ with $\phi(0) = 0$. It is clear that it cannot be optimal to pay any dividend before time T. Thus, we just have $c_t = 0$. In a similar way as for the case with optimal consumption, we obtain the *Hamilton–Jacobi–Bellman equation*

$$\sup_\theta\{[(1-\theta)\delta + m\theta]xV_x(t,x) + \tfrac{1}{2}\sigma^2\theta^2x^2V_{xx}(t,x) + V_t(t,x)\} = 0\,.$$

We look for a solution $V : [0,T]\times \mathbb{R}_+ \to \mathbb{R}_+$ that is twice continuously differentiable, strictly increasing and concave in x, and continuously differentiable in t with $V(t,0) = 0$ and $V(T,x) = \phi(x)$. The optimal strategy again is given by

$$\theta^*(t,x) = -\frac{(m-\delta)V_x(t,x)}{\sigma^2 x V_{xx}(t,x)}\,.$$

The verification theorem can be proved in a similar way as before.

Theorem 3.4. *Suppose that there is a solution $f(t,x)$ to the Hamilton–Jacobi–Bellman equation that is twice continuously differentiable increasing in x and continuously differentiable in t with boundary condition $f(T,x) = \phi(x)$. Then $V(t,x) \le f(t,x)$. If $\theta^*(t,x)$ defined by (3.2) is bounded and $f(t,0) = 0$ for all t, then $V(t,x) = f(t,x)$. Moreover, $\{\theta^*(t,X_t^*)\}$ is an optimal strategy.* □

Example 3.5. Let $\phi(x) = x^\alpha$. Letting $f(t,x) = g(t)x^\alpha$ for a continuously differentiable function $g : [0,T] \to \mathbb{R}_+$ with $g(T) = 1$, we obtain the equation

$$\Bigl(\delta + \frac{(m-\delta)^2}{2(1-\alpha)\sigma^2}\Bigr)\alpha x^\alpha g(t) + g'(t)x^\alpha = 0\,.$$

Thus, $g(t) = e^{-(1-\alpha)r^*(T-t)}$, fulfilling the boundary condition $g(T) = 1$. The value r^* is defined as in Example 3.3. ■

Example 3.6. Let us now consider a case where $V(t, x)$ is not bounded from below. Let $\phi(x) = \log x$. It is therefore impossible to choose $\phi(0) = 0$. This means that we cannot apply Theorem 3.4 directly. We first note that $\alpha^{-1}(x^\alpha - 1)$ tends to $\log x$ as $\alpha \to 0$. Thus, the problem from Example 3.5 is connected. We therefore guess that the solution is the limit of $\alpha^{-1}(V_\alpha(t, x) - 1)$, i.e.,

$$V(t, x) = \log x + \Big(\delta + \frac{(m-\delta)^2}{2\sigma^2}\Big)(T - t)\,.$$

Direct verification shows that $V(t, x)$ indeed solves the Hamilton–Jacobi–Bellman equation with the correct boundary condition.

In order to prove a verification theorem, we need to make an assumption about the possible strategies. We suppose that $\mathbb{E}[\int_0^T \theta_s^2 \,\mathrm{d}s] < \infty$. We need to prove part of the verification theorem because we no longer have that the solution is positive. Define $\tau_\varepsilon^n = \inf\{t : X_t \notin [\varepsilon, n]\}$. In the same way as in the proof of Theorem 3.2, one finds that

$$\mathbb{E}[V(\tau_\varepsilon^n \wedge T, X_{\tau_\varepsilon^n \wedge T})] \le V(t, x)\,.$$

We want to let $n \to \infty$ and then $\varepsilon \to 0$. Because $T - (\tau_\varepsilon^n \wedge T)$ is bounded, we see that the time term tends to zero. By Fatou's lemma,

$$\lim_{\varepsilon \to 0} \mathbb{E}[\log X_{\tau_\varepsilon \wedge T}] \le \lim_{\varepsilon \to 0} \lim_{n \to \infty} \mathbb{E}[\log X_{\tau_\varepsilon^n \wedge T}] \le V(t, x)\,,$$

where $\tau_\varepsilon = \inf\{t : X_t < \varepsilon\}$.

If $\tau_0 \le T$, then $\log X_T = \log X_{\tau_\varepsilon} = -\infty$. Hence, if $\mathbb{P}[\tau_0 \le T] > 0$, then $\mathbb{E}[\log X_T] = -\infty < V(t, x)$. We can therefore assume that $\tau_0 > T$. We want to show that $\mathbb{E}[\log X_T] \le \lim_{\varepsilon \to 0} \mathbb{E}[\log X_{\tau_\varepsilon \wedge T}]$. By Itô's formula,

$$\log X_{\tau_\varepsilon \wedge T} - \log X_T = \int_{\tau_\varepsilon \wedge T}^T (\tfrac{1}{2}\sigma^2\theta_t^2 - (m-\delta)\theta_t - \delta)\,\mathrm{d}t - \int_{\tau_\varepsilon \wedge T}^T \sigma\theta_t \,\mathrm{d}W_t\,.$$

By our assumption, the stochastic integral is quadratically integrable, and therefore uniformly integrable. The integrand of the other integral is bounded from below. Thus, $\lim_{\varepsilon \to 0} \mathbb{E}[\log X_{\tau_\varepsilon \wedge T} - \log X_T] \ge 0$. Therefore,

$$\mathbb{E}[\log X_T] \le V(t, x)\,.$$

For the constant strategy $\theta^* = (m - \delta)/\sigma^2$ we find that

$$\mathbb{E}[\log X_T^* \mid X_t^*] = \log X_t^* + \int_t^T \Big(\delta + \frac{(m-\delta)^2}{2\sigma^2}\Big)\,\mathrm{d}s = V(t, X_t^*)\,.$$

This proves the verification theorem. ■

3.1.2 Single Life Insurance Contract

In this section we will not talk about the existence of solutions to the Hamilton–Jacobi–Bellman equations. In this sense the section is incomplete and not applicable until one shows the existence of a solution. It may therefore be considered as an exercise in modelling and in deriving the Hamilton–Jacobi–Bellman equation. If one uses a specific model for the mortality rate μ_{x+t} defined below, it may be possible to use results from analysis in order to prove that a solution to the Hamilton–Jacobi–Bellman equation exists. Another possibility to circumvent the problem is to apply general results from stochastic control theory, such as what can be found in [61]. Even though a verification theorem without proving the existence of a solution is problematic, we have decided to consider the problem of this section. The reader thus has an example, how to deal with jump diffusion processes. The author also thinks that in practical applications one often deals with numerical solutions only. The latter is, of course, problematic because the solution could be wrong.

We next suppose that the insurer has a life insurance contract in his portfolio. The goal is to maximise the final expected utility. We assume that the equity market and the death of the insurer are independent. This seems to be reasonable. However, one can think of situations where this assumption is not fulfilled. If the insured works or invests in the stock market, he could get a heart attack because of movements in the financial market. Or death could be caused by a catastrophe that also causes a collapse of the financial markets.

As above, let $\phi(x)$ be a utility function defined on $[0, \infty)$ with $\phi(0) = 0$. In addition to the market above, the agent has insured a person whose age is y at time zero. The force of mortality is denoted by μ_{y+t}. That is, ${}_tp_y = \mathbb{P}[T_y > t] = \exp\{-\int_0^t \mu_{y+s}\,\mathrm{d}s\}$, where T_y is the time of death of the insured. In particular, T_y is an absolutely continuous random variable with density $\mu_{y+t}\,{}_tp_y$. The initial capital is denoted by x and includes the premia paid until time zero. The problem is now that the insurer always needs to have enough money in the portfolio to cover the liability; otherwise, the supervisor would close the company. This implies that one cannot invest into the risky asset whenever the lower bound (the minimal amount needed in the portfolio to cover the liability) is reached. Our derivation of the Hamilton–Jacobi–Bellman equation shows that the equation still should hold whenever one is away from the boundary.

Pure Endowment Insurance

We start by considering the case where the insurer has promised to pay one monetary unit whenever the insured is still alive at time T. We consider the case of a single premium, i.e., there are no further premia to be paid. We measure money here in terms of the insured sum. That means that the utility function $\phi(x)$ may have to be adjusted. If we consider the case $\phi(x) = x^\alpha$, the monetary unit does not play any rôle. In order for the liability to be secured the boundary condition $X_t \geq \mathrm{e}^{-\delta(T-t)}$ has to be fulfilled. If the boundary is reached, it is no longer possible to invest, because otherwise $X_T \geq 1$ would not be fulfilled with probability 1.

Denote by $V^{\mathrm{M}}(t,x)$ the value of the classical Merton problem. If the boundary $X_t = \mathrm{e}^{-\delta(T-t)}$ is reached, we can calculate the solution. As long as the insured is alive, no investment into the risky asset is possible. Thus, $X_t = \mathrm{e}^{-\delta(T-t)}$. If the insured dies, the only restriction left is $X_t \geq 0$; thus, the value is $V^{\mathrm{M}}(t, \mathrm{e}^{-\delta(T-t)})$. If $X_{T-} = 1$, then the value 1 has to be paid and the expected utility left is zero. This gives the value

$$V(t, \mathrm{e}^{-\delta(T-t)}) = \int_t^T V^{\mathrm{M}}(s, \mathrm{e}^{-\delta(T-s)}) \mu_{y+s\,s}p_y \,\mathrm{d}s \;. \tag{3.4}$$

Hence, we have turned the restriction into a boundary condition.

Note that because there are no liabilities in the classical Merton problem, we have $V(t,x) < V^{\mathrm{M}}(t,x)$. The boundary condition (3.4) corresponds to the strategy $\theta_t = 0$ on (t, T_y) and to the classical optimal strategy after that point. Thus we know the optimal strategy if we start at the boundary.

Let us next motivate the Hamilton–Jacobi–Bellman equation. We again do the calculations for $t = 0$. Suppose that $x > \mathrm{e}^{-\delta T}$. Let $\tau = \inf\{t : X_t = \mathrm{e}^{-\delta(T-t)}\}$ and T_y denote the time of death. Consider a constant strategy $\theta_t = \theta$ on $(0, \tau \wedge T_y \wedge h)$. In the same way as before, we conclude that

$$V(0,x) \geq \mathbb{E}[V(\tau \wedge T_y \wedge h, X_{\tau\wedge T_y \wedge h})] \;,$$

where we, slightly abusing the notation, interpret $V(T_y, z)$ as $V^{\mathrm{M}}(T_y, z)$. From Itô's formula, we find that

$$\begin{aligned} \mathbb{E}\Bigl[\int_0^{\tau\wedge T_y\wedge h} &\{[(1-\theta)\delta + m\theta]X_t V_x(t,X_t) + \tfrac12 \sigma^2\theta^2 X_t^2 V_{xx}(t,X_t) + V_t(t,X_t)\}\,\mathrm{d}t \\ &+ \int_0^{\tau\wedge T_y\wedge h} \sigma\theta X_t V_x(t,X_t)\,\mathrm{d}W_t \\ &+ (V^{\mathrm{M}}(T_y, X_{T_y}) - V(T_y, X_{T_y}))\mathbb{1}_{T_y \leq \tau\wedge h}\Bigr] \leq 0 \;. \end{aligned}$$

Suppose that the stochastic integral is a martingale. We then divide by h and let $h \to 0$. In the limit the set $\{\tau < h\}$ does not contribute to the limit. On $\{T_y \leq h\}$, because $X_{T_y} \to x$ as $h \to 0$ and $h^{-1}\mathbb{P}[T_y \leq h] \to \mu_y$, we get the limit

$$h^{-1}\mathbb{E}[(V^{\mathrm{M}}(T_y, X_{T_y}) - V(T_y, X_{T_y}))\mathbb{1}_{T_y \leq \tau\wedge h}] \to \mu_y(V^{\mathrm{M}}(0,x) - V(0,x)) \;.$$

This leads to the inequality

$$\begin{aligned} [(1-\theta)\delta + m\theta]x V_x(0,x) + \tfrac12\sigma^2\theta^2 x^2 V_{xx}(0,x) + V_t(0,x) \\ + \mu_y(V^{\mathrm{M}}(0,x) - V(0,x)) \leq 0 \;. \end{aligned}$$

We have therefore motivated the *Hamilton–Jacobi–Bellman equation*

$$\sup_\theta\{[(1-\theta)\delta+m\theta]xV_x(t,x)+\tfrac{1}{2}\sigma^2\theta^2x^2V_{xx}(t,x)+V_t(t,x) + \mu_{y+t}(V^{\rm M}(t,x)-V(t,x))\}=0\;. \tag{3.5}$$

We are looking for a solution $V: B\to \mathbb{R}_+$ that is twice continuously differentiable and strictly increasing in x and continuously differentiable in t. The boundary conditions are $V(T,x)=\phi(x-1)$ and (3.4). The set B is the set $\{(t,x)\in[0,T]\times\mathbb{R}: x\ge e^{-\delta(T-t)}\}$.

We next prove the verification theorem.

Theorem 3.7. *Suppose that there is a solution $f(t,x)$ to Equation* (3.5) *that is twice continuously differentiable and increasing in x and continuously differentiable in t satisfying the boundary conditions above. Then $f(t,x)\ge V(t,x)$. If the strategy* (3.2) *is bounded, then $f(t,x)=V(t,x)$ and an optimal strategy is given by* (3.2).

Proof. From the boundary conditions we conclude that

$$f(t,x)\ge f(t,e^{-\delta(T-t)})>0\;.$$

Thus, $f(t,x)$ is bounded from below. Let θ be an arbitrary strategy. By Itô's formula,

$$\begin{aligned}
&f^{(\rm M)}(\tau\wedge T_y\wedge T, X_{\tau\wedge T_y\wedge T})\\
&\quad= f(t,x)+\int_t^{\tau\wedge T_y\wedge T}\sigma\theta_sX_sf_x(s,X_s)\,dW_s\\
&\qquad+\int_t^{\tau\wedge T_y\wedge T}\{[(1-\theta_s)\delta+m\theta_s]X_sf_x(s,X_s)+\tfrac{1}{2}\sigma^2\theta_s^2X_s^2f_{xx}(s,X_s)\\
&\qquad+f_t(s,X_s)\}\,ds+(V^{\rm M}(T_y,X_{T_y})-f(T_y,X_{T_y}))1\!\!{\rm I}_{T_y\le\tau\wedge T}\;.
\end{aligned}$$

We write $f^{(\rm M)}$ on the left-hand side in order to indicate that the function is $V^{\rm M}$ if $T_y\le\tau\wedge T$. It can be verified directly that the process

$$(V^{\rm M}(T_y,X_{T_y})-f(T_y,X_{T_y}))1\!\!{\rm I}_{T_y\le\tau\wedge t}-\int_0^{T_y\wedge\tau\wedge t}\mu_{y+s}(V^{\rm M}(s,X_s)-f(s,X_s))\,ds$$

is a martingale. Taking expected values and using (3.5) shows that

$$\mathbb{E}[f^{(\rm M)}(\tau\wedge T_y\wedge T, X_{\tau\wedge T_y\wedge T})]\le f(t,x)+\mathbb{E}\Big[\int_t^{\tau\wedge T_y\wedge T}\sigma\theta_sX_sf_x(s,X_s)\,dW_s\Big]\;.$$

Replacing T by a localisation sequence of the stochastic integral and using Fatou's lemma shows that

$$\mathbb{E}[f^{(\mathrm{M})}(\tau \wedge T_y \wedge T, X_{\tau \wedge T_y \wedge T})] \le f(t,x) .$$

On the left-hand side we have the value of the strategy θ. Thus, $V(t,x) \le f(t,x)$.

Let $\tau_\varepsilon = \inf\{t : X_t^* < \mathrm{e}^{-\delta(T-t)} + \varepsilon\}$. Then $f_x(s, X_s^*)$ is bounded on $s \in [t, \tau_\varepsilon \wedge T_y \wedge T]$. Thus, the stochastic integral is a martingale and

$$\mathbb{E}[f^{(\mathrm{M})}(\tau_\varepsilon \wedge T_y \wedge T, X^*_{\tau_\varepsilon \wedge T_y \wedge T})] = f(t,x) .$$

Equality holds because of Equation (3.5). That $f(t,x) = V(t,x)$ follows now as in the proof of Theorem 3.2. □

The jump term makes it hard to solve the equation. On the one hand, we have time dependence via the mortality μ_{y+t}. On the other hand, V^{M} is connected to the utility function $\phi(x)$, the boundary condition to the utility function $\phi(x-1)$. For example, if $\phi(x) = x^\alpha$, then the solution cannot be of the form $g(t)x^\alpha$ because of the boundary condition. Neither can it be of the form $g(t)(x-1)^\alpha$, because x^α appears in the equation. In particular, the optimal strategy will depend on the capital X_t.

Term Insurance

Next we consider an insurance contract where one unit is paid at the time of death provided that the insured dies in $(0,T]$. Because death can happen immediately, the process has to fulfil $X_t \ge 1$. The problem is thus different from the pure endowment case. If the boundary is reached, one can no longer invest. But interest ensures that one immediately leaves the boundary. We therefore cannot as in the pure endowment case immediately determine the value at the boundary.

The derivation of the Hamilton–Jacobi–Bellman equation at points away from the boundary is not influenced by the properties at the boundary. Thus,

$$\begin{aligned}\sup_\theta\{[(1-\theta)\delta + m\theta]xV_x(t,x) + \tfrac{1}{2}\sigma^2\theta^2x^2V_{xx}(t,x) + V_t(t,x)\\ + \mu_{y+t}(V^{\mathrm{M}}(t,x-1) - V(t,x))\} = 0 . \quad (3.6)\end{aligned}$$

We are looking for a continuous solution $V : [0,T] \times [1,\infty) \to \mathbb{R}_+$ that is twice continuously differentiable and strictly increasing in x, and continuously differentiable in t on the interior of $[0,T] \times [1,\infty)$. Because the process $\{X_t\}$ does not leave $[1,\infty)$, we expect that $\theta^*(t,x)$ tends to zero as $x \to 1$. On the boundary the function should therefore fulfil

$$\delta V_x(t,1) + V_t(t,1) - \mu_{y+t}V(t,1) = 0 . \quad (3.7)$$

Because $V(t,x)$ should be concave in x, it is not possible that $V_x(t,1) = 0$. Thus, $V_{xx}(t,1) = -\infty$. We see that the function is not twice differentiable on the boundary. That means that we have to change the proof of the verification theorem slightly. For technical reasons we only allow strategies such that $\theta_t \to 0$ as $X_t \to 1$. This is also needed in order for the solution to the Hamilton–Jacobi–Bellman equation to be well defined.

Theorem 3.8. *Suppose that there is a solution $f(t,x)$ to Equation (3.6) that is twice continuously differentiable and increasing in x, and continuously differentiable in t, satisfying the boundary conditions $f(T,x) = \phi(x)$ and (3.7). Then $f(t,x) \ge V(t,x)$. If the strategy (3.2) is bounded, then $f(t,x) = V(t,x)$ and an optimal strategy is given by (3.2).*

Proof. Proving the result completely would be beyond the scope of the present book. We therefore give a heuristic argument.

Note that by continuity $\theta^*(t,x)$ tends to zero as $x \downarrow 1$. Suppose that $x > 1$, and choose ε such that $x > 1 + \varepsilon$. Note that $f(x,t)$ is bounded from below by continuity. Let $\tau_\varepsilon = \inf\{t : X_t < 1 + \varepsilon\}$. Then in the same way as for Theorem 3.7, it follows that

$$\mathbb{E}[f^{(\mathrm{M})}(\tau_\varepsilon \wedge T_y \wedge T, X_{\tau_\varepsilon \wedge T_y \wedge T})] \le f(t,x) .$$

The left-hand side is bounded. Thus, we can let $\varepsilon \to 0$. This shows that

$$\mathbb{E}[f^{(\mathrm{M})}(\tau \wedge T_y \wedge T, X_{\tau \wedge T_y \wedge T})] \le f(t,x) .$$

Until time τ the process $\{f(t, X_t)\}$ is a supermartingale. The boundary condition implies that it is a supermartingale on $[0,T]$. This follows in the same way as for Markov processes; see, for instance, [57]. Thus, the value of the strategy is bounded by $f(t,x)$.

Replacing θ by θ^* yields

$$\mathbb{E}[f^{(\mathrm{M})}(\tau^* \wedge T_y \wedge T, X^*_{\tau^* \wedge T_y \wedge T})] = f(t,x) .$$

The proof for the arbitrary strategy also applies, but the process is a martingale instead of a supermartingale. Thus, $f(t,x)$ is the value of the strategy θ^*. By continuity, this also holds on the boundary. □

Bibliographical Remarks

Merton's problem goes back to Merton [131, 132]. Extensions have been discussed by many authors. Some references are Lehoczky, Sethi, and Shreve [123, 124], Aase [1], Jacka [110], Karatzas, Lehoczky, Sethi, and Shreve [115], Karatzas, Lehoczky, and Shreve [116], Sethi and Taksar [166], Fleming and Zariphopoulou [62], Gabih, Richter, and Wunderlich [68], Bäuerle and Rieder [16], Gabih and Wunderlich [69] and Gabih, Grecksch, and Wunderlich [67]. In some of these papers an alternative approach is used, the martingale method. Richard [150] considered a life insurance contract, but in contrast to Section 3.1.2, he optimised the consumption and the utility at the (stochastic) time of death of the policyholder.

3.2 Optimal Dividends and Bonus Payments

3.2.1 Utility Maximisation of Dividends

An insurer is interested to optimise dividend payments in a life insurance contract. A dividend in a life insurance contract means a payment to the insured. We model the dividend payments and the surplus process as in Appendix F. Thus, the state of the contract is given by a Markov process $\{J_t\}$ on a finite state space $\mathcal{J} = \{0, 1, \ldots, J\}$. State 0 means "dead", and no further dividends will be paid. We can let $J_0 = 1$ be the initial state. By $\mu_{ij}(t)$ we denote the intensity for changing from state i to state j, i.e.,

$$\mathbb{P}[J_s = j \mid J_t = i] = \Bigl(\exp\Bigl\{\int_t^s (\mu_{hk}(v))_{\{hk\in\mathcal{J}\}}\,dv\Bigr\}\Bigr)_{ij}$$

for $s \ge t$ where $(\mu_{hk}(v))_{\{hk\in\mathcal{J}\}}$ denotes the matrix with off-diagonal elements $\mu_{hk}(v)$ and diagonal elements $\mu_{hh}(v) = -\sum_{k\neq h}\mu_{hk}(v)$. The integration is taken element-wise. This can be interpreted to mean we have $\mathbb{P}[J_{t+v} = j \mid J_t = i] = \mu_{ij}v + o(v)$ as $v \downarrow 0$ for $i \neq j$.

By N_t^{ij} we denote the number of jumps of $\{J_s\}$ from state i to state j until time t. The dividends are

$$D_t = \sum_{i=1}^J \int_0^t \mathbb{1}_{J_s=i} d_s^i\,ds + \sum_{i=1}^J\sum_{j=0}^J d_s^{ij}\,dN_s^{ij} + \sum_{i=1}^J \Delta D_T^i \mathbb{1}_{J_T=i,t\ge T}\;.$$

d_t^i is the rate at which dividends are paid to the insured if the contract is in state i. d_t^{ij} is a lump sum to be paid upon transition from state i to state j. We only consider the case where $d_t^i \ge 0$ and $d_t^{ij} \ge 0$. Thus, we exclude the case where the insured pays a continuous premium to the insurer. Finally, Δ_T^i is a payment to be paid upon termination of the contract if the insured is in state i at time T.

A dividend strategy is a collection of positive adapted stochastic processes, $\{d_t^i\}$, $\{d_t^{ij}\}$, and a random variable $\Delta D_T^i \ge 0$. Moreover, the insurer can choose an adapted investment strategy $\{\theta_t\}$. The surplus process then fulfils

$$dX_t = \{[(1-\theta_t)\delta + m\theta_t]X_t - d_t^{J_t}\}dt + \sigma\theta_t X_t\,dW_t - d_t^{J_{t-}J_t}\,dN_t^{J_{t-}J_t}\;.$$

Here we use the short notation $dN_s^{J_{s-}J_s} = \sum_{i\neq j}\mathbb{1}_{J_s-=i,J_s=j}\,dN_s^{ij}$. The final payment is $\Delta_T^{J_T} = X_{T-}$, i.e., $X_T = 0$. A positive adapted strategy $\{(d_t, \theta_t)\}$ is *admissible* if the process $\{X_t\}$ is well defined and $\Delta_T^{J_T} \ge 0$. Here we write d_t for the vector $(d_t^1, \ldots, d_t^J, d_t^{1,0}, \ldots, d_t^{J,J_1})^\top$ The filtration $\{\mathcal{F}_t\}$ is the natural filtration generated by the process $\{(J_t, W_t)\}$. The remaining value of a strategy is defined as

$$V^{d,\theta}(t,i,x) = \mathbb{E}\Big[\int_t^T \phi^c(J_s, d_s^{J_s}, s)\,\mathrm{d}s + \int_t^T \phi^d(J_{s-}, J_s, d_s^{J_{s-}J_s}, s)\,\mathrm{d}N_s^{J_{s-}J_s} + \phi^T(J_T, \Delta D_T^{J_T})\,\Big|\, J_t = i, X_s = x\Big]\,.$$

The functions $\phi^c(i,d,t)$, $\phi^d(i,j,d,t)$, and $\phi^T(i,D)$ are utility functions, i.e., strictly increasing and strictly concave functions. For simplicity we norm the utility functions such that $\phi^c(i,0,t) = 0$, $\phi^d(i,j,0,t) = 0$, and $\phi^T(i,0) = 0$. This is possible if the value in 0 is finite. The value function becomes

$$V(t,i,x) = \sup_{d,\theta} V^{d,\theta}(t,i,x)\,,$$

where the supremum is taken over all admissible strategies. More specifically, we will be interested in $V(0,1,x)$. Note that it is possible that a pension has to be paid also after time T. We consider this problem here by letting $\phi^T(i, \Delta D_T^i)$ denote the value of the dividends after time T and ΔD_T^i denote the value of the future pension.

Remark 3.9. We add here just the utility of a dividend rate and the utility of lump sums. This seems to be strange at first sight because it is not immediately clear how to compare these two versions of dividends. If the dividend rate is now constant over a unit time interval, one could argue that dividend and utility integrated over the time interval are completely comparable with utilities of lump sums. However, to make the argument usable, one should consider a process where dividends are accumulated over unit time intervals and their utilities are added. That means lump sum payments and continuous dividends should be added before taking their utility. We therefore need to keep in mind that our approach just approximates the customers "happiness." ■

The problem we consider here is thus similar to the classical Merton problem considered in Section 3.1.1. But we additionally have lump sum payments. The payment at the maturity date T just leads to the boundary condition $V(T,i,x) = \phi^T(i,x)$. This is because any reasonable dividend/bonus scheme should pay all surplus at the expiring date.

Proceeding as in Section 3.1.1 yields the *Hamilton–Jacobi–Bellman equation*

$$\sup_{\theta,d^c,d^d}\Big\{\phi^c(i,d^c,t) + \{[(1-\theta)\delta + m\theta]x - d^c\}V_x(t,i,x) + \tfrac{1}{2}\sigma^2\theta^2x^2V_{xx}(t,i,x) + V_t(t,i,x) + \sum_{j=0}^{J}\mu_{ij}(t)[V(t,j,x-d^j) + \phi^d(i,j,d^j,t) - V(t,i,x)]\Big\} = 0\,. \tag{3.8}$$

The function $V : [0,T] \times \mathcal{J} \times \mathbb{R}_+ \to \mathbb{R}_+$ should be increasing, twice continuously differentiable in x and continuously differentiable in t with boundary conditions $V(t,i,0) = 0$ and $V(T,i,x) = \phi^T(i,x)$. Here we make the convention that $\phi^d(i,i,d,t) = 0$. Because $\mu_{ii}(t) < 0$, the optimal value for d^i is $d^i = 0$.

Solving for the supremum, we see that

$$\theta^*(t,i,x) = -\frac{(m-\delta)V_x(t,i,x)}{\sigma^2 x V_{xx}(t,i,x)},$$

and the dividend strategy is the solution to $\phi_d^c(i, d^{c*}(t,i,x), t) = V_x(t,i,x)$ and $\phi_d^d(i,j,d^{d*}(t,i,j,x),t) = V_x(t,j,x-d^{d*}(t,i,j,x))$, provided that a solution exists. As in Lemma 3.1, it follows that $V(t,i,x)$ is concave in x. If no solution $d^{c*}(t,i,x)$ exists, then $d^{c*}(t,i,x) = 0$. If no solution $d^{d*}(t,i,j,x)$ exists, then $d^{ij*}(t,i,j,x) = 0$. The supremum is unique by strict concavity.

We next prove the verification theorem.

Theorem 3.10. *Suppose that there exists a solution $f(t,i,x)$ to (3.8) that is twice continuously differentiable with respect to x and continuously differentiable with respect to t, increasing in x and with boundary conditions $f(T,i,x) = \phi^T(i,x)$. Then $V(t,i,x) \le f(t,i,x)$. Suppose, moreover, that $\theta^*(t,x,i)$ is bounded and $f(t,i,0) = 0$. Then $V(t,i,x) = f(t,i,x)$. An optimal strategy is given by $\theta^*(t,x,i)$, $d^{c*}(t,i,x)$, $d^{d*}(t,i,j,x)$ obtained above, and $\Delta D_T^i(x) = x$.*

Proof. As a continuous function on a compact interval, the function $[0,T] \to \mathbb{R}, t \mapsto f(t,i,0)$ is bounded from below. Because the function is increasing in x, we obtain that $f(t,i,x)$ is bounded from below. Consider first an arbitrary strategy and fix $t \in [0,T)$. According to [26, p. 27], the process M' defined as

$$\begin{aligned} M'_s = &\sum_{t<v\le s: J_v \ne J_{v-}} f(v,J_v,X_v) + \phi^d(J_{v-},J_v,d_v^{J_{v-}J_v},v) - f(v,J_{v-},X_{v-}) \\ &- \int_t^s \sum_{j=0}^J \mu_{J_v,j}(v)[f(v,j,X_v - d_v^{J_v j}) + \phi^d(J_v,j,d_v^{J_v j},v) \\ &- f(v,i,X_v)]\,\mathrm{d}v \end{aligned}$$

is a martingale. Let S be a point where $\{J_t\}$ jumps, and $T = \inf\{t > S : J_t \ne J_S\}$. From Itô's formula, we get that

$$\begin{aligned} &f(T,J_{T-},X_{T-}) - f(S,J_S,X_S) \\ &\quad = \int_S^T (\{[(1-\theta_v)\delta + m\theta_v]X_v - d_v^{J_v}\} f_x(v,J_v,X_v) \\ &\qquad + \tfrac12 \sigma^2\theta_v^2 X_v^2 f_{xx}(v,J_v,X_v) + f_t(v,J_v,X_v))\,\mathrm{d}v \\ &\qquad + \int_S^T \sigma\theta_v X_v f_x(v,J_v,X_v)\,\mathrm{d}W_t\,. \end{aligned}$$

Therefore, we get that

$$f(s, J_s, X_s) + \sum_{t<v\le s: J_v \ne J_{v-}} \phi^d(J_{v-}, J_v, d_v^{J_{v-} J_v}, v)$$
$$- \int_t^s \Big(\{[(1-\theta_v)\delta + m\theta_v]X_v - d_v^{J_v}\} f_x(v, J_v, X_v) + \tfrac{1}{2}\sigma^2\theta_v^2 X_v^2 f_{xx}(v, J_v, X_v)$$
$$+ f_t(v, J_v, X_v)$$
$$+ \sum_{j=0}^{J} \mu_{J_v j}(v)(f(v, j, X_v - d_v^{J_v,j}) + \phi^d(J_v, j, d^{J_v j}, v) - f(v, J_v, X_v))\Big) \,\mathrm{d}v$$

is a local martingale. Let $\{T_n\}$ be a localisation sequence. Using (3.8), we find that

$$f(s \wedge T_n, J_{s\wedge T_n}, X_{s\wedge T_n}) + \sum_{t<v\le s\wedge T_n: J_v \ne J_{v-}} \phi^d(J_{v-}, J_v, d_v^{J_{v-} J_v}, v)$$
$$+ \int_t^{s\wedge T_n} \phi^c(J_v, d_v^{J_v}, v)\,\mathrm{d}v$$

is a supermartingale. From the martingale property we obtain

$$\begin{aligned} f(t, i, x) \ge \mathbb{E}\Big[& f(T \wedge T_n, J_{T\wedge T_n}, X_{T\wedge T_n}) \\ & + \sum_{t<v\le T\wedge T_n: J_v \ne J_{v-}} \phi^d(J_{v-}, J_v, d_v^{J_{v-} J_v}, v) \\ & + \int_t^{T\wedge T_n} \phi^c(J_v, d_v^{J_v}, v)\,\mathrm{d}v\Big] \,. \end{aligned}$$

Fatou's lemma and monotone convergence imply that

$$\begin{aligned} f(t, i, x) &\ge \mathbb{E}\Big[\phi^T(J_T, X_T) + \sum_{t<v\le T: J_v \ne J_{v-}} \phi^d(J_{v-}, J_v, d_v^{J_{v-} J_v}, v) \\ &\qquad + \int_t^T \phi^c(J_v, d_v^{J_v}, v)\,\mathrm{d}v\Big] \\ &= V^{d,\theta}(t, i, x) \,. \end{aligned}$$

This shows the first part of the theorem. In a similar way as in the proof of Theorem 3.2, the remaining part of the theorem can be shown. □

Let us now specify the utility functions. We let

$$\begin{aligned} \phi^c(i, d, t) &= a(t, i)^{1-\alpha} d^\alpha \,, \\ \phi^d(i, j, d, t) &= a(t, i, j)^{1-\alpha} d^\alpha \,, \\ \phi^T(i, D) &= A(i)^{1-\alpha} D^\alpha \,, \end{aligned}$$

where $\alpha \in (0,1)$, $a(t,i), a(t,i,j) \geq 0$, and $A(i) > 0$. Note that the capital independent parts of the utility are written in a convenient way but are arbitrary functions. We try the function $V(t,i,x) = g(t,i)^{1-\alpha}x^\alpha$. The optimal strategy becomes

$$\theta^*(t,i,x) = \frac{m-\delta}{(1-\alpha)\sigma^2}\,,$$

$$d^{c*}(t,i,x) = \frac{a(t,i)}{g(t,i)}x\,,$$

$$d^{d*}(t,i,j,x) = \frac{a(t,i,j)}{g(t,j)+a(t,i,j)}x\,.$$

The functions $a(t,i)$ and $a(t,i,j)$ are therefore some sort of shadow contract. The function $g(t,i)/x$ has the rôle of determining how many contracts are in force at time t.

Equation (3.8) reads

$$\begin{aligned} &a(t,i)^{1-\alpha}\frac{a(t,i)^\alpha}{g(t,i)^\alpha}x^\alpha + \Bigl\{\delta - \frac{a(t,i)}{g(t,i)} + \frac{(m-\delta)^2}{2\sigma^2(1-\alpha)}\Bigr\}g(t,i)^{1-\alpha}\alpha x^\alpha \\ &+ (1-\alpha)g(t,i)^{-\alpha}g'(t,i)x^\alpha + \sum_{j=0,j\neq i}^J \mu_{ij}(t)\Bigl\{g(t,j)^{1-\alpha}\Bigl(\frac{g(t,j)}{g(t,j)+a(t,i,j)}\Bigr)^\alpha \\ &+ a(t,i,j)^{1-\alpha}\frac{a(t,i,j)^\alpha}{(g(t,j)+a(t,i,j))^\alpha} - g(t,i)^{1-\alpha}\Bigr\}x^\alpha = 0\,, \end{aligned}$$

where $g'(t,i)$ denotes the derivative with respect to t. This yields

$$\begin{aligned} &\alpha\Bigl\{\delta + \frac{(m-\delta)^2}{2\sigma^2(1-\alpha)}\Bigr\}g(t,i) + (1-\alpha)a(t,i) + (1-\alpha)g'(t,i) \\ &+ \sum_{j=0,j\neq i}^J \mu_{ij}(t)\Bigl((g(t,j)+a(t,i,j))^{1-\alpha}g(t,i)^\alpha - g(t,i)\Bigr) = 0\,. \end{aligned} \tag{3.9}$$

The boundary condition is $g(T,i) = A(i)$. Equation (3.9) has to be solved numerically. We do not show that there is a solution $g(t,i)$ to (3.9). However, we have a system of coupled ordinary differential equations. Under some conditions on the functions $a(t,i)$, $a(t,i,j)$, and $A(i)$, the theory of ordinary differential equations yields the existence of a solution. It should be remarked that the surplus process X is not involved in solving $g(t,i)$. The function $g(t,i)$ is completely determined through $\{J_t\}$.

From the equation one can see that the function could be obtained in the following way. We can express $g(t,i)$ as

$$g(t,i) = \mathbb{E}^*\Bigl[\int_t^T e^{r^*(s-t)}\,\mathrm{d}A_s \Bigm| J_t = i\Bigr]\,. \tag{3.10}$$

The measure $\mathbb{E}^*$ is the measure under which the transition matrix is $\boldsymbol{\mu}(t)/(1-\alpha)$. The "inflation" rate is

$$r^* = \frac{\alpha}{1-\alpha}\Big\{\delta + \frac{(m-\delta)^2}{2\sigma^2(1-\alpha)}\Big\}\,.$$

The process $\{A_t\}$ is the payment process where a dividend at rate $a(t, J_t)$ is paid. Upon transition from i to j, the amount

$$(g(t,j) + a(t,i,j))^{1-\alpha} g(t,i)^\alpha - g(t,j)$$

is paid. The reader should, however, note that this interpretation does not help to solve Equation (3.9) explicitly. The equation remains nonlinear. But in order to show the existence of a solution, one has to verify that (3.10) is continuously differentiable. Then the usual arguments used in Markov process theory will show that $g(t,i)$ solves (3.9).

3.2.2 Utility Maximisation of Bonus

The dividend scheme found in Section 3.2.1 strongly depends on the return on investment. The dividend is a linear function of X_t, the surplus at time t. A person looking for life insurance coverage usually is interested in a certain guarantee. It is therefore not very popular if "bad" investment leads to lower dividends. A quite common form of bonus is a scheme under which new "payment processes" are bought.

Suppose that the dividend process is given by (F.1), where the payments are deterministic. For simplicity we operate with a single premium, i.e., a premium is paid at time zero only. The value is given by (F.2), where the value is calculated under the technical basis, that is, with interest rate $\bar{\delta}$ and under a measure $\overline{\mathbb{P}}$ for the state process $\{J_t\}$. We denote the corresponding technical value by $\overline{V}(t,i)$. The "market" value is denoted by $\widehat{V}(t,i)$, and the corresponding force of interest is δ. That is, the real interest rate is δ and the value is calculated under some equivalent pricing measure $\mathbb{P}^*$; see Appendix E. Here we use $\widehat{V}(t,i)$ because we still want to use $V(t,i,x)$ for the value function of the optimisation problem.

Let $K(t)$ be the number of payment processes bought up to time t. $K(0)$ is then the number of payment processes guaranteed. Instead of considering the real dividends that are paid, it is reasonable to consider the purchase of more payment processes as dividend payoff, even though only the guaranteed sum is increased. The change of the considered dividend stream is then given by $\overline{V}(t, J_t)\, \mathrm{d}K(t)$.

The purchase of additional payments is made for the technical price, which is larger than the market value. Thus, the reserve also increases, $(\overline{V}(t,J_t) - \widehat{V}(t,J_t))\, \mathrm{d}K(t)$. Therefore the change in the reserve becomes

$$-\overline{V}(t,J_t)\,\mathrm{d}K(t)+(\overline{V}(t,J_t)-\widehat{V}(t,J_t))\,\mathrm{d}K(t)=-\widehat{V}(t,J_t)\,\mathrm{d}K(t)\,.$$

This is the market value of the payment streams bought.

The surplus process now becomes

$$\mathrm{d}X_t^{K\theta}=[(1-\theta_t)\delta+\theta_t m]X_t^{K\theta}\,\mathrm{d}t+\sigma\theta_t X_t^{K\theta}\,\mathrm{d}W_t-\widehat{V}(t,J_t)\,\mathrm{d}K(t)\,,$$

where again $\{\theta_t\}$ is an investment strategy. The value of the bonus strategy is defined as

$$V^{K\theta}(t,i,x,\kappa)=\mathbb{E}\Bigl[\int_t^T \phi^c(J_s,K(t),t)\,\mathrm{d}s+\phi^T(J_T,K(T))$$
$$+\int_t^T \phi^d(J_{s-},J_s,K(t-),s)\,\mathrm{d}N_s^{J_{s-}J_s}\Bigm| J_t=i,X_t=x,K_t=\kappa\Bigr]\,.$$

The functions $\phi^c(i,K,t)$, $\phi^d(i,j,K,t)$, and $\phi^T(i,K)$ are utility functions. It is natural to use $K(t)$ to measure the utility. The higher $K(t)$ is, the higher the payment stream will be. The value function of the problem becomes $V(t,i,x,\kappa)=\sup_{K,\theta}V^{K\theta}(t,i,x,\kappa)$. The supremum is taken over all adapted strategies $\{(K(t),\theta_t)\}$, where $K(t)$ is increasing such that $\{X_t^{K,\theta}\}$ is well defined.

Let us now motivate the Hamilton–Jacobi–Bellman equation. To simplify the problem for the moment, we only allow absolutely continuous processes $K(t)=1+\int_0^t k(s)\,\mathrm{d}s$, where $0\le k(s)\le k_0$. Let $\theta\in\mathbb{R}$ and $k\in[0,k_0]$. We start at t with $X_t=x>0$, $K(t)=\kappa$, and $J_t=i$. If $X_t=0$, then only $\theta=k=0$ is allowed. We let $\tau=\inf\{s>t: X_t=0\}$. We choose $h>0$, and the constant strategy $(k_s,\theta_s)=(k,\theta)$ in $[t,\tau\wedge(t+h))$. From time $\tau^*=\tau\wedge\tau_t\wedge(t+h)$ on we choose an ε-optimal strategy. We let $\tau_t=\inf\{s>t: J_s\ne J_t\}$. Then, as in Section 3.1.1, we find that

$$V(t,i,x,\kappa)\ge\mathbb{E}\Bigl[\int_t^{\tau^*}\phi^c(i,\kappa+ks,s)\,\mathrm{d}s+V(\tau^*,J_{\tau^*},X_{\tau^*},\kappa+k\tau^*)$$
$$\Bigm| J_t=i,X_t=x,K(t)=\kappa\Bigr]\,.$$

This yields the inequality

$$\begin{aligned}0\ge{}&\phi^c(i,\kappa,t)+\{[(1-\theta)\delta+\theta m]x-\widehat{V}(t,i)k\}V_x(t,i,x,\kappa)\\&+\tfrac12\sigma^2\theta^2x^2V_{xx}(t,i,x,\kappa)+V_t(t,i,x,\kappa)+kV_\kappa(t,i,x,\kappa)\\&+\sum_{j=0}^J\mu_{ij}(t)(V(t,j,x,\kappa)+\phi^d(i,j,\kappa,t)-V(t,i,x))\,.\end{aligned}$$

The corresponding *Hamilton–Jacobi–Bellman equation* for the restricted problem is

$$\begin{aligned}0 = \sup_{k,\theta}\Bigl\{&\phi^c(i,\kappa,t) + \{[(1-\theta)\delta + \theta m]x - \widehat{V}(t,i)k\}V_x(t,i,x,\kappa)\\&+ \tfrac{1}{2}\sigma^2\theta^2x^2V_{xx}(t,i,x,\kappa) + V_t(t,i,x,\kappa) + kV_\kappa(t,i,x,\kappa)\\&+\sum_{j=0}^J \mu_{ij}(t)(V(t,j,x,\kappa) + \phi^d(i,j,\kappa,t) - V(t,i,x,\kappa))\Bigr\}\;.\end{aligned}$$

The optimal investment strategy is again given by

$$\theta^*(t,i,x,\kappa) = -\frac{(m-\delta)V_x(t,i,x,\kappa)}{\sigma^2 x V_{xx}(t,i,x,\kappa)}\;. \tag{3.11}$$

The optimal strategy for $k(t)$ is

$$k^*(t) = k_0 1\!\!\mathrm{I}_{V_\kappa(t,i,x,\kappa)\ge \widehat{V}(t,i)V_x(t,i,x,\kappa)}\;.$$

In a similar way as in Section 2.4.2, one can motivate the equation

$$\begin{aligned}0 = \max\Bigl\{\sup_\theta\Bigl\{&\phi^c(i,\kappa,t) + \{[(1-\theta)\delta + \theta m]x\}V_x(t,i,x,\kappa)\\&+ \tfrac{1}{2}\sigma^2\theta^2x^2V_{xx}(t,i,x,\kappa) + V_t(t,i,x,\kappa)\\&+\sum_{j=0}^J \mu_{ij}(t)(V(t,j,x,\kappa) + \phi^d(i,j,\kappa,t))\Bigr\},\\&V_\kappa(t,i,x,\kappa) - \widehat{V}(t,i)V_x(t,i,x,\kappa)\Bigr\}\;.\end{aligned}$$

This is the Hamilton–Jacobi–Bellman equation for the problem considered. We are looking for a function $V : [0,T] \times \mathcal{J} \times \mathbb{R}^2 \to \mathbb{R}$ that is increasing and twice continuously differentiable in x and continuously differentiable in t. The boundary condition is $V(T,i,x,\kappa) = \phi^T(i,\kappa + x/\widehat{V}(T,i))$. The boundary condition follows because it cannot be optimal to have $X_T > 0$. If $\widehat{V}(T,i) = 0$, then one would expect that $X_t \to 0$ as $t \uparrow T$.

The optimal investment strategy is given by (3.11). The optimal $K(t)$ jumps at the jump times of $\{J_t\}$ if the process after the jump is in the interior of the area where $V_\kappa(t,i,x,\kappa) = \widehat{V}(t,i)V_x(t,i,x,\kappa)$. The process $\{X_t\}$ is reflected at the boundary of the area where $V_\kappa(t,i,x,\kappa) = \widehat{V}(t,i)V_x(t,i,x,\kappa)$. The process $K(t)$ increases accordingly. Note that the process may behave in an unexpected way because the area $V_\kappa(t,i,x,\kappa) = \widehat{V}(t,i)V_x(t,i,x,\kappa)$ changes with time t. Because of this behaviour at the boundary, we do not prove a verification theorem here. One will have to prove a verification theorem for each choice of utility functions separately.

Bibliographical Remarks

The results of this section are taken from Steffensen [175]; see also Møller and Steffensen [135]. Some generalisations can also be found in Nielsen [138] and [137]; see also [139]. Similar problems are considered in Steffensen [174] and [176].

3.3 Optimal Control of a Pension Fund

A pension fund has some outgo for benefits and earns contributions. The contribution rate c_t can be determined by the fund manager, who also determines the investment strategy. Investment again is possible into a riskfree bond and a risky asset. As before, we model these financial instruments by the Black–Scholes model (E.3) and (E.1). Because we will introduce another Brownian motion below, we denote the Brownian motion driving the risky asset by W^{I} and the volatility by σ_{I}.

The outgo for benefits is modelled by a diffusion approximation

$$bt + \sigma_{\mathrm{B}} W_t^{\mathrm{B}} .$$

b is the rate at which the pension fund expects to pay the benefits, and σ_{B} is a parameter modelling the fluctuations of the outgo. These two parameters will be dependent on the region in which the pension fund operates. It also depends on the specific composition of the portfolio. For a large portfolio one would expect smaller variability and therefore a smaller value of σ_{B}. It is further assumed that W^{I} and W^{B} are independent. This should almost hold if the pension fund is large.

As modelled here, the benefits do not increase with inflation. It is therefore natural to assume that the fund size is measured in terms of a benefit unit. Because one expects benefits to increase with inflation, it is natural to let $\delta = 0$. We will, however, in the following let δ be arbitrary, but the reader should keep in mind that δ is not the return from a riskless bond but the return of the inflation-adjusted riskless bond.

In this section we let $\{\mathcal{F}_t\}$ denote the natural filtration of the process $\{(W_t^{\mathrm{I}}, W_t^{\mathrm{B}})\}$. Let $\{\theta_t\}$ denote the proportion of the fund size invested into the risky asset. By $\{c_t\}$ we denote the contribution rate of the insured to the pension fund. Then the fund size follows the stochastic differential equation

$$\mathrm{d}X_t^{\theta,c} = [(1-\theta_t)\delta + m\theta_t]X_t^{\theta,c}\,\mathrm{d}t + \sigma_{\mathrm{I}}\theta_t X_t^{\theta,c}\,\mathrm{d}W_t^{\mathrm{I}} + (c_t - b)\,\mathrm{d}t - \sigma_{\mathrm{B}}\,\mathrm{d}W_t^{\mathrm{B}} .$$

We allow only adapted cadlag strategies $\{(\theta_t, c_t)\}$ such that the stochastic integrals are well defined. We could make further constraints such as, for, example $\min\{|\theta_t|, |\theta_t X_t^{\theta,c}|\} \le K$ for some $K > 0$. That is, for large capital there is a limit to how many times the wealth can be invested. In addition, there is an upper bound for how much wealth can be invested. We call the allowed (constrained or unconstrained) strategies *admissible*.

For securing the benefits the fund size should be larger than a predefined value x_p, say. On the other hand, the contributors would like to have c_t as small as possible. In order to measure the deviation from the ideal values, we introduce the loss function $L(c, x)$, measuring the "loss" if the contribution rate is c and the fund size is x. We denote the accumulated loss of the strategy $\{(\theta_t, c_t)\}$ by

$$V^{\theta,c}(x) = \mathbb{E}\Bigl[\int_0^\infty e^{-\beta t} L(c_t, X_t^{\theta,c})\, dt \Bigm| X_0^{\theta,c} = x\Bigr] .$$

We use the discounting of the loss $e^{-\beta t}$ in order to give more weight to the near future. For simplicity we consider an infinite horizon. We do not restrict to positive $\{X_t\}$ either. Because the benefits are modelled by a Brownian motion, there is always a positive probability that $\{X_t^{\theta,c}\}$ becomes negative. Thus, we need $L(c, x) < \infty$ for all x in order for the problem to be well defined.

The value function is $V(x) = \inf_{\theta,c} V^{\theta,c}(x)$, where we take the infimum over all admissible strategies. Recall that we consider losses. Hence, we need to minimise the value. In principle, this is not a value but costs. Hence we should define the value $-V^{\theta,c}(t, x)$. For simplicity of notation, we disregard the sign and still call it value.

3.3.1 No Constraints

We allow all adapted cadlag strategies $\{(\theta_t, c_t)\}$ such that the process $X^{\theta,c}$ is well defined. We start by motivating the Hamilton–Jacobi–Bellman equation. Let $\theta \in \mathbb{R}$, $c \ge 0$, and $h > 0$. Consider the strategy

$$(\theta_t, c_t) = \begin{cases} (\theta, c) , & \text{if } t < h, \\ (\theta^\varepsilon_{t-h}, c^\varepsilon_{t-h}) , & \text{if } t \ge h. \end{cases}$$

$\{(\theta^\varepsilon_t, c^\varepsilon_t)\}$ is an ε-optimal strategy for initial fund size X_h. As before, we discard from discussing measurability because we only look for candidate value functions at the moment. We find that

$$\begin{aligned} V(x) \le V^{\theta,c}(x) &= \mathbb{E}\Bigl[\int_0^h e^{-\beta t} L(c, X_t)\, dt + e^{-\beta h} V^\varepsilon(X_h)\Bigr] \\ &\le \mathbb{E}\Bigl[\int_0^h e^{-\beta t} L(c, X_t)\, dt + e^{-\beta h} V(X_h)\Bigr] + \varepsilon . \end{aligned}$$

We can let ε tend to zero. By Itô's formula,

$$\begin{aligned} V(X_h) = V(x) &+ \int_0^h \sigma_{\mathrm{I}} \theta X_t V'(X_t)\, dW_t^{\mathrm{I}} + \int_0^h \sigma_{\mathrm{B}} V'(X_t)\, dW_t^{\mathrm{B}} \\ &+ \int_0^h (\{[(1-\theta)\delta + m\theta] X_t + c - b\} V'(X_t) \\ &\qquad + \tfrac12 (\sigma_{\mathrm{I}}^2 \theta^2 X_t^2 + \sigma_{\mathrm{B}}^2) V''(X_t))\, dt , \end{aligned}$$

provided that $V(x)$ is twice continuously differentiable. Assuming that the stochastic integrals are martingales, plugging this in yields,

$$\begin{aligned} 0 \le \mathbb{E}\Bigl[\int_0^h &\{e^{-\beta t} L(c, X_t) + e^{-\beta h} (\tfrac12 (\sigma_{\mathrm{I}}^2 \theta^2 X_t^2 + \sigma_{\mathrm{B}}^2) V''(X_t) \\ &+ \{[(1-\theta)\delta + m\theta] X_t + c - b\} V'(X_t))\}\, dt\Bigr] \\ &- (1 - e^{-\beta h}) V(x) . \end{aligned}$$

Dividing by h and letting $h \downarrow 0$ gives, under the assumption that the limit and integration can be interchanged,

$$L(c,x) + \tfrac{1}{2}(\sigma_{\rm I}^2\theta^2x^2 + \sigma_{\rm B}^2)V''(x) \\ + \{[(1-\theta)\delta + m\theta]x + c - b\}V'(x) - \beta V(x) \geq 0\,.$$

This motivates the *Hamilton–Jacobi–Bellman equation*

$$\inf_{\theta,c}\{L(c,x) + \tfrac{1}{2}(\sigma_{\rm I}^2\theta^2x^2 + \sigma_{\rm B}^2)V''(x) \\ + \{[(1-\theta)\delta + m\theta]x + c - b\}V'(x) - \beta V(x)\} = 0\,. \tag{3.12}$$

The function $V : \mathbb{R} \to \mathbb{R}$ should be twice continuously differentiable. The left-hand side is quadratic in θ. In order for the infimum to be finite, we must have $V''(x) \geq 0$. If $V''(x) = 0$, the infimum is only finite if $V'(x) = 0$. This means that $V(x)$ does not depend on x. If this holds on some interval, then $L(c,x)$ does not depend on x. Because we want to punish if X_t deviates from a target value, this should not be the case. It is therefore natural to suppose that $V(x)$ is strictly convex.

The value at which the infimum for θ is taken is

$$\theta^*(x) = -\frac{(m-\delta)V'(x)}{\sigma_{\rm I}^2 xV''(x)}\,.$$

Not surprisingly, for the investment there will be a drift in the direction of the minimum of $V(x)$.

In order to calculate the c for which the minimum is taken, we differentiate with respect to c and set to zero: $L_c(c,x) + V'(x) = 0$. In order that there is a unique point for which the infimum is taken, we need to assume that $L_c(c,x)$ is monotone in c. If the infimum is taken in $(0,\infty)$, the loss function $L(c,x)$ must be convex in c. We therefore now assume that $L(c,x)$ is strictly convex in c. Moreover, if we suppose that $L_c(c,x) \to \infty$ as $c \to \infty$, we will get that there is a unique $c^*(x) \in [0,\infty)$ at which the infimum is taken.

It is hard to prove a verification theorem in general without any boundedness conditions. We will therefore just treat some examples.

Example 3.11. Consider the quadratic loss function

$$L(c,x) = (c-\hat{c})^2 + 2\rho(c-\hat{c})(x-\hat{x}) + (k+\rho^2)(x-\hat{x})^2\,, \tag{3.13}$$

where $k > 0$. That is, we have some target values $\hat{c}$ and $\hat{x}$, and deviation from these target values is punished. For the optimal c, we get

$$c^*(x) = \hat{c} - \rho(x-\hat{x}) - \tfrac{1}{2}V'(x)\,.$$

Equation (3.12) then reads

$$-\tfrac{1}{4}f'^2(x) + k(x-\hat{x})^2 - \frac{(m-\delta)^2 f'^2(x)}{2\sigma_{\mathrm{I}}^2 f''(x)} + \tfrac{1}{2}\sigma_{\mathrm{B}}^2 f''(x)$$
$$+ (\delta x + \hat{c} - \rho(x-\hat{x}) - b)f'(x) - \beta f(x) = 0\,.$$

We try a solution of the form $f(x) = Ax^2 + Bx + C$. Comparison of the terms yields the three equations

$$-A^2 + k - \frac{(m-\delta)^2 A}{\sigma_{\mathrm{I}}^2} + 2(\delta-\rho)A - \beta A = 0\,,$$

$$-AB - 2k\hat{x} - \frac{(m-\delta)^2 B}{\sigma_{\mathrm{I}}^2} + (\delta-\rho)B + 2(\hat{c}+\rho\hat{x}-b)A - \beta B = 0\,,$$

$$-\frac{B^2}{4} + k\hat{x}^2 - \frac{(m-\delta)^2 B^2}{4A\sigma_{\mathrm{I}}^2} + \sigma_{\mathrm{B}}^2 A + (\hat{c}+\rho\hat{x}-b)B - \beta C = 0\,.$$

Solving this system yields

$$A = -\frac{(m-\delta)^2}{2\sigma_{\mathrm{I}}^2} + (\delta-\rho) - \tfrac{1}{2}\beta + \sqrt{\Bigl(-\frac{(m-\delta)^2}{2\sigma_{\mathrm{I}}^2} + (\delta-\rho) - \tfrac{1}{2}\beta\Bigr)^2 + k}\,,$$

$$B = \frac{2[(\hat{c}+\rho\hat{x}-b)A - k\hat{x}]}{A + (m-\delta)^2\sigma_{\mathrm{I}}^{-2} + \beta + \rho - \delta}\,,$$

$$C = \beta^{-1}\Bigl[-\frac{B^2}{4} + k\hat{x}^2 - \frac{(m-\delta)^2 B^2}{4A\sigma_{\mathrm{I}}^2} + \sigma_{\mathrm{B}}^2 A + (\hat{c}+\rho\hat{x}-b)B\Bigr]\,.$$

Note that the other solution for A is strictly negative and therefore is not the solution we are looking for. Note also that the denominator in the expression for B is strictly positive. We find the optimal strategy

$$\theta^*(x) = -\frac{(m-\delta)(Ax + B/2)}{\sigma_{\mathrm{I}}^2 Ax}\,, \qquad c^*(x) = \hat{c} - \rho(x-\hat{x}) - Ax - B/2\,.$$

We note that $f(x)$ is minimal at $x = -B/(2A)$. The investment strategy is chosen in such a way that we take a short position in the risky asset if the reserve lies above the minimum and a long position if the reserve lies below. We could interpret this as we are throwing away money if the reserve lies above the minimising value.

We need to verify that we really found the solution. Let $\{(\theta_t, c_t)\}$ be an arbitrary adapted cadlag strategy. By Itô's formula ,

$$\begin{aligned}
\mathrm{e}^{-\beta t} f(X_t) = f(x) &+ \int_0^t \mathrm{e}^{-\beta s}\sigma_{\mathrm{I}}\theta_s X_s f'(X_s)\,\mathrm{d}W_t^{\mathrm{I}} \\
&+ \int_0^t \mathrm{e}^{-\beta s}\sigma_{\mathrm{B}} f'(X_s)\,\mathrm{d}W_s^{\mathrm{B}} \\
&+ \int_0^t \bigl(f'(X_s) + \{[(1-\theta_s)\delta + m\theta_s]X_s + c_s - b\}f'(X_s) \\
&+ \tfrac{1}{2}(\sigma_{\mathrm{I}}^2\theta_s^2 X_s^2 + \sigma_{\mathrm{B}}^2)f''(X_s) - \beta f(X_s)\bigr)\,\mathrm{e}^{-\beta s}\,\mathrm{d}s\,.
\end{aligned}$$

We first want to restrict the strategies further. We first note that

$$L(c,x) \geq k(x-\hat{x})^2 \geq kx^2/2 - k\hat{x}^2$$

and

$$2Ax^2 - f(x) \geq -C - \frac{B^2}{4A}\,.$$

Suppose that $\limsup_{t\to\infty} \mathbb{E}[\mathrm{e}^{-\beta t} f(X_t^{\theta,c})] > 0$. Then there exists $\varepsilon > 0$ such that $\mathbb{E}[(X_t^{\theta,c})^2] > \varepsilon \mathrm{e}^{\beta t}$ for $t \in D$ for an unbounded set D with infinite measure (at least if ε is small enough). Note that the drift term in the stochastic differential equation for $\{(X_t^{\theta,c})^2\}$ is bounded from below (Itô's formula). We then conclude that

$$V^{\theta,c}(x) \geq \int_D \mathrm{e}^{-\beta t}(k\varepsilon \mathrm{e}^{\beta t}/2 - k\hat{x}^2)\,\mathrm{d}t = \infty\,.$$

We can therefore restrict to strategies such that $\lim_{t\to\infty} \mathbb{E}[\mathrm{e}^{-\beta t} f(X_t^{\theta,c})] = 0$. Because $\mathbb{E}[(\mathrm{e}^{-\beta t} f'(X_t))^2] \leq n\mathbb{E}[\mathrm{e}^{-\beta t} f(X_t)]\mathrm{e}^{-\beta t}$ for some n, we find that the stochastic integrals above are martingales. Using (3.12), we conclude that

$$f(x) \leq \mathbb{E}[\mathrm{e}^{-\beta t} f(X_t)] + \mathbb{E}\Bigl[\int_0^t \mathrm{e}^{-\beta s} L(c_s, X_s)\,\mathrm{d}s\Bigr]\,.$$

Letting $t \to \infty$ shows that $f(x) \leq V^{\theta,c}(x)$, and therefore $f(x) \leq V(x)$.

For the strategy $\{(\theta_t^*, c_t^*)\}$, we get the martingale property of the stochastic integrals directly and

$$f(x) = \mathbb{E}[\mathrm{e}^{-\beta t} f(X_t^*)] + \mathbb{E}\Bigl[\int_0^t \mathrm{e}^{-\beta s} L(c_s^*, X_s^*)\,\mathrm{d}s\Bigr]\,.$$

The process is mean-reverting. Therefore, $\mathrm{e}^{-\beta t} f(X_t^*) \to 0$. By Fatou's lemma and monotone convergence, we obtain $f(x) \geq V^*(x) \geq V(x)$.

The reader should be aware that if we added some constraint on the strategies, such as $c_t \geq 0$ or $0 \leq \theta_t \leq 1$, the solution could not longer be quadratic. Indeed, in the area where $\{(\theta^*(x), c^*(x))\}$ fulfils the constraint, the solution must necessarily coincide with $f(x) = Ax^2 + Bx + C$ obtained above. The probability of leaving this area is strictly positive. Thus, the solution to the constrained problem must be strictly larger than $f(x)$. ∎

Example 3.12. Consider the loss function $L(c,x) = \mathrm{e}^{\alpha c - \gamma x}$ for $\alpha, \gamma > 0$. For this choice large values of c and low values of x are punished. For matching the target values the reader might prefer the loss function $\mathrm{e}^{\alpha(c-\hat{c})-\gamma(x-\hat{x})}$. But this only results in a constant multiplied to the value function. We therefore choose the present form. We try a solution of the form $f(x) = \mathrm{e}^{a-gx}$. The optimal strategy becomes

$$\theta^*(x) = \frac{m-\delta}{g\sigma_\mathrm{I}^2 x}\,, \qquad c^*(x) = \alpha^{-1}[a - (g-\gamma)x + \log g/\alpha]\,.$$

We would expect that $c^*(x)$ is decreasing in x. We therefore expect $g \ge \gamma$. For the investment strategy we observe that the amount invested $\theta^*(x)x$ becomes constant.

Plugging our guess into the Hamilton–Jacobi–Bellman equation (3.12) and dividing by $f(x)$ yields

$$\frac{g}{\alpha} - \frac{(m-\delta)^2}{2\sigma_\mathrm{I}^2} + \sigma_\mathrm{B}^2 g^2 - \Bigl(\Bigl(\delta + \frac{\gamma - g}{\alpha}\Bigr)x + \frac{a}{\alpha} - b + \alpha^{-1}\log g/\alpha\Bigr)g - \beta = 0\,.$$

We see that $g = \gamma + \delta\alpha > \gamma$ and

$$a = 1 + \alpha\Bigl(\sigma_\mathrm{B}^2(\gamma+\delta\alpha) + b - \frac{(m-\delta)^2 + 2\sigma_\mathrm{I}^2\beta}{2\sigma_\mathrm{I}^2(\gamma+\delta\alpha)}\Bigr) - \log\frac{\gamma+\delta\alpha}{\alpha}\,.$$

The optimal strategy becomes

$$x\theta^*(x) = \frac{m-\delta}{\sigma_\mathrm{I}^2(\beta+\delta\alpha)}\,, \quad c^*(x) = \alpha^{-1} + \sigma_\mathrm{B}^2(\gamma+\delta\alpha) + b - \frac{(m-\delta)^2 + 2\sigma_\mathrm{I}^2\beta}{2\sigma_\mathrm{I}^2(\gamma+\delta\alpha)} - \delta x\,.$$

We observe that the contribution rate is some constant minus the expected income from the investments under the risk-neutral probability measure; see Appendix E..

The optimal reserve process is a Brownian motion with drift

$$\alpha^{-1} + \sigma_\mathrm{B}^2(\gamma+\delta\alpha) + \frac{(m-\delta)^2 - 2\sigma_\mathrm{I}^2\beta}{2\sigma_\mathrm{I}^2(\gamma+\delta\alpha)}$$

and diffusion parameter

$$\frac{(m-\delta)^2}{\sigma_\mathrm{I}^2(\gamma+\delta\alpha)^2} + \sigma_\mathrm{B}^2\,.$$

Thus, there is a value β_0 such that for $\beta < \beta_0$ the process drifts away to infinity, whereas for $\beta > \beta_0$ the process drifts to $-\infty$. In any case the solution is not stationary. This is not what one would intend to have. This means that one should not use the exponential loss function.

It is understandable that the surplus tends to infinity. This is because it is not punished if x is large. Large x also allows one to keep the contribution rate low. In reality, however, the pension fund would be punished with high taxes and therefore would not be interested in keeping the surplus high. If the surplus tended to $-\infty$, the supervisor would close the pension fund. Thus, one would not allow the reserves to become small. But for our setup, if the discounting is "too" large, then it is better to have a small loss at the beginning (contribution rate "too" low) and a high loss later will no longer matter.

That our solution really is the value function follows similarly as in Example 3.11. The verification is left to the reader. ∎

3.3.2 Fixed θ

Let us now consider the situation where θ is fixed by the pension fund. This is the case, for example, if the contributors have fixed the average return on their capital. Thus we only minimise over $\{c_t\}$. The *Hamilton–Jacobi–Bellman equation* is then

$$\inf_c \{L(c,x) + \tfrac{1}{2}(\sigma_{\rm I}^2\theta^2x^2 + \sigma_{\rm B}^2)V''(x) + \{[(1-\theta)\delta + m\theta]x + c - b\}V'(x) - \beta V(x)\} = 0 \,. \quad (3.14)$$

Again $V : \mathbb{R} \to \mathbb{R}$ should be twice continuously differentiable. The optimal strategy $c^*(x)$ is again determined by the equation $L_c(c^*(x), x) + V'(x) = 0$.

Example 3.13. Let us again consider the loss function (3.13). The Hamilton–Jacobi–Bellman equation reads

$$-\tfrac{1}{4}f'^2(x) + k(x-\hat{x})^2 + \tfrac{1}{2}(\sigma_{\rm I}^2\theta^2x^2 + \sigma_{\rm B}^2)f''(x) + \{[(1-\theta)\delta + m\theta - \rho]x + \hat{c} + \rho\hat{x} - b\}f'(x) - \beta f(x) = 0 \,.$$

We try a solution of the form $f(x) = Ax^2 + Bx + C$. Plugging in this guess yields the equations

$$-A^2 + k + \sigma_{\rm I}^2\theta^2 A + 2[(1-\theta)\delta + m\theta - \rho]A - \beta A = 0 \,,$$
$$-AB - 2k\hat{x} + [(1-\theta)\delta + m\theta - \rho]B + 2(\hat{c} + \rho\hat{x} - b)A - \beta B = 0 \,,$$
$$-\frac{B^2}{4} + k\hat{x}^2 + \sigma_{\rm B}^2 A + (\hat{c} + \rho\hat{x} - b)B - \beta C = 0 \,.$$

The solution is

$$A = (1-\theta)\delta + m\theta - \rho - \beta/2 + \tfrac{1}{2}\sigma_{\rm I}^2\theta^2 + \sqrt{[(1-\theta)\delta + m\theta - \rho - \beta/2 + \tfrac{1}{2}\sigma_{\rm I}^2\theta^2]^2 + k} \,,$$
$$B = \frac{2[(\hat{c} + \rho\hat{x} - b)A - k\hat{x}]}{A - (1-\theta)\delta - m\theta + \rho + \beta} \,,$$
$$C = \beta^{-1}[-\tfrac{1}{4}B^2 + k\hat{x}^2 + \sigma_{\rm B}^2 A + (\hat{c} + \rho\hat{x} - b)B] \,.$$

That we really found the value function follows similarly as in Example 3.11.

The optimal strategy is

$$c^*(x) = \hat{c} - \rho(x - \hat{x}) - Ax - B/2 \,.$$

Suppose that we now want to minimise over θ. The solution will of course depend on x because the parameter θ appears in A, B, and C. An idea could be to minimise the expected loss for large x, that is, we should minimise A. Let $Q = (1-\theta)\delta + m\theta - \rho - \beta/2 + \frac{1}{2}\sigma_{\rm I}^2\theta^2$. Then $A = Q + \sqrt{Q^2 + k}$. The derivative of A with respect to θ is then

$$\frac{\mathrm{d}A}{\mathrm{d}\theta} = \frac{\mathrm{d}A}{\mathrm{d}Q}\frac{\mathrm{d}Q}{\mathrm{d}\theta} = \Bigl(1 + \frac{Q}{\sqrt{Q^2+k}}\Bigr)(m - \delta + \sigma_{\mathrm{I}}^2\theta)\;.$$

Equating to zero yields $\theta^* = -(m-\delta)/\sigma_{\mathrm{I}}^2$ because $\mathrm{d}A/\mathrm{d}Q > 0$. $\mathrm{d}A/\mathrm{d}\theta$ is negative if $\theta < \theta^*$ and positive if $\theta > \theta^*$. Thus, θ^* is indeed the minimum. The strategy is thus to take a long position in the stock if the capital is negative and a short position if the capital is positive. This strategy seems weird because we reduce the expected profit compared with the riskless return if our capital is positive.

Note that the optimal strategy of Example 3.11 tends to θ^* as $|x| \to \infty$. Calculating A and B, we find that these two parameters coincide with the parameters in Example 3.11. The difference of the value function to the value of the optimal unconstrained problem is therefore just a constant,

$$\frac{(m-\delta)^2 B^2}{4A\beta\sigma_{\mathrm{I}}^2}\;.$$

■

3.3.3 Fixed c

Let us now suppose that we are free to choose the investment strategy, but the contribution rate $c_t = \hat{c}$ is fixed. This could be the case if the contribution rate is fixed by the contract. It would not make sense to choose another contribution rate than $\hat{c}$, because this is the target value. The loss function is a function of x only, $L(x)$. We find the *Hamilton–Jacobi–Bellman equation*

$$\begin{aligned}\inf_{\theta}\{L(x) &+ \tfrac{1}{2}(\sigma_{\mathrm{I}}^2\theta^2x^2 + \sigma_{\mathrm{B}}^2)V''(x) \\ &+ \{[(1-\theta)\delta + m\theta]x + \hat{c} - b\}V'(x) - \beta V(x)\} = 0\;. \qquad (3.15)\end{aligned}$$

We are looking for a twice continuously differentiable function $V : \mathbb{R} \to \mathbb{R}$. The optimal investment strategy again becomes

$$\theta^*(x) = -\frac{(m-\delta)V'(x)}{\sigma_{\mathrm{I}}^2 x V''(x)}\;.$$

Example 3.14. We again take the example considered before. We let $L(x) = (x-\hat{x})^2$. Note that the constant $k + \rho^2$ does not play any rôle. The equation to solve becomes

$$(x-\hat{x})^2 + \frac{\sigma_{\mathrm{B}}^2}{2}f''(x) + (\delta x + \hat{c} - b)f'(x) - \beta f(x) - \frac{(m-\delta)^2 f'^2(x)}{2\sigma_{\mathrm{I}}^2 f''(x)} = 0\;.$$

We try the function $f(x) = Ax^2 + Bx + C$. Ordering the terms, we find the equations

$$1 + 2\delta A - \beta A - \frac{(m-\delta)^2}{\sigma_{\rm I}^2} A = 0 \,,$$

$$-2\hat{x} + 2(\hat{c}-b)A + \delta B - \beta B - \frac{(m-\delta)^2}{\sigma_{\rm I}^2} B = 0 \,,$$

$$\hat{x}^2 + \sigma_{\rm B}^2 A + (\hat{c}-b)B - \beta C - \frac{(m-\delta)^2 B^2}{4\sigma_{\rm I}^2 A} = 0 \,.$$

If $\beta + (m-\delta)^2/\sigma_{\rm I}^2 - 2\delta > 0$, the solution becomes

$$A = \frac{\sigma_{\rm I}^2}{(m-\delta)^2 + (\beta - 2\delta)\sigma_{\rm I}^2} \,,$$

$$B = \frac{2\hat{\sigma}_{\rm I}^2((\hat{c}-b)A - \hat{x})}{(m-\delta)^2 + (\beta - \delta)\sigma_{\rm I}^2} \,,$$

$$C = \beta^{-1}\Bigl(\hat{x}^2 + \sigma_{\rm B}^2 A + (\hat{c}-b)B - \frac{(m-\delta)^2 B^2}{4\sigma_{\rm I}^2 A}\Bigr) \,.$$

In order for $f(x)$ to be the solution to the problem, we need $A > 0$. The condition $\beta + (m-\delta)^2/\sigma_{\rm I}^2 - 2\delta > 0$ is therefore necessary. Because $\beta > 0$ is necessary anyway, the condition is fulfilled if $(m-\delta)^2 \ge 2\delta\sigma_{\rm I}^2$.

In the same way as in Example 3.11, it follows that $V(x) = f(x)$. From the definition of $V(x)$ it follows that $V(x)$ is decreasing in β. Suppose that $(m-\delta)^2 < 2\delta\sigma_{\rm I}^2$. Letting $\beta \downarrow (2\delta - (m-\delta)^2/\sigma_{\rm I}^2)$, the function $V(x)$ tends to infinity. We therefore can conclude that $V(x) = \infty$ for $\beta \le 2\delta - (m-\delta)^2/\sigma_{\rm I}^2$.

If $0 < \beta \le 2\delta - (m-\delta)^2/\sigma_{\rm I}^2$, we see that it is important also to be able to control c_t. Without this possibility the reserve may grow or decrease too fast. At least if the reserve becomes large the customers would prefer a lower contribution rate to the "poor" investment strategy

$$\theta^*(x) = -\frac{(m-\delta)(2Ax + B)}{2\sigma_{\rm I}^2 Ax}$$

the fund manager chooses in order to decrease the reserve.

3.3.4 Power Loss Function and $\sigma_{\rm B} = 0$

This example shows that one always has to verify that the solution to the Hamilton–Jacobi–Bellman equation really is the value function.

We assume that the benefit stream is deterministic bt. Let $0 < \gamma < 1$. We use the loss function

$$L(c,x) = \begin{cases} -\frac{1}{\gamma}(\hat{c}-c)^\gamma \,, & \text{for } c \le \hat{c}, \\ \infty \,, & \text{for } c > \hat{c}. \end{cases}$$

The loss function does not depend on x. Thus, in principle, the solution is $c = -\infty$, and θ would not matter. In order to prevent this, we restrict to strategies for which $X_t^{\theta,c} \geq x^*$ for some fixed $x^* \in \mathbb{R}$. This restriction is similar to the restriction in the Black–Scholes model that the wealth needs to be positive in order to exclude doubling-type strategies that would lead to an arbitrage. The set of admissible strategies are then the adapted cadlag processes $\{(\theta_t, c_t)\}$ with $c_t \leq \hat{c}$ such that $\{X_t^{\theta,c}\}$ is well defined and $X_t^{\theta,c} \geq x^*$ for all $t \geq 0$.

The corresponding *Hamilton–Jacobi–Bellman equation* is

$$\inf_{\theta,c}\Bigl\{-\frac{1}{\gamma}(\hat{c}-c)^\gamma+\tfrac{1}{2}\sigma_I^2\theta^2x^2V''(x)+\{[(1-\theta)\delta+m\theta]x+c-b\}V'(x)-\beta V(x)\Bigr\}=0\,, \tag{3.16}$$

where $V : [x^*, \infty) \to \mathbb{R}$ is twice continuously differentiable. Because θ_t should tend to zero as X_t approaches x^*, we also add the boundary condition

$$-\frac{1-\gamma}{\gamma}(-V'(x^*))^{-\gamma/(1-\gamma)} + (\delta x^* + \hat{c} - b)V'(x^*) - \beta V(x^*) = 0\,.$$

Because the drift close to x^* cannot be strictly negative, a solution is only possible if

$$b + (-V'(x^*))^{-1/(1-\gamma)} \geq \delta x^* + \hat{c}\,,$$

which yields an additional condition.

Taking the derivative with respect to c and equating to zero yields $c^*(x) = \hat{c} - (-V'(x))^{-1/(1-\gamma)}$. This makes sense because the value function should be decreasing in x. Differentiating with respect to θ and equating to zero yields

$$\theta^*(x) = -\frac{(m-\delta)V'(x)}{\sigma_I^2 x V''}\,.$$

Thus, $V(x)$ has to solve the equation

$$-\frac{1-\gamma}{\gamma}(-V'(x))^{-\gamma/(1-\gamma)} - \frac{(m-\delta)^2V'^2(x)}{2\sigma_I^2V''(x)} + (\delta x + \hat{c} - b)V'(x) - \beta V(x) = 0\,.$$

Motivated by Merton's problem (Section 3.1.1), we try a solution of the form $V(x) = -k(x-\hat{x})^\gamma$, where k and $\hat{x}$ are constants. Note that the value is well defined only if $x \geq \hat{x}$. We find that

$$\begin{aligned}-(1-\gamma)(\gamma k^\gamma)^{-1/(1-\gamma)}(x-\hat{x})^\gamma &- \frac{(m-\delta)^2k\gamma(x-\hat{x})^\gamma}{2\sigma_I^2(1-\gamma)} \\ &- [\delta x + \hat{c} - b]\gamma k(x-\hat{x})^{\gamma-1} + \beta k(x-\hat{x})^\gamma = 0\,.\end{aligned}$$

This can be a solution only if $\hat{x} = (b-\hat{c})/\delta$. Dividing by $k(x-\hat{x})^\gamma$ yields

$$-(1-\gamma)(\gamma k)^{-1/(1-\gamma)} - \frac{(m-\delta)^2\gamma}{2\sigma_I^2(1-\gamma)} - \delta\gamma + \beta = 0\,.$$

Letting

$$c_1 = (1-\gamma)^{-1}\Big(\beta - \delta\gamma - \frac{(m-\delta)^2\gamma}{2\sigma_I^2(1-\gamma)}\Big),$$

we find that $k = \gamma^{-1}c_1^{\gamma-1}$. This is well defined only if $c_1 > 0$. The optimal strategy becomes

$$c^*(x) = \hat{c} - c_1(x-\hat{x}), \qquad \theta^*(x) = \frac{(m-\delta)(x-\hat{x})}{\sigma_I^2(1-\gamma)x}.$$

We need to verify that indeed $X_t^* \geq x^*$. Let $Y_t = X_t^* - \hat{x}$. Then

$$\mathrm{d}Y_t = \Big(\delta - c_1 + \frac{(m-\delta)^2}{\sigma_I^2(1-\gamma)}\Big)Y_t\,\mathrm{d}t + \frac{m-\delta}{\sigma_I(1-\gamma)}Y_t\,\mathrm{d}W_t^I.$$

We see that Y is a geometric Brownian motion and therefore strictly positive. This geometric Brownian motion tends to zero if

$$0 > \delta - c_1 + \frac{(m-\delta)^2}{\sigma_I^2(1-\gamma)} - \tfrac{1}{2}\Big(\frac{m-\delta}{\sigma_I(1-\gamma)}\Big)^2 = \frac{\delta-\beta}{1-\gamma} + \frac{(m-\delta)^2}{2\sigma_I^2(1-\gamma)},$$

that is, if $2\sigma_I^2(\beta-\delta) > (m-\delta)^2$. If $2\sigma_I^2(\beta-\delta) < (m-\delta)^2$, then Y tends to infinity. If $2\sigma_I^2(\beta-\delta) = (m-\delta)^2$, then $\limsup_{t\to\infty} Y_t = \infty$ and $\liminf_{t\to\infty} Y_t = 0$.

Because any value in $(\hat{x},\infty)$ can be reached, we conclude that $\hat{x} \geq x^*$. In order to prove a verification theorem, we first consider processes with $X_t \geq \hat{x}$ only. We only need to consider strategies $\{c_t\}$ for which $c_t \leq \hat{c}$. Then $f(X_t) = -k(X_t-\hat{x})^\gamma$ is well defined. From Itô's formula, we get

$$\begin{aligned}
\mathrm{e}^{-\beta t}f(X_t) = f(x) &+ \int_0^t \mathrm{e}^{-\beta s}\sigma_I\theta_s X_s f'(X_s)\,\mathrm{d}W_t^I \\
&+ \int_0^t \big(f'(X_s) + \{[(1-\theta_s)\delta + m\theta_s]X_s + c_s - b\}f'(X_s) \\
&\qquad + \tfrac{1}{2}\sigma_I^2\theta_s^2X_s^2f''(X_s) - \beta f(X_s)\big)\mathrm{e}^{-\beta s}\,\mathrm{d}s.
\end{aligned}$$

Using (3.16), we conclude that

$$f(x) \leq \mathbb{E}[\mathrm{e}^{-\beta(t\wedge T_n)}f(X_{t\wedge T_n})] + \mathbb{E}\Big[\int_0^{t\wedge T_n}\mathrm{e}^{-\beta s}L(c_s,X_s)\,\mathrm{d}s\Big]$$

for some localisation sequence $\{T_n\}$. Letting $n\to\infty$, it follows from Fatou's lemma $[f(x)\leq 0]$ and monotone convergence $[L(c,x)\leq 0]$ that

$$f(x) \leq \mathbb{E}[\mathrm{e}^{-\beta t}f(X_t)] + \mathbb{E}\Big[\int_0^t \mathrm{e}^{-\beta s}L(c_s,X_s)\,\mathrm{d}s\Big] \leq \mathbb{E}\Big[\int_0^t \mathrm{e}^{-\beta s}L(c_s,X_s)\,\mathrm{d}s\Big].$$

Letting $t\to\infty$ shows that $f(x) \leq V^{\theta,c}(x)$, and therefore $f(x)\leq V(x)$.

Using the strategy (θ^*, c^*) and the fact that $\{Y_t = X_t^* - \hat{x}\}$ is a geometric Brownian motion, we find that the stochastic integrals are martingales, and therefore

$$f(x) = \mathbb{E}[\mathrm{e}^{-\beta t} f(X_t^*)] + \mathbb{E}\Bigl[\int_0^t \mathrm{e}^{-\beta s} L(c_s^*, X_s^*)\,\mathrm{d}s\Bigr]\,.$$

By Itô's formula,

$$Y_t^\gamma = \int_0^t \gamma\Bigl(\delta - c_1 + \frac{(m-\delta)^2}{2\sigma_I^2(1-\gamma)}\Bigr) Y_s^\gamma\,\mathrm{d}s + \int_0^t \frac{\gamma(m-\delta)}{\sigma_I(1-\gamma)} Y_s^\gamma\,\mathrm{d}W_s^I\,.$$

Taking the expected values, we find that

$$\mathbb{E}[Y_t^\gamma] = \int_0^t \gamma\Bigl(\delta - c_1 + \frac{(m-\delta)^2}{2\sigma_I^2(1-\gamma)}\Bigr)\mathbb{E}[Y_s^\gamma]\,\mathrm{d}s$$

because the stochastic integral is a martingale. In order for $\mathbb{E}[\mathrm{e}^{-\beta t} f(X_t)]$ to tend to zero, we need

$$\beta > \gamma\Bigl(\delta - c_1 + \frac{(m-\delta)^2}{2\sigma_I^2(1-\gamma)}\Bigr)\,,$$

or equivalently

$$\beta > \gamma\Bigl(\delta + \frac{(m-\delta)^2}{2\sigma_I^2(1-\gamma)}\Bigr)\,.$$

This is fulfilled because we have assumed that $c_1 > 0$.

This result is strange. One would therefore expect that the value function as a function of β has a jump at

$$\beta = \gamma\Bigl(\delta + \frac{(m-\delta)^2}{2\sigma_I^2(1-\gamma)}\Bigr)\,.$$

Suppose now that we have $x^* < \hat{x}$ and $\delta = 0$. Considering the wealth process $X_t' = X_t + \hat{x} - x^*$, we have $X_t' \ge \hat{x}$. Choosing the "optimal" investment process for initial capital $x + \hat{x} - x^*$, i.e., $\theta(x) = m(x-x^*)/(\sigma_I^2(1-\gamma)x)$ for the process $\{X_t\}$, the value is the same as for the value of $\{X_t'\}$, that is, $-k(x-x^*)^\gamma$. We see that the value function is larger than the value calculated above. Thus, we only found the solution if $\hat{x} = x^*$.

The ansatz we have tried is therefore not valid for the general solution. Because it is hard to solve the Hamilton–Jacobi–Bellman equation in this case, we do not give a solution to the general case here.

Bibliographical Remarks

The results in this section are taken from Cairns [31]. This paper also contains a discussion of additional topics of interest. Similar topics are discussed in Boulier et al. [25], Cairns and Parker [32], and Gerber and Shiu [79].

4

Asymptotics of Controlled Risk Processes

In this chapter we return to the problem of minimising the probability of ultimate ruin discussed in Section 2.3. We are interested in asymptotic properties of the ruin probability of the risk process under optimal control as well as the asymptotic behaviour of the optimal controls. We will see that analysis of the Hamilton–Jacobi–Bellman equation combined with methods from ruin theory will yield the results. We will only treat the Cramér–Lundberg case. Because the optimal policy in the diffusion case is explicitly known, it is possible to get the corresponding results just by standard methods.

4.1 Maximising the Adjustment Coefficient

We consider a classical risk model where $\sum_{i=1}^{N_t} Y_i$ is a compound Poisson process. We use the notation of Appendix D.1.

We start here with a problem connected to the problems of Sections 2.3.1 to 2.3.3. Suppose that we want to minimise the ruin probability through a decision that cannot be altered in the future. The strategy has to be constant, and the decision has to be made at time zero. This decision should also still be correct after some time, where the surplus has changed. One therefore looks for a criterion for an optimal decision that does not change with the initial capital.

A simple idea is to consider the adjustment coefficient. If the adjustment coefficient is maximised, then the ruin probability is asymptotically minimised. That is, the rate at which the ruin probability tends to zero is maximised. This seems to be the correct decision because, if ruin does not occur, the surplus will converge to infinity. We assume in this section that $r_\infty = \inf\{r : M_Y(r) < \infty\} > 0$. Otherwise, the considerations in this section would not make sense because the adjustment coefficient would not exist.

4.1.1 Optimal Reinsurance

As a first problem consider the problem of optimal choice of reinsurance. The surplus of a portfolio is modelled as a classical risk model; see Section D.1. Suppose that the insurer can choose reinsurance from a set of reinsurance possibilities indexed by $b \in \mathcal{U} \subset [-\infty, \infty]$. We consider the one-dimensional case only for convenience. The multidimensional case can be treated similarly. $\mathcal{U}$ is assumed to be a closed interval.

We denote by $s(Y;b)$ the self-insurance function. Examples are proportional reinsurance $s(Y;b) = bY$ or excess of loss reinsurance $s(Y;b) = Y \wedge b$. For possible reinsurance forms, see Section D.5.

For the reinsurance a premium at rate $c - c(b)$ has to be paid, i.e., the premium rate left to the insurer is $c(b)$. We denote by $M(r;b) = \mathbb{E}[e^{rs(Y;b)}]$ the moment-generating function of the part of the claim the insurer has to pay if the reinsurance level b is chosen. Then the adjustment coefficient $R(b)$ fulfils

$$\lambda(M(r;b) - 1) - c(b)r = 0 ; \tag{4.1}$$

see also Section D.1.2. In order for $R(b) \neq \infty$ for all b, we have to assume that $c(b_0) < 0$, where b_0 stands for full reinsurance. We also have to assume that there is b_1 such that the net profit condition $c(b_1) > \lambda\mathbb{E}[s(Y;b_1)]$ is fulfilled. Otherwise, no adjustment coefficient would exist.

Lemma 4.1 (Waters [185]). *Suppose that both $M(r;b)$ and $c(b)$ are twice differentiable (with respect to r and b). Suppose, moreover, that*

$$\lambda\mathbb{E}\Bigl[\frac{\partial^2 s(Y;b)}{\partial b^2}\Bigr] - c''(b) \geq 0 . \tag{4.2}$$

Then $R(b)$ is unimodal.

Remark 4.2. In the examples given a larger b means less reinsurance. The function $\lambda\mathbb{E}[s(Y;b)] - c(b)$ should therefore be decreasing, meaning that less reinsurance means less expected loss. Condition (4.2) implies that the expected loss is a convex function in b. For the reinsurer this means that the expected profit is convex. This seems plausible because if the insurer takes over a part of the large risk, he should profit more than by taking over a part of the small risk. The latter property would also follow from a utility approach. ■

Proof. In order for the adjustment coefficient to exist, the net profit condition $c(b) > \lambda M'(0;b)$ has to be fulfilled. Formally taking the derivative with respect to b in (4.1) gives

$$R'(b)\Bigl\{\lambda\frac{\partial M(R(b);b)}{\partial r} - c(b)\Bigr\} + \lambda\frac{\partial M(R(b);b)}{\partial b} - c'(b)R(b) = 0 .$$

We denote the partial derivatives by $M_r(r;b)$ and $M_b(r;b)$, respectively. Because the left-hand side of (4.1) is a convex function in r, we have that $\lambda M_r(R(b);b) - c(b) > 0$. By the implicit function theorem, $R'(b)$ exists and

$$R'(b) = -\frac{\lambda M_b(R(b);b) - c'(b)R(b)}{\lambda M_r(R(b);b) - c(b)}\;.$$

Let b^* be an argument such that $R'(b^*) = 0$. Note that $\lambda M_b(R(b^*);b^*) - c'(b^*)R(b^*) = 0$. The second derivative at b^* becomes

$$R''(b^*) = -\frac{\lambda M_{bb}(R(b^*);b^*) - c''(b^*)R(b^*)}{\lambda M_r(R(b^*);b^*) - c(b^*)}\;.$$

The denominator is strictly positive. We have

$$M_{bb}(r;b) = \mathbb{E}\Bigl[r e^{rs(Y;b)}\Bigl\{\frac{\partial^2 s(Y;b)}{\partial b^2} + r\Bigl(\frac{\partial s(Y;b)}{\partial b}\Bigr)^2\Bigr\}\Bigr]\;.$$

This is strictly bounded from below by

$$r\mathbb{E}\Bigl[\frac{\partial^2 s(Y;b)}{\partial b^2}\Bigr]\;.$$

By (4.2), $R''(b^*) < 0$. This must be the case at any point where $R'(b) = 0$. Thus, $R(b)$ is unimodal. □

One should note that the result does not imply that there is a value b^* such that $R'(b^*) = 0$. It only says that if such a value exists, then it is the maximum. If $R'(b) \neq 0$ for all b, then $R(b)$ is monotone and the optimal value is at the boundary. This can happen, for instance, in the case of proportional reinsurance, where it is optimal not to underwrite a reinsurance contract.

Consider now the case of proportional reinsurance. Then $\mathbb{E}[s(Y;b)] = b\mathbb{E}[Y]$ is linear in b. Thus, (4.2) is fulfilled for all concave $c(b)$. The derivative of $R(b)$ at $b = 1$ becomes

$$R'(1) = -R(1)\frac{\lambda\mathbb{E}[Y\exp\{R(1)Y\}] - c'(1)}{\lambda\mathbb{E}[Y\exp\{R(1)Y\}] - c}\;.$$

Thus, we get a simple criterion to determine whether or not $b^* \in (0,1)$. If $c'(1) < \lambda\mathbb{E}[Y\exp\{R(1)Y\}]$, then $b^* \in (0,1)$.

Let us consider some of the most popular premium principles. The definitions of the premium principles can be found in Section D.4.

Expected value principle

Here $c(b) = c - (1-b)(1+\theta)\lambda\mathbb{E}[Y]$, where θ is the safety loading. This is also linear in b. Therefore, (4.2) is fulfilled. Here $c'(1) = (1+\theta)\lambda\mathbb{E}[Y]$. Thus, reinsurance is taken if $(1+\theta)\mathbb{E}[Y] < \mathbb{E}[Y\exp\{R(1)Y\}]$, i.e., the reinsurer's safety loading is not too large. At first sight it looks like the premium rate c does not matter. It should be noted that $R(1)$ depends on c.

Variance principle

Here $c(b) = c - (1-b)\lambda\mathbb{E}[Y] - \alpha(1-b)^2\lambda\mathbb{E}[Y^2]$. This is concave and (4.2) is fulfilled. Moreover, $c'(1) = \lambda\mathbb{E}[Y] < \lambda\mathbb{E}[Y\exp\{R(1)Y\}$. Thus, the cedent always takes reinsurance.

Modified variance principle

Here $c(b) = c - (1-b)\lambda\mathbb{E}[Y] - \alpha(1-b)\mathbb{E}[Y^2]/E[Y]$ is linear in b and (4.2) is fulfilled. Reinsurance is chosen if α is not too large.

Standard deviation principle

Here $c(b) = c - (1-b)(\lambda\mathbb{E}[Y] + \alpha\sqrt{\lambda\mathbb{E}[Y^2]})$ is linear in b and (4.2) is fulfilled. Reinsurance is chosen if α is not too large.

Exponential principle

Here $c(b) = c - \alpha^{-1}\lambda(M_Y(\alpha(1-b)) - 1)$, where $M_Y(r) = \mathbb{E}[e^{rY}]$ denotes the moment-generating function of the claim sizes. This is concave and (4.2) is fulfilled. We get $c'(1) = \lambda M_Y'(0) = \lambda\mathbb{E}[Y]$. Thus, $b^* \in (0,1)$, i.e., the cedent will always choose reinsurance.

Calculation of the maximal value is numerically not easy. The function $R(b)$ is only determined implicitly through (4.1). That means that at each point where $R(b)$ is needed, Equation (4.1) has to be solved. This is quite time-consuming, unless $R(b)$ is known explicitly.

In some situations $R(b)$ and the maximiser b^* can be calculated in a simple way. Consider the case of proportional reinsurance. Suppose that the reinsurance premium is calculated by the expected value principle, $c(b) = c - (1-b)(1+\theta)\lambda\mu$ with $\mu = \mathbb{E}[Y]$. In order for the problem not to be trivial, we need $c(0) < 0$. Let $\underline{b}$ be the argument for which $c(\underline{b}) = 0$. Then we only need to consider $b \in (\underline{b}, \infty)$. Here we allow $b > 1$ for simplicity of the considerations. If $b^* > 1$, the optimal retention level will just be 1.

The equation to solve is

$$\lambda(M_Y(bR(b)) - 1) - (c - (1-b)(1+\theta)\lambda\mu)R(b) = 0\;.$$

Let $r(b) = bR(b)$. Note that $r(b)$ is continuous. Then

$$\lambda(M_Y(r(b)) - 1) - (1+\theta)\lambda\mu r(b) - (c - (1+\theta)\lambda\mu)\frac{r(b)}{b} = 0\;.$$

Because $R(b)$ is unimodal, the function $r(b)$ is strictly increasing on some interval $(\underline{b}, b^* + \varepsilon)$ for some $\varepsilon > 0$. Thus, $r(b)$ is invertible, and

$$b(r) = \frac{((1+\theta)\lambda\mu - c)r}{(1+\theta)\lambda\mu r - \lambda(M_Y(r) - 1)}\;.$$

Therefore, as a function of r,

$$R(b(r)) = \frac{r}{b(r)} = \frac{(1+\theta)\lambda\mu r - \lambda(M_Y(r) - 1)}{(1+\theta)\lambda\mu - c}\;.$$

This function is strictly concave. Thus, there is a unique maximum. Because there is a (local) maximum at $r = b^* R(b^*)$, the equation to solve is

$$M_Y'(r) = (1+\theta)\mu\;. \tag{4.3}$$

If a solution exists, we denote it by r^*. If (4.3) does not have a solution, then we let $r^* = \sup\{r : M_Y(r) < \infty\}$. Note that only the claim size distribution and the safety loading of the reinsurer are needed to calculate r^*.

The value b^* then becomes

$$b^* = b(r^*) = \frac{((1+\theta)\lambda\mu - c)r^*}{(1+\theta)\lambda\mu r^* - \lambda(M_Y(r^*) - 1)}\;.$$

The maximal adjustment coefficient then is $R(b^*) = r^*/b^*$.

Example 4.3. Suppose that the claim sizes are *gamma-distributed* with parameters $\gamma, \beta > 0$, i.e.,

$$\mathbb{P}[Y \le y] = \int_0^y \frac{\beta^\gamma}{\Gamma(\gamma)} z^{\gamma-1} \mathrm{e}^{-\beta z}\,\mathrm{d}z\;.$$

The mean value is γ/β and the moment-generating function is

$$M_Y(r) = \Bigl(\frac{\beta}{\beta - r}\Bigr)^\gamma\;.$$

Equation (4.3) has the solution $r^* = \beta(1 - (1+\theta)^{-1/(\gamma+1)})$. We obtain

$$b^* = \frac{\gamma(\theta - \eta)(1 - (1+\theta)^{-1/(\gamma+1)})}{\gamma\theta - (\gamma+1)((1+\theta)^{\gamma/(\gamma+1)} - 1)}\;,$$

where we write $c = (1+\eta)\lambda\mu = (1+\eta)\lambda\gamma/\beta$. If $b^* \le 1$, the adjustment coefficient becomes

$$R(b^*) = \frac{r^*}{b^*} = \beta\frac{\gamma\theta - (\gamma+1)((1+\theta)^{\gamma/(\gamma+1)} - 1)}{\gamma(\theta - \eta)}\;.$$

Note that we obtain a closed expression for R^*, even though in general $R(b)$ cannot be obtained in closed form. ■

4.1.2 Optimal Investment

Suppose that it is possible to invest the amount A of the portfolio into a risk asset modelled as (E.1). Investment is even possible if $A > X_t$, i.e., the insurer takes a short position. The risk process then becomes

$$X_t^A = x + (c + Am)t - \sum_{i=1}^{N_t} Y_i + A\sigma W_t\,.$$

Note that no net profit condition is needed, because A can be chosen such that $c + Am > \lambda\mu$.

By (D.9) the adjustment coefficient $R(A)$ is defined through

$$\lambda(M_Y(r) - 1) - (c + mA)r + \frac{A^2 r^2 \sigma^2}{2} = 0\,. \tag{4.4}$$

Lemma 4.4. *The function $R(A)$ is infinitely often differentiable and unimodal.*

Proof. Formally taking the derivative with respect to A gives

$$(\lambda M_Y'(R(A)) - (c + mA) + A^2\sigma^2 R(A))R'(A) - mR(A) + A\sigma^2 R(A)^2 = 0\,.$$

Because $\lambda M_Y'(R(A)) - (c + mA) + A^2\sigma^2 R(A) > 0$ by the convexity of (4.4), the implicit function theorem yields

$$R'(A) = \frac{mR(A) - A\sigma^2 R(A)^2}{\lambda M_Y'(R(A)) - (c + mA) + \sigma^2 A^2 R(A)}\,.$$

In particular, the right-hand side is differentiable and therefore $R(A)$ is twice differentiable. By induction, $R(A)$ is infinitely often differentiable. Moreover, at any A^* at which $R'(A^*) = 0$, we have $\sigma^2 A^* R(A^*) = m$, and the second derivative becomes

$$R''(A^*) = -\frac{\sigma^2 R(A^*)^2}{\lambda M_Y'(R(A^*)) - (c + mA^*) + {A^*}^2\sigma^2 R(A^*)} < 0\,.$$

Thus, any point with $R'(A) = 0$ is a local maximum. That is, $R(A)$ is a unimodal function. □

Let $R^* = R(A^*)$. Then R^* fulfils

$$\lambda(M_Y(R^*) - 1) - cR^* - \frac{m^2}{2\sigma^2} = 0\,. \tag{4.5}$$

Because the left-hand side is a convex function in R^*, there is at most one solution $R^* > 0$. If no such solution exists, we have $R^* = r_\infty$. The optimal A^* then becomes $A^* = m/(R^*\sigma^2)$.

4.1.3 Optimal Reinsurance and Investment

Suppose now that both investment and reinsurance are allowed. The adjustment coefficient $R(A,b)$ is then determined through

$$\lambda(M(r;b)-1)-(c(b)+mA)r+\frac{A^2r^2\sigma^2}{2}=0\;.$$

The maximisation of the adjustment coefficient turns out not to be a harder problem than in the previous sections, because A and b turn up in different terms.

Lemma 4.5. *Suppose that both $M(r;b)$ and $c(b)$ are twice differentiable (with respect to r and b). Suppose, moreover, that (4.2) holds. Then there is a unique local maximum of $R(A,b)$.*

Proof. As for the previous cases, we obtain the partial derivatives

$$R_A(A,b)=\frac{mR(A,b)-AR(A,b)^2\sigma^2}{\lambda M_Y'(R(A,b))-(c(b)+mA)+A^2R(A,b)\sigma^2}$$

and

$$R_b(A,b)=-\frac{\lambda M_b(R(A,b);b)-c'(b)R(A,b)}{\lambda M_r(R(A,b);b)-(c(b)+mA)+A^2R(A,b)\sigma^2}\;.$$

Let (A^*,b^*) be a point where $R_A(A^*,b^*)=R_b(A^*,b^*)=0$. The second-order partial derivatives at this point become $R_{Ab}(A^*,b^*)=0$,

$$R_{AA}(A^*,b^*)=-\frac{R(A^*,b^*)^2\sigma^2}{\lambda M_Y'(R(A^*,b^*))-(c(b^*)+mA^*)+{A^*}^2R(A^*,b^*)\sigma^2}<0\;,$$

$$R_{bb}(A^*,b^*)=-\frac{\lambda M_{bb}(R(A^*,b^*);b^*)-c''(b^*)R(A^*,b^*)}{\lambda M_Y'(R(A^*,b^*))-(c(b^*)+mA^*)+{A^*}^2R(A^*,b^*)\sigma^2}<0\;,$$

where $R_{bb}(A^*,b^*)<0$ follows by the same argument as in the proof of Lemma 4.1. Thus, the second derivative of $R(a,B)$ is negative definite at (A^*,b^*). Therefore, it is a local maximum. By the mountain pass lemma, there can be at most one local maximum. Consider now a local maximum at (A^*,b^*) on the boundary. From $R_A(A^*,b^*)$ we conclude that $A^*\in\mathbb{R}$ and that $R_A(A^*,b^*)=0$. In particular, $\sigma^2A^*R(A^*,b^*)=m$ and b^* must be on the boundary. Thus, $R(A^*,b^*)$ fulfils

$$\lambda(M(r;b)-1)-c(b)r-\frac{m^2}{2\sigma^2}=0\;. \tag{4.6}$$

Hence, following the arguments used in the proof of Lemma 4.1, $R(b)$ is unimodal, i.e., increasing in b on $\mathcal{U}$. □

Now let us again consider proportional reinsurance with $c(b) = c-(1-b)(1+\theta)\lambda\mu$. Then the optimal strategy can be obtained in the following way. Let r^* denote the solution to (4.3), if such a solution exists, and $r^* = r_\infty$ otherwise. Then

$$b^* = \min\Bigl\{\frac{((1+\theta)\lambda\mu - c)r^*}{(1+\theta)\lambda\mu r^* + m^2/(2\sigma^2) - \lambda(M_Y(r^*)-1)}, 1\Bigr\}\;.$$

If $b^* < 1$, then $R(A^*, b^*) = r^*/b^*$; otherwise, $R(A^*, 1)$ has to be calculated from $\lambda(M_Y(r)-1)-cr-m^2/(2\sigma^2) = 0$. Finally, $A^* = m/(\sigma^2 R(A^*, b^*))$. Note that one chooses more reinsurance than in the case without investment. That means it is optimal for the insurer to choose a higher risk on the financial market in order to create more income and a smaller risk in the insurance business.

Bibliographical Remarks

The idea of maximising the adjustment coefficient goes back to Waters [185]. The method using (4.3) in the case of proportional reinsurance was introduced by Hald and Schmidli [90].

4.2 Cramér–Lundberg Approximations for Controlled Classical Risk Models

Consider the controlled risk models of Section 2.3. From far away the processes look like diffusion processes if one scales the axes appropriately. Concerning the ruin probabilities for a diffusion approximation, we found in Section 2.2 that $\psi^*(x) = \mathrm{e}^{-Rx}$ for some $R > 0$. Moreover, the optimal strategies are constant. One would therefore expect that the Cramér–Lundberg approximation $\psi^*(x) \sim C\mathrm{e}^{-Rx}$ also holds for the optimally controlled classical risk model and that the optimal strategies converge as $x \to \infty$. We will show this under a light-tail condition for the claim size distributions. In addition, we will also discuss the large claim case in Section 4.3.

The method we use will be the same in Sections 4.2.1 to 4.2.4. Because the technical problems to circumvent are different, we will consider all three cases separately. However, where the proofs without additional technical problems can be done in the same way, we will not repeat it.

4.2.1 Optimal Proportional Reinsurance

We start by defining the adjustment coefficient. In the setup of Section 4.1, let $R(b)$ be the adjustment coefficient if a fixed retention level b is used, i.e., $R(b)$ is the maximal solution to

$$\theta(r; b) := \lambda(M_Y(br) - 1) - c(b)r = 0 .$$

The adjustment coefficient for our problems is then $R = \sup_{b\in[0,1]} R(b)$. By the assumptions made in Section 2.3.1, the function $\theta(r; b)$ is continuous in b. Moreover, the function $\theta(r; b)$ is strictly convex in r; see also Section D.1.2. Because $\theta(0; b) = 0$ and by the definition of $R(b)$ also $\theta(R(b); b) = 0$, we get that $\theta(R; b) \ge 0$. If $r < R$, then there is a b such that $R(b) > r$. By concavity, we have $\theta(r; b) < 0$. This shows that we could define R as the strictly positive solution to

$$\inf_{b\in[0,1]} \theta(r; b) = \inf_{b\in[0,1]} \{\lambda(M_Y(br) - 1) - c(b)r\} = 0 . \tag{4.7}$$

Because $\theta(R; b)$ is continuous in b and $[0, 1]$ is compact, there exists a b^* for which $\theta(R; b^*) = 0$, i.e., for which the infimum is taken. In particular, if the conditions of Lemma 4.1 are fulfilled, then b^* is unique.

In Section 2.3.1 we found the value function $\delta(x)$ and an optimal strategy $b(X_t)$ of feedback form for which the minimal ruin probability is attained. We now denote by $\psi(x)$ the ruin probability under the optimal strategy. We start by showing an upper bound.

Proposition 4.6. *The minimal ruin probability is bounded by $\psi(x) < \mathrm{e}^{-Rx}$.*

Proof. Suppose that we follow the constant strategy $b_t = b^*$. Then by Lundberg's inequality, $\psi(x) \le \psi^{b^*}(x) < \mathrm{e}^{-Rx}$; see Section D.1.2. □

We now assume that $\mathbb{E}[\mathrm{e}^{RY_1}] < \infty$.

For the considerations below we do not stop the process at ruin. We therefore have to define the strategy when $X_t < 0$. We let $b(x) = 1$ if $x < 0$. Note that $c(1) = c$.

Consider the process M defined as

$$M_t = \exp\Bigl\{-R(X_t - x) - \int_0^t \theta(R; b(X_s))\,\mathrm{d}s\Bigr\} .$$

Lemma 4.7. *The process M is a strictly positive martingale with mean value 1.*

Proof. The proof is quite technical. We therefore recommended skipping the proof upon first reading. That M is strictly positive is evident. With

$$X_t = x + \int_0^t c(b(X_s))\,\mathrm{d}s - \sum_{i=1}^{N_t} b(X_{T_i-})Y_i ,$$

the martingale reads

$$M_t = \exp\Bigl\{\sum_{i=1}^{N_t} b(X_{T_i-})RY_i - \lambda \int_0^t (M_Y(b(X_s)R) - 1)\,\mathrm{d}s\Bigr\}\,.$$

We next show that $\{M_{T_n \wedge t}\}$ is a martingale. By the Markov property of M, we only have to show that $\mathbb{E}[M_{T_n \wedge t}] = 1$. Clearly, $\mathbb{E}[M_{T_0 \wedge t}] = \mathbb{E}[M_0] = 1$. Suppose that $\mathbb{E}[M_{T_{n-1} \wedge t}] = 1$. Then we condition on $\mathcal{F}_{T_{n-1}}$ and show that $\mathbb{E}[M_{T_n \wedge t} \mid \mathcal{F}_{T_{n-1}}] = M_{T_{n-1} \wedge t}$. The assertion is trivial if $T_{n-1} > t$. Suppose therefore that $T_{n-1} \le t$. Given $\mathcal{F}_{T_{n-1}}$, the process $\{X_t\}$ is deterministic on $[T_{n-1}, T_n)$. We split into the event $\{T_n > t\}$ and $\{T_n \le t\}$. The expression to calculate becomes

$$\begin{aligned}
&\mathrm{e}^{-\lambda(t-T_{i-1})} \exp\Bigl\{-\lambda \int_{T_{i-1}}^t (M_Y(b(X_s)R) - 1)\,\mathrm{d}s\Bigr\} \\
&\qquad + \int_{T_{i-1}}^t M_Y(b(X_{s-})R) \exp\Bigl\{-\lambda \int_{T_{i-1}}^s (M_Y(b(X_v)R) - 1)\,\mathrm{d}v\Bigr\} \\
&\qquad \times \lambda \mathrm{e}^{-\lambda(s-T_{i-1})}\,\mathrm{d}s \\
&= \exp\Bigl\{-\lambda \int_{T_{i-1}}^t M_Y(b(X_s)R)\,\mathrm{d}s\Bigr\} \\
&\qquad + \int_{T_{i-1}}^t \lambda M_Y(b(X_{s-})R) \exp\Bigl\{-\lambda \int_{T_{i-1}}^s M_Y(b(X_v)R)\,\mathrm{d}v\Bigr\}\,\mathrm{d}s \\
&= 1\,,
\end{aligned}$$

where we used integration by parts. Thus, $\mathbb{E}[M_{T_n \wedge t} \mid \mathcal{F}_{T_{n-1}}] = M_{T_{n-1} \wedge t}$ and by induction $\mathbb{E}[M_{T_n \wedge t}] = 1$. The assertion is proved if we can show that $\{M_{T_n \wedge t}\}$ is uniformly integrable for each t. But

$$\exp\Bigl\{\sum_{i=1}^{N_t} Rb(X_{T_i-})Y_i\Bigr\} \le \exp\Bigl\{\sum_{i=1}^{N_t} RY_i\Bigr\}\,,$$

which is integrable by our assumption that $M_Y(R) < \infty$. □

We can now use the martingale to change the measure. Let $\mathbb{P}^*[A] = \mathbb{E}[M_t; A]$ for $A \in \mathcal{F}_t$. The measure is independent of t. Let us start by finding the law of X under $\mathbb{P}^*$.

Lemma 4.8. *Under the measure $\mathbb{P}^*$, the process X is a piecewise deterministic Markov process with jump intensity $\lambda^*(x) = \lambda M_Y(b(x)R)$ and claim size distribution $G^*_x(y) = M_Y(b(x)R)^{-1} \int_0^y \mathrm{e}^{Rb(x)y}\,\mathrm{d}G(y)$. The premium rate is $c(b(x))$.*

Proof. Also, this proof is technical; hence, we recommend skipping it upon first reading. Let B be a Borel set. By Lemma C.1, we have

$$\begin{aligned}
\mathbb{P}^*[X_{t+s} \in B \mid \mathcal{F}_t] &= \mathbb{E}[M_t^{-1} M_{t+s}; X_{t+s} \in B \mid \mathcal{F}_t] \\
&= \mathbb{E}[M_t^{-1} M_{t+s}; X_{t+s} \in B \mid X_t]\,.
\end{aligned}$$

Thus, X remains a Markov process under $\mathbb{P}^*$. Because its paths are deterministic between jumps, it is a piecewise deterministic Markov process. We therefore need to calculate the distribution of T_1 and of Y_1. For this purpose, denote by x_s the deterministic path on $[0, T_1)$. For the distribution of T_1, we obtain

$$\begin{aligned}\mathbb{P}^*[T_1 > t] = \mathbb{E}[M_t; T_1 > t] &= \mathrm{e}^{-\lambda t} \exp\Bigl\{-\int_0^t \lambda(M_Y(b(x_s)R) - 1)\,\mathrm{d}s\Bigr\} \\ &= \exp\Bigl\{-\int_0^t \lambda M_Y(b(x_s)R)\,\mathrm{d}s\Bigr\}\;.\end{aligned}$$

This yields the intensity. Let us now consider the first claim size

$$\begin{aligned}\mathbb{P}^*[T_1 \le t, Y_1 \in B] &= \mathbb{E}[M_t; T_1 \le t, Y_1 \in B] = \mathbb{E}[\mathbb{E}[M_t; T_1 \le t, Y_1 \in B \mid \mathcal{F}_{T_1}]] \\ &= \mathbb{E}[M_{T_1}; T_1 \le t, Y_1 \in B] \\ &= \int_0^t \int_0^\infty \exp\Bigl\{Rb(x_s)y - \lambda \int_0^s (M_Y(b(x_v)R) - 1)\,\mathrm{d}v\Bigr\} \\ &\qquad \times \mathbb{1}_B(y)\,\mathrm{d}G(y)\lambda \mathrm{e}^{-\lambda s}\,\mathrm{d}s \\ &= \int_0^t \int_0^\infty \mathbb{1}_B(y)\,\mathrm{d}G^*(y)\lambda^*(x_s) \exp\Bigl\{-\int_0^s \lambda^*(x_v)\,\mathrm{d}v\Bigr\}\,\mathrm{d}s\;.\end{aligned}$$

This gives the claim size distribution. □

Let us now consider the drift of the process under $\mathbb{P}^*$. The infinitesimal drift becomes

$$c(b(x)) - \lambda^*(x) \int_0^\infty y\,\mathrm{d}G_x^*(y) = -\theta'(R; b(x))\;,$$

where $\theta'(r; b)$ denotes the derivative with respect to the first variable. Because of the convexity of the function $\theta(r; b)$, we get $\theta'(R; b(x)) > 0$. That is, the drift is strictly negative. We therefore find that $\tau < \infty$ under $\mathbb{P}^*$.

The ruin probability can now be expressed using the law $\mathbb{P}^*$,

$$\psi(x) = \mathbb{E}^*\Bigl[\exp\Bigl\{RX_\tau + \int_0^\tau \theta(R; b(X_s))\,\mathrm{d}s\Bigr\}\Bigr]\mathrm{e}^{-Rx}\;.$$

We next prove a lower bound for $\psi(x)$.

Proposition 4.9. *Let*

$$C_- = \inf_z \frac{1}{\mathbb{E}[\mathrm{e}^{R(Y-z)} \mid Y > z]}\;, \tag{4.8}$$

where z is taken over the set where $\mathbb{P}[Y > z] > 0$. Then $\psi(x) \ge C_- \mathrm{e}^{-Rx}$.

Proof. Clearly, $\psi(x) \ge \mathbb{E}^*[\mathrm{e}^{RX_\tau}]\mathrm{e}^{-Rx}$. Suppose that we know $X_{\tau-} = z$. Then

$$\begin{aligned}
\mathbb{E}^*[\mathrm{e}^{RX_\tau} \mid X_{\tau-} = z] &= \mathbb{E}^*[\mathrm{e}^{R(z-b(z)Y)} \mid b(z)Y > z] \\
&= \mathbb{E}^*[\mathrm{e}^{b(z)R(z/b(z)-Y)} \mid Y > z/b(z)] \\
&= \frac{1}{\mathbb{E}[\mathrm{e}^{b(z)R(Y-z/b(z))} \mid Y > z/b(z)]} \\
&\ge \inf_b \frac{1}{\mathbb{E}[\mathrm{e}^{bR(Y-z/b(z))} \mid Y > z/b(z)]}\;.
\end{aligned}$$

Thus,

$$\begin{aligned}
\mathbb{E}^*[\mathrm{e}^{RX_\tau}] &\ge \inf_{b,z} \frac{1}{\mathbb{E}[\mathrm{e}^{bR(Y-z/b(z))} \mid Y > z/b(z)]} \ge \inf_{b,z} \frac{1}{\mathbb{E}[\mathrm{e}^{bR(Y-z)} \mid Y > z]} \\
&\ge \inf_z \frac{1}{\mathbb{E}[\mathrm{e}^{R(Y-z)} \mid Y > z]}\;.
\end{aligned}$$

This is the expression claimed. □

Under weak conditions we obtain that $C_- > 0$; see Lemma D.1. Note that C_- is the constant also obtained in the classical model; see (D.4).

We have that $\psi(x)$ also solves (2.14). We first rewrite the equation. Fubini's theorem yields

$$\begin{aligned}
\int_0^{x/b(x)} &\psi(x - b(x)y)\,\mathrm{d}G(y) \\
&= \psi(0)G(x/b(x)) + \int_0^{x/b(x)} \int_0^{x-b(x)y} \psi'(z)\,\mathrm{d}z\,\mathrm{d}G(y) \\
&= \psi(0)G(x/b(x)) + \int_0^x \int_0^{(x-z)/b(x)} \mathrm{d}G(y)\psi'(z)\,\mathrm{d}z \\
&= \psi(0)G(x/b(x)) + \int_0^x \psi'(z)G((x-z)/b(x))\,\mathrm{d}z\;.
\end{aligned}$$

Hence, the Hamilton–Jacobi–Bellman equation can be written as

$$\begin{aligned}
c(b(x))\psi'(x) + \lambda\Bigl[&(1 - G(x/b(x))) - \psi(0)(1 - G(x/b(x))) \\
&- \int_0^x \psi'(z)(1 - G((x-z)/b(x)))\,\mathrm{d}z\Bigr] = 0\;.
\end{aligned}$$

Because we are interested in a Cramér–Lundberg approximation, we multiply $\psi(x)$ by e^{Rx} and consider $f(x) = \psi(x)\mathrm{e}^{Rx}$. We already proved that $f(x) < 1$. Then $f(x)$ fulfils

$$\begin{aligned}
c(b(x))(f'(x) - Rf(x)) + \lambda\Bigl[&\delta(0)(1 - G(x/b(x)))\mathrm{e}^{Rx} \\
&+ \int_0^x (Rf(z) - f'(z))(1 - G((x-z)/b(x)))\mathrm{e}^{R(x-z)}\,\mathrm{d}z\Bigr] = 0\;.
\end{aligned}$$

Because $\psi(x)$ is strictly decreasing, we have that $f'(x) < Rf(x)$, giving that $f'(x)$ is bounded from above. Let $g(x) = Rf(x) - f'(x) = -\psi'(x)\mathrm{e}^{Rx}$. Then the equation reads

$$\lambda\Big[\delta(0)(1 - G(x/b(x)))\mathrm{e}^{Rx} + \int_0^x g(x-z)(1 - G(z/b(x)))\mathrm{e}^{Rz}\,\mathrm{d}z\Big] - c(b(x))g(x) = 0\,.$$

$b(x)$ is the argument at which the left-hand side becomes minimal. If we therefore replace $b(x)$ by b^*, we obtain the inequality

$$\lambda\Big[\delta(0)(1 - G(x/b^*))\mathrm{e}^{Rx} + \int_0^x g(x-z)(1 - G(z/b^*))\mathrm{e}^{Rz}\,\mathrm{d}z\Big] - c(b^*)g(x) \ge 0\,.$$

Note that

$$\begin{aligned} M_Y(b^*R) &= 1 + \int_0^\infty \int_0^{b^*y} R\mathrm{e}^{Rz}\,\mathrm{d}z\,\mathrm{d}G(y) \\ &= 1 + R\int_0^\infty \int_{z/b^*}^\infty \mathrm{d}G(y)\mathrm{e}^{Rz}\,\mathrm{d}z \\ &= 1 + R\int_0^\infty (1 - G(z/b^*))\mathrm{e}^{Rz}\,\mathrm{d}z\,. \end{aligned}$$

Thus, by the definition of b^*,

$$\lambda\int_0^\infty (1 - G(z/b^*))\mathrm{e}^{Rz}\,\mathrm{d}z - c(b^*) = 0\,.$$

Using this we can express $c(b^*)$ in the above inequality and obtain

$$\begin{aligned} &\int_0^x (g(x-z) - g(x))(1 - G(z/b^*))\mathrm{e}^{Rz}\,\mathrm{d}z \\ &\qquad \ge \int_x^\infty (1 - G(z/b^*))\mathrm{e}^{Rz}\,\mathrm{d}z g(x) - \delta(0)(1 - G(x/b^*))\mathrm{e}^{Rx}\,. \end{aligned} \tag{4.9}$$

We observe that the second term on the right-hand side converges to zero as $x \to \infty$; see Lemma A.12.

The functions $f(x)$ and $g(x)$ have the following properties.

Lemma 4.10. i) *$g(x)$ is bounded. In particular, $f'(x)$ is bounded.*

ii) *Let $\zeta = \limsup_{x\to\infty} g(x)/R$. Then $\limsup_{x\to\infty} f(x) = \zeta$. In particular, $\zeta > 0$ if $C_- > 0$.*

iii) *For any $\beta > 0$, $x_0 > 0$, and $\varepsilon > 0$ there is an $x \ge x_0$ such that $f(y) > \zeta - \varepsilon$ for $y \in [x - \beta, x]$.*

Proof. i) Suppose that $g(x)$ is unbounded. Let $x_n = \inf\{x : g(x) = 2nR\}$. Choose $0 < \varepsilon < 1$. If n is large enough, then the right-hand side of (4.9) becomes larger than $-\varepsilon$ for $x = x_n$. Let $x' = \inf\{x : (1 - G(x/b^*))e^{Rx} \le \varepsilon\}$. We can assume that $x' \le x_n$. Then

$$-nR\varepsilon \int_0^{x'} \mathbb{I}_{g(x_n-z)\le Rn}\,dz \ge \int_0^{x'} (g(x_n-z)-g(x_n))(1-G(z/b^*))e^{Rz}\,dz \ge -\varepsilon\,.$$

Thus, $\int_0^{x'} \mathbb{I}_{g(x_n-z)\le Rn}\,dz \le 1/(nR)$. If $g(x) > nR$, then $f'(x) = Rf(x) - g(x) < -(n-1)R$ by Proposition 4.6. This yields

$$\begin{aligned}
f(x_n) &= f(x_n - x') + \int_{x_n-x'}^{x_n} f'(x)\,dx \\
&< 1 + R\int_0^{x'} \mathbb{I}_{g(x-z)\le Rn}\,dz - (n-1)R\int_0^{x'} \mathbb{I}_{g(x-z)>Rn}\,dz \\
&\le 1 + 1/n - (n-1)R(x' - 1/(nR)) = 2 - (n-1)Rx'\,.
\end{aligned}$$

This becomes negative for n large enough, which is a contradiction. That $f'(x)$ is bounded follows from $f'(x) = Rf(x) - g(x)$. In particular, we have shown that the right-hand side of (4.9) converges to zero.

ii) Let $\eta = \limsup_{x\to\infty} f(x)$. Choose $\varepsilon > 0$ and x_0 such that $g(x) < R\zeta + \varepsilon$ for $x \ge x_0$. There exists $x' \ge x_0$ such that $f(x') > \eta - \varepsilon/R$ and $f'(x') < \varepsilon$. Then

$$R\eta - \varepsilon < Rf(x') = g(x') + f'(x') < R\zeta + 2\varepsilon\,.$$

Because ε is arbitrary, we get $\eta \le \zeta$.

Choose $\beta > 0$ (large) and $\varepsilon, \delta > 0$ (small). We want to show that there is an x_0 such that $\int_{x_0-\beta}^{x_0} \mathbb{I}_{g(z)<R\zeta-\varepsilon}\,dz < \delta$. Suppose first that $G(\beta) < 1$. Let $\xi = \delta\varepsilon/4$ and $\xi' = (2\int_0^\infty (1 - G(z/b^*))e^{Rz}\,dz)^{-1}\xi^2$. We can assume that $\xi' < \varepsilon/2$ and $(1 - G(z/b^*))e^{Rz} > \xi$ for $z \le \beta$. Choose x_1 such that

$$\int_0^\beta (g(x_1 - z) - g(x_1))(1 - G(z/b^*))e^{Rz}\,dz \ge -\xi^2$$

and $g(x) < R\zeta + \xi'$ for all $x \ge x_1 - \beta$ and $g(x_1) > R\zeta - \xi'$. From (4.9) we conclude that

$$\begin{aligned}
&-\tfrac{1}{2}\varepsilon\xi \int_0^\beta \mathbb{I}_{g(x_1-z)<R\zeta-\varepsilon}\,dz + \xi^2 \\
&\quad\ge -\tfrac{1}{2}\varepsilon \int_0^\beta \mathbb{I}_{g(x_1-z)<R\zeta-\varepsilon}(1 - G(z/b^*))e^{Rz}\,dz \\
&\qquad + 2\xi' \int_0^\beta (1 - G(z/b^*))e^{Rz}\,dz \\
&\quad> \int_0^\beta (g(x_1 - z) - g(x_1))(1 - G(z/b^*))e^{Rz}\,dz \\
&\quad\ge -\xi^2\,.
\end{aligned}$$

Thus,

$$\int_0^\beta \mathbb{I}_{g(x_1-z)<R\zeta-\varepsilon}\,\mathrm{d}z < 4\xi/\varepsilon = \delta\,.$$

This proves the claim if $G(\beta) < 1$. Suppose that we have proved in the case $G((2/3)^n\beta) < 1$ that for any $\varepsilon, \delta > 0$ there are ξ' and x_1 such that whenever $g(x) > R\zeta - \xi'$ for some $x > x_1 - \beta$, then $\int_0^\beta \mathbb{I}_{g(x-z)<R\zeta-\varepsilon}\,\mathrm{d}z < \delta$. Suppose now that $G((2/3)^{n+1}\beta) < 1$. Let $\tilde\beta = 2/3\beta$, $\tilde\delta = \delta/2\mathrm{m}$ and $\tilde\varepsilon = \xi'$. It is no loss of generality to assume that $\tilde\delta < \beta/3$. Then there are $\tilde x_1$ and $\tilde\xi'$ such that the induction assumption holds for the tilde quantities. Let $x_1 = 3\tilde x_1/2$. Choose $x > x_1$ such that $g(x) > R\zeta - \tilde\xi'$. There must be a $y \in [x - \tilde\beta, x - \tilde\beta/2]$ such that $g(y) > R\zeta - \tilde\varepsilon$. Thus, $\int_0^{2\beta/3} \mathbb{I}_{g(y-z)<R\zeta-\varepsilon}\,\mathrm{d}z < \tilde\delta = \delta/2$. This proves the claim by induction.

Suppose now that $\eta < \zeta$. Let $\varepsilon = R(\zeta - \eta)/3$. For x large enough, $f(x) < \eta + \varepsilon/R$. If $g(x) > R\zeta - \varepsilon$, then $f'(x) = Rf(x) - g(x) < -\varepsilon$. For any β we can find an x such that $\int_0^\beta \mathbb{I}_{g(x-z)<R\zeta-\varepsilon}\,\mathrm{d}z < \delta$. Thus,

$$f(x) = f(x-\beta) + \int_{x-\beta}^x f'(z)\,\mathrm{d}z < \eta + \varepsilon/R + \delta R - (\beta - \delta)\varepsilon\,.$$

For β large enough, we obtain $f(x) < 0$. This is a contradiction.

iii) Fix δ to be determined later. By the proof of part ii) there is an x large enough such that $\int_{x-(n+1)\beta}^x \mathbb{I}_{g(z)\le R(\zeta-\varepsilon/3)}\,\mathrm{d}z < \delta$. Suppose that $g(z) > R(\zeta - \varepsilon/3)$ and $f(z) < \zeta - 2\varepsilon/3$. Then $f'(z) = Rf(z) - g(z) < -\varepsilon R/3$. Suppose that $f(z) < \zeta - 2\varepsilon/3$ for all $z \in [y, x]$, then

$$f(y) = f(x) - \int_y^x f'(z)\,\mathrm{d}z > -R\delta + (x - y - \delta)R\varepsilon/3\,.$$

This proves that $f(y) \ge \zeta - 2\varepsilon/3$ for some $y > x - \delta - 3(\zeta - 2\varepsilon/3 + \delta R)/(\varepsilon R)$. Without loss of generality, we can assume that $x - y < n\beta$. If $f(z) < \zeta - \varepsilon/3$ and $g(z) \ge R(\zeta - \varepsilon/3)$, then $f'(z) < 0$. This shows that

$$f(v) = f(y) - \int_v^y f'(z)\,\mathrm{d}z \ge \zeta - 2\varepsilon/3 - R\delta\,.$$

If, therefore, $\delta < \varepsilon/(3R)$, then $f(v) > \zeta - \varepsilon$ for all $v \in [x - (n+1)\beta, y]$, which is an interval of at least length β. □

We can now prove the main result of this section.

Theorem 4.11. *Suppose that $C_- > 0$. Then $\lim_{x\to\infty} \psi(x)\mathrm{e}^{Rx} = \zeta > 0$, where ζ is defined above.*

Proof. Fix $\beta > 0$ and $\varepsilon > 0$. Then there exists x_0 such that $f(y) > \zeta - \varepsilon$ for all $y \in [x_0 - \beta, x_0]$. Suppose that the initial capital x is larger than $2x_0$. We let $T = \inf\{t : X_t < x_0\}$. By the change of measure formula,

$$\begin{aligned}
f(x) &= \mathbb{E}^*\Bigl[\exp\Bigl\{RX_\tau + \int_0^\tau \theta(R;b(X_s))\,\mathrm{d}s\Bigr\}\Bigr] \\
&= \mathbb{E}^*\Bigl[\mathbb{E}^*\Bigl[\exp\Bigl\{RX_\tau + \int_0^\tau \theta(R;b(X_s))\,\mathrm{d}s\Bigr\}\Bigm| \mathcal{F}_T\Bigr]\Bigr] \\
&= \mathbb{E}^*\Bigl[\mathbb{E}^*\Bigl[\exp\Bigl\{RX_\tau + \int_T^\tau \theta(R;b(X_s))\,\mathrm{d}s\Bigr\}\Bigm| X_T\Bigr] \\
&\qquad \times \exp\Bigl\{\int_0^T \theta(R;b(X_s))\,\mathrm{d}s\Bigr\}\Bigr] \\
&\ge \mathbb{E}^*[f(X_T)] \ge (\zeta-\varepsilon)\mathbb{P}^*[x_0 - X_T \le \beta]\,.
\end{aligned}$$

By a similar argument as the one used in the proof of Proposition 4.9, the expression $\mathbb{P}^*[x_0 - X_T \le \beta]$ can be made uniformly close to 1. Thus, $\liminf_{x\to\infty} f(x) \ge \zeta$. By Lemma 4.10, the result is proved. □

Remark 4.12. The condition $C_- > 0$ is only needed for technical reasons. The result could also be proved if one could establish that under $\mathbb{P}^*$ the only way the process X can reach the ruin state is by going with a probability arbitrarily close to 1 via a finite set $[0, A]$, say. This looks very much like results from large deviation theory. However, a typical large deviation result would establish that X goes to ruin via a set of the form $[0, Ax]$. Thus, it still would be possible that ruin is reached by a very large jump, avoiding a finite set with a probability bounded away from zero. ■

After proving the Cramér–Lundberg approximation, it is possible to determine the strategy for large capital x.

Corollary 4.13. *Suppose that* $C_- > 0$. *Then* $\lim_{x\to\infty} f'(x) = 0$. *If, moreover,* b^* *is unique, then* $\lim_{x\to\infty} b(x) = b^*$.

Proof. Clearly, we have $\liminf_{x\to\infty} f'(x) = R\zeta - \limsup_{x\to\infty} g(x) = 0$. Now consider the Hamilton–Jacobi–Bellman equation (2.14), where $\psi(x)$ is replaced by $f(x)\mathrm{e}^{-Rx}$:

$$\begin{aligned}
\lambda\Bigl[\int_0^{x/b(x)} f(x - b(x)y)\mathrm{e}^{Rb(x)y}\,\mathrm{d}G(y) - f(x)\Bigr] - c(b(x))Rf(x) \\
+ c(b(x))f'(x) + \lambda\mathrm{e}^{Rx}(1 - G(x/b(x))) = 0\,.
\end{aligned}$$

Let $\varepsilon > 0$. Because $\int_x^\infty \mathrm{e}^{Ry}\,\mathrm{d}G(y)$ converges to zero as $x \to \infty$, we obtain that

$$\lambda\Bigl|\int_0^{x/b(x)} f(x - b(x)y)\mathrm{e}^{Rb(x)y}\,\mathrm{d}G(y) - f(x) - \zeta(M_Y(Rb(x)) - 1)\Bigr| < \varepsilon/4$$

for x large enough. Finally, $c(b(x))R|f(x) - \zeta| < \varepsilon/4$ for x large enough. We conclude that

$$c(b(x))f'(x) < -\zeta\theta(R;b(x)) + \varepsilon/2\,.$$

Let $\underline{b} = \sup\{b : c(b) \le 0\}$. Because $f'(x)$ is bounded and $\theta(R; \underline{b}) > 0$, we get that $c(b(x))$ cannot be arbitrarily small. That means that $c(b(x))$ is bounded away from zero. We conclude therefore that $\limsup_{x\to\infty} f'(x) \le 0$. Thus, $\lim_{x\to\infty} f'(x) = 0$. If x is large enough, we get $\lambda e^{Rx}(1 - G(x/b(x))) < \varepsilon/4$ and $c(b(x))|f'(x)| < \varepsilon/4$. Thus, we have

$$0 \le \zeta\theta(R; b(x)) < \varepsilon\,.$$

This proves that $\lim_{x\to\infty} \theta(R; b(x)) = 0$. If b^* is unique, this is only possible if $\lim_{x\to\infty} b(x) = b^*$. □

Remark 4.14. From $\lim_{x\to\infty} f'(x) = 0$ it follows that $\lim_{x\to\infty} g(x) = R\zeta$. From the definition of $g(x)$ this is $\lim_{x\to\infty} -\psi'(x)e^{Rx} = R\zeta$. Hence, $\psi'(x) \sim -R\zeta e^{-Rx}$; that is, we can interchange the limit and derivative. ■

4.2.2 Optimal Excess of Loss Reinsurance

We consider now excess of loss reinsurance, $s(y, a) = \min\{y, a\}$. Let

$$M(r; a) = \int_0^a e^{rx}\, dG(x) + e^{ra}(1 - G(a)) = 1 + r\int_0^a (1 - G(z))e^{rz}\, dz\,.$$

The function $\theta(r; a)$ is then defined by $\theta(r; a) := \lambda(M(r; a) - 1) - c(a)r$, where $a \in [0, \infty]$ is the deductible. The solution $r = R(a)$ to $\theta(r; a) = 0$ is the adjustment coefficient corresponding to the fixed strategy $a_t = a$. Denote by a^* a value for which $R(a^*)$ is maximal. The Lundberg exponent is then $R = R(a^*)$ and the solution to

$$\inf_{a\ge 0} \theta(r; a) = 0\,.$$

We can rewrite the equation as

$$\inf_{a\ge 0}\Bigl\{\lambda \int_0^a (1 - G(z))e^{rz}\, dz - c(a)\Bigr\} = 0\,.$$

Because $c(0) < 0$, one can see that $a^* > 0$. As in Section 4.2.1, we assume that $M_Y(R) < \infty$.

We next define the martingale M as

$$M_t = \exp\Bigl\{-R(X_t - x) - \int_0^t \theta(R; a(X_s))\, ds\Bigr\}\,. \tag{4.10}$$

Similarly as in Section 4.2.1, we obtain the following

Lemma 4.15. *The process M is a strictly positive martingale with mean value 1.* □

In the same way as before, we define the new measure $\mathbb{P}^*$. A similar proof as in Section 4.2.1 shows the law of X under $\mathbb{P}^*$.

Lemma 4.16. *Under the measure $\mathbb{P}^*$ the process X is a piecewise deterministic Markov process with jump intensity $\lambda^*(x) = \lambda M(R; a(x))$ and claim size distribution $G_x^*(y) = M(R; a(x))^{-1} \int_0^y \mathrm{e}^{R(z \wedge a(x))} \,\mathrm{d}G(z)$. The premium rate is $c(a(x))$.* □

The upper and lower Lundberg bounds can now be easily derived.

Proposition 4.17. *Let C_- be defined by* (4.8). *Then*

$$C_- \mathrm{e}^{-Rx} \le \psi(x) < \mathrm{e}^{-Rx} .$$

Proof. The upper bound is obtained from the constant strategy $a_t = a^*$. The lower bound is obtained by conditioning on $\{X_{\tau-} = z\}$. This implies that $a(X_{\tau-}) > z$ and $Y > z$. The quantity of interest $\mathbb{E}^*[\mathrm{e}^{RX_\tau} \mid X_{\tau-} = z]$ is minimised if $a(X_{\tau-}) = \infty$. But then we obtain the expression from the classical case, which is C_-. □

In order to prove the Cramér–Lundberg approximation, we proceed as in Section 4.2.1. Equation (2.14) then reads

$$c(a(x))\psi'(x) + \lambda \Bigl[\delta(0)(1 - G(x))\mathbb{1}_{a(x)=\infty} - \int_0^{x \wedge a(x)} \psi'(x - z)(1 - G(z)) \,\mathrm{d}z\Bigr] = 0 .$$

Note that $a(x) = \infty$ if $a(x) > x$. It cannot be optimal to take reinsurance if ruin occurs even if the reinsurer has to pay part of the claim size.

We again write $f(x) = \psi(x)\mathrm{e}^{Rx}$ and $g(x) = Rf(x) - f'(x) = -\psi'(x)\mathrm{e}^{Rx}$. The Hamilton–Jacobi–Bellman equation then reads

$$\lambda\Bigl[\delta(0)(1 - G(x))\mathrm{e}^{Rx}\mathbb{1}_{a(x)=\infty} + \int_0^{x \wedge a(x)} g(x - z)\mathrm{e}^{Rz}(1 - G(z)) \,\mathrm{d}z\Bigr] - c(a(x))g(x) = 0 . \tag{4.11}$$

Replacing $a(x)$ by a^* yields the inequality

$$\lambda\Bigl[\delta(0)(1 - G(x))\mathrm{e}^{Rx}\mathbb{1}_{a^* > x} + \int_0^{x \wedge a^*} g(x - z)\mathrm{e}^{Rz}(1 - G(z)) \,\mathrm{d}z\Bigr] - c(a^*)g(x) \ge 0 .$$

Noting that the definition of a^* reads

$$\lambda \int_0^{a^*} (1 - G(z))\mathrm{e}^{Rz} \,\mathrm{d}z = c(a^*) ,$$

we get

$$\int_0^{x\wedge a^*} (g(x-z)-g(x))\mathrm{e}^{Rz}(1-G(z))\,\mathrm{d}z$$
$$\geq \Bigl[\int_x^{a^*} (1-G(z))\mathrm{e}^{Rz}\,\mathrm{d}z g(x) - \delta(0)(1-G(x))\mathrm{e}^{Rx}\Bigr] 1\!\!\mathrm{I}_{a^*>x}\,. \qquad (4.12)$$

Repeating the proof of Lemma 4.10 gives us the analogous results.

Lemma 4.18. i) *$g(x)$ is bounded. In particular, $f'(x)$ is bounded.*
ii) *Let $\zeta = \limsup_{x\to\infty} g(x)/R$. Then $\limsup_{x\to\infty} f(x) = \zeta$. In particular, $\zeta > 0$ if $C_- > 0$.*
iii) *For any $\beta > 0$, $x_0 > 0$, and $\varepsilon > 0$, there is an $x \geq x_0$ such that $f(y) > \zeta - \varepsilon$ for $y \in [x-\beta, x]$.* □

The same proof as for Theorem 4.11 applies for the next result.

Theorem 4.19. *Suppose that $C_- > 0$. Then $\lim_{x\to\infty} \psi(x)\mathrm{e}^{Rx} = \zeta$.* □

A similar proof as for Corollary 4.13 yields convergence of the optimal strategy.

Corollary 4.20. *Suppose that $C_- > 0$. Then $\lim_{x\to\infty} f'(x) = 0$. If, moreover, a^* is unique, then $\lim_{x\to\infty} a(x) = a^*$.* □

In particular, we also have $\psi'(x) \sim -R\zeta\mathrm{e}^{-Rx}$, so the limit and derivative can be interchanged.

4.2.3 Optimal Investment

We now turn to the case where no reinsurance is possible but the surplus can be invested in a risky asset. From Section 4.1.2 we know that the adjustment coefficient $R(A)$ is unimodal. There is a value A^* such that $R = \sup_A R(A) = R(A^*)$. In particular, we get that $M_Y(R) < \infty$. We have to assume here that the Lundberg coefficient is well defined. We therefore assume that

$$\lambda(M_Y(R)-1) - cR - \frac{m^2}{2\sigma^2} = 0\,.$$

Because Equation (4.4) determining $R(A)$ is strictly convex, the right-hand side is positive at $r = R$ for any A. It is zero only at $A = A^*$. We thus can express the equation as

$$\inf_{A\in\mathbb{R}} \{\lambda(M_Y(r)-1) - (c+mA)r + \tfrac{1}{2}A^2\sigma^2 r^2\} = 0\,. \qquad (4.13)$$

Note that there are three solutions, and R is the strictly positive one.

By choosing the strategy $A_t = A^*$, we obtain the following

Proposition 4.21. *The minimal ruin probability is bounded by $\psi(x) < \mathrm{e}^{-Rx}$.*

Proof. If $A_t = A^*$, then X is a perturbed risk model. Thus, Lundberg's inequality follows from (D.10). □

We next define the martingale M. Let

$$\theta(r; A) = \lambda(M_Y(r) - 1) - (c + mA)r + \tfrac{1}{2}\sigma^2 A^2 r^2 .$$

For negative values we define $A(x) = 0$, $x \le 0$. Using (4.4), we can write

$$\theta(R; A) = \tfrac{1}{2}(\sigma R A - m/\sigma)^2 .$$

We define the process M as

$$M_t = \exp\Bigl\{-R(X_t - x) - \int_0^t \theta(R; A(X_s))\,\mathrm{d}s\Bigr\} .$$

We now show that M is a martingale.

Lemma 4.22. *The process M is a martingale.*

Proof. The process M can be written as

$$\exp\Bigl\{R\sum_{i=1}^{N_t} Y_i - \lambda t(M_Y(R) - 1) - R\int_0^t A(X_s)\,\mathrm{d}W_s - \tfrac{1}{2}R^2\int_0^t A^2(X_s)\,\mathrm{d}s\Bigr\} .$$

This is a Markov process. Therefore, it is enough to prove that its expected value is 1. Let $\tau_n = \inf\{t : X_t > n\}$. Because $A(x)$ is bounded on $[0, n]$, by conditioning on N and $\{Y_i\}$ we get that

$$\mathbb{E}[M_{\tau_n \wedge t} \mid N, \{Y_k\}] = \exp\Bigl\{R\sum_{i=1}^{N_{\tau_n \wedge t}} Y_i - \lambda(\tau_n \wedge t)(M_Y(R) - 1)\Bigr\} ;$$

see Proposition A.5. We have $X_{\tau_n} > 0$. Because $\theta(R; A) \ge 0$, we find that $M_{\tau_n} < \mathrm{e}^{-Rx}$. Thus, the bounded convergence theorem shows that $\mathbb{E}[M_{\tau_n}; \tau_n \le t \mid N, \{Y_k\}]$ converges to zero. Furthermore, we have that

$$M_t = \lim_{n\to\infty} M_t \mathbb{1}_{\tau_n > t} = \lim_{n\to\infty} M_{\tau_n \wedge t} \mathbb{1}_{\tau_n > t}$$

is monotone in n. Thus, by the monotone convergence theorem,

$$\begin{aligned}
\mathbb{E}[M_t \mid N, \{Y_k\}] &= \mathbb{E}[\lim_{n\to\infty} M_{\tau_n \wedge t}; \tau_n > t \mid N, \{Y_k\}] \\
&= \lim_{n\to\infty} \mathbb{E}[M_{\tau_n \wedge t}; \tau_n > t \mid N, \{Y_k\}] \\
&= \lim_{n\to\infty} \mathbb{E}[M_{\tau_n \wedge t} \mid N, \{Y_k\}] \\
&= \lim_{n\to\infty} \exp\Bigl\{R\sum_{i=1}^{N_{\tau_n \wedge t}} Y_i - \lambda(\tau_n \wedge t)(M_Y(R) - 1)\Bigr\} \\
&= \exp\Bigl\{R\sum_{i=1}^{N_t} Y_i - \lambda t(M_Y(R) - 1)\Bigr\} .
\end{aligned}$$

Because the process $\{\exp\{\sum_{i=1}^{N_t} Y_i - \lambda t(M_Y(r)-1)\}\}$ is a martingale, it follows that $\mathbb{E}[M_t] = 1$. Thus, M is a martingale. □

We again define the new measure $\mathbb{P}^*[A] = \mathbb{E}[M_t; A]$ for $A \in \mathcal{F}_t$. Then X follows the following law under $\mathbb{P}^*$.

Lemma 4.23. *Under $\mathbb{P}^*$ the process X is a jump diffusion process. The claim intensity is $\lambda^* = \lambda M_Y(R)$, the claim size distribution is $G^*(y) = (M_Y(R))^{-1} \int_0^y e^{Rz}\, dG(z)$, the drift of the diffusion is $c_x^* = c + A(x)m - \sigma^2 A^2(x) R$, and the diffusion rate is $\sigma_x^* = \sigma A(x)$.*

Proof. As in the proof of Lemma 4.8, X remains a Markov process. Let us first consider the law of the jump times. Let $t \geq 0$. Then

$$\begin{aligned}
\mathbb{P}^*[T_1 > t] &= \mathbb{E}[M_t; T_1 > t] \\
&= \mathbb{E}\Big[\exp\Big\{-R\sigma \int_0^t A(X_s)\, dW_s - \tfrac{1}{2}\sigma^2 R^2 \int_0^t A^2(X_s)\, ds \\
&\qquad - \lambda(M_Y(R)-1)t\Big\}; T_1 > t\Big] \\
&= \exp\{-\lambda(M_Y(R)-1)t\}\mathbb{P}[T_1 > t] = \exp\{-\lambda M_Y(R) t\}\,.
\end{aligned}$$

Thus, the claim arrival process is a Poisson process with rate λ^*. We next consider the claim size distribution. By conditioning on $\mathcal{F}_{T_1}$, we have that

$$\begin{aligned}
\mathbb{P}^*[Y_1 \leq y, T_1 \leq t] &= \mathbb{E}[M_{T_1}; Y_1 \leq y, T_1 \leq t] \\
&= \mathbb{E}\Big[\exp\Big\{-R\sigma \int_0^{T_1} A(X_s)\, dW_s - \tfrac{1}{2}\sigma^2 R^2 \int_0^{T_1} A^2(X_s)\, ds \\
&\qquad + RY_1 - \lambda(M_Y(R)-1)t\Big\}; Y_1 \leq y, T_1 \leq t\Big] \\
&= \mathbb{E}\Big[\exp\Big\{RY_1 - \lambda(M_Y(R)-1)t\Big\}; Y_1 \leq y, T_1 \leq t\Big] \\
&= \int_0^t \lambda e^{-\lambda M_Y(R)s}\, dt \int_0^y e^{Rz}\, dG(z) \\
&= (1 - e^{-\lambda M_Y(R)})(M_Y(R))^{-1} \int_0^y e^{Rz}\, dG(z)\,.
\end{aligned}$$

Thus, the jump sizes are independent of the jump times and have the claimed distribution.

Let $\mathcal{B}$ be a Borel set of the space of cadlag functions on $[0,t]$. Then in the same way as above,

$$\begin{aligned}
\mathbb{P}^*[T_1 > t, X \in \mathcal{B}] &= e^{-\lambda M_Y(R)t} \\
&\times \mathbb{E}\Big[\exp\Big\{-R\sigma \int_0^{T_1} A(X_s)\, dW_s - \tfrac{1}{2}\sigma^2 R^2 \int_0^{T_1} A^2(X_s)\, ds\Big\}; \mathcal{B}\Big]\,.
\end{aligned}$$

Thus, the law of the diffusion process between jumps follows from Girsanov's theorem. □

Let us calculate the infinitesimal drift of the process under $\mathbb{P}^*$. We get

$$c + A(x)m - \sigma^2 A^2(x)R - \lambda M_Y(R)M_Y'(R)/M_Y(R) = -\theta'(R; A(x)) .$$

Because $r \mapsto \theta(r; A)$ is a strictly convex function and R is to the right of $R(A(x))$, we have that $\theta'(R; A(x)) > 0$. Thus, the process has a negative drift and therefore tends to $-\infty$ under $\mathbb{P}^*$. In particular, we get $\mathbb{P}^*[\tau < \infty] = 1$. The ruin probability can then be written as

$$\psi(x) = \mathbb{E}^*\Bigl[\exp\Bigl\{RX_\tau + \int_0^\tau \theta(R; A(X_s))\,\mathrm{d}s\Bigr\}\Bigr]\mathrm{e}^{-Rx} .$$

We next find the lower bound.

Proposition 4.24. *Let C_- be defined by* (4.8)*. Then $\psi(x) \ge C_-\mathrm{e}^{-Rx}$.*

Proof. As in the proof of Proposition 4.9, we find that $\psi(x)\mathrm{e}^{Rx} \ge \mathbb{E}^*[\mathrm{e}^{RX_\tau}]$. As in the proof of Proposition 4.9, the latter expression is bounded from below by C_-. □

The function $\delta(x) = 1 - \psi(x)$ solves (2.18). Thus, $\psi(x)$ fulfils

$$-\frac{m^2}{2\sigma^2}\frac{\psi'(x)^2}{\psi''(x)} + c\psi'(x) + \lambda\Bigl[\int_0^x \psi(x-y)\,\mathrm{d}G(y) + 1 - G(x) - \psi(x)\Bigr] = 0 . \quad (4.14)$$

Using integration by parts, we can write

$$-\frac{m^2}{2\sigma^2}\frac{\psi'(x)^2}{\psi''(x)} + c\psi'(x) - \lambda\int_0^x \psi'(x-y)(1-G(y))\,\mathrm{d}y + \lambda(1-G(x))\delta(0) = 0 . \quad (4.15)$$

Let $f(x) = \psi(x)\mathrm{e}^{Rx}$ and $g(x) = Rf(x) - f'(x) = -\psi'(x)\mathrm{e}^{Rx}$. Because $\psi(x)$ is strictly decreasing and strictly convex, we have $g(x) > 0$ and $g'(x) < Rg(x)$. The Hamilton–Jacobi–Bellman equation can be written as

$$-\frac{m^2}{2\sigma^2}\frac{(Rf(x) - f'(x))^2}{R^2 f(x) - 2Rf'(x) + f''(x)} - c(Rf(x) - f'(x)) + \lambda\Bigl[\int_0^x f(x-y)\mathrm{e}^{Ry}\,\mathrm{d}G(y) + (1-G(x))\mathrm{e}^{Rx} - f(x)\Bigr] = 0 , \quad (4.16)$$

or

$$-\frac{m^2}{2\sigma^2}\frac{g(x)^2}{Rg(x) - g'(x)} - cg(x) + \lambda\int_0^x g(x-y)(1-G(y))\mathrm{e}^{Ry}\,\mathrm{d}y + \lambda(1-G(x))\mathrm{e}^{Rx}\delta(0) = 0 . \quad (4.17)$$

Because $f'(x) < Rf(x)$, the derivative $f'(x)$ is bounded from above. Looking at a local minimum of $f'(x)$, i.e., $f''(x) = 0$, it follows that $f'(x)$ also is bounded from below. In particular, $g(x)$ is bounded, and $g'(x)$ is bounded from above.

Multiplying (4.5) by $g(x)$, we have that

$$\Bigl(\lambda\int_0^\infty (1-G(z))\mathrm{e}^{Rz}\,\mathrm{d}z - c - \frac{m^2}{2\sigma^2 R}\Bigr)g(x) = 0\,.$$

Taking the difference to (4.17) yields

$$-\frac{m^2}{2\sigma^2 R}\frac{g'(x)g(x)}{Rg(x)-g'(x)} + \lambda\int_0^x (g(x-y)-g(x))(1-G(y))\mathrm{e}^{Ry}\,\mathrm{d}y$$
$$-\lambda g(x)\int_x^\infty (1-G(z))\mathrm{e}^{Rz}\,\mathrm{d}z + \lambda(1-G(x))\mathrm{e}^{Rx}\delta(0) = 0\,. \quad (4.18)$$

We let $\zeta = R^{-1}\limsup_{x\to\infty} g(x)$. Note that $\zeta > 0$, because otherwise $f'(x) = Rf(x) - g(x)$ would be bounded away from zero for large x and $f(x)$ could not be bounded. The functions $f(x)$ and $g(x)$ have the following properties.

Lemma 4.25. i) *For any* $\varepsilon, \beta, x_0 > 0$, *there is* $x \ge x_0$ *such that* $g(z) > R\zeta - \varepsilon$ *for* $z \in [x-\beta, x]$.
ii) *We have* $\limsup_{x\to\infty} f(x) = \zeta$.
iii) *For any* $\varepsilon, \beta, x_0 > 0$, *there is* $x \ge x_0$ *such that* $f(z) > \zeta - \varepsilon$ *for* $z \in [x-\beta, x]$.

Proof. i) If $g(x)$ is ultimately monotone, then $g(x)$ is converging and the assertion is trivial. We therefore assume that $g(x)$ is not ultimately monotone. Suppose first that $G(\beta) < 1$. Choose $0 < \chi < \varepsilon/3$ to be determined later. Let $\eta = \inf_{y\le\beta}\{\mathrm{e}^{Ry}(1-G(y))\}$. We can then find $x \ge x_0$ such that $g(x) > R\zeta - \chi$, $g'(x) \ge 0$, $g(z) < R\zeta + \chi$ for all $z \ge x - \beta$, and

$$\phi(x) := 2R\zeta\lambda\int_x^\infty \mathrm{e}^{Ry}(1-G(y))\,\mathrm{d}y + \lambda\delta(0)\mathrm{e}^{Rx}(1-G(x)) < \chi\,.$$

Then

$$\lambda\int_0^x (g(x-y)-g(x))\mathrm{e}^{Ry}(1-G(y))\,\mathrm{d}y > -\chi\,.$$

Suppose now that $g(x-y) \le R\zeta - \varepsilon$ for some $y \le \beta$. We have that $g'(z) \le Rg(z) \le R(R\zeta + \chi)$. Thus, $g(z) < R\zeta - 2\varepsilon/3$ for all $z \in [x-y, x-y+\varepsilon/(3R(R\zeta+\chi))]$. That means we can estimate the integral above:

$$\lambda 2\chi\int_0^\infty \mathrm{e}^{Ry}(1-G(y))\,\mathrm{d}y - \frac{\lambda\eta\varepsilon^2}{9R(R\zeta+\chi)} > -\chi\,.$$

If one chooses χ small enough, one obtains a contradiction. This proves the result if $G(\beta) < 1$.

Suppose that the result is proven in the sense of the construction above for $\tilde{\beta} = 2\beta/3$. There exists $\tilde{\chi}$ such that whenever $g(x) > R\zeta - \tilde{\chi}$, $g'(x) \ge 0$, $g(z) < R\zeta + \tilde{\chi}$ for all $z \ge x - 2\beta$, and $\phi(x) < \tilde{\chi}$, then $g(z) > R\zeta - \varepsilon$ for all $z \in [x - 2\beta/3, x]$. Now choose $x \ge x_0 + \beta$ such that $g'(x) \ge 0$, $g(z) > R\zeta - \tilde{\chi}$ for all $z \in [x - 2\beta/3, x]$, $g(z) < R\zeta + \tilde{\chi}$ for $z \ge x - 3\beta$, and $\phi(z) < \tilde{\chi}$ for $z > x - \beta$. If there is $\tilde{x} \in [x - 2\beta/3, x - \beta/3)$ such that $g'(\tilde{x}) \ge 0$, the result follows from the proof above. If there is no such $\tilde{x}$, then $g(z)$ is decreasing on $[x - 2\beta/3, x - \beta/3]$. If $g(z)$ is also decreasing on $[x - \beta, x - 2\beta/3]$, the result follows. Otherwise, $z_0 := \sup\{z \le x - 2\beta/3 : g(z) = 0\}$ exists, and the result follows from the considerations above. This proves the assertion if $G(2\beta/3) < 1$. In the same way the result follows if $G((2/3)^n\beta) < 1$ by induction.
ii) Denote by $\eta = \limsup_{u\to\infty} f(x)$. Note there must be points where $f'(x)$ is arbitrarily close to zero and $f(x)$ is close to η. Because $g(x) = Rf(x) - f'(x)$, we must have $\eta \le \zeta$. Suppose that $\eta < \zeta$. By part i) there exists x such that $g(z) > (\eta + 2\zeta)R/3$ and $f(z) < (2\eta + \zeta)/3$ for $z \in [x - 3/(R(\zeta - \eta)), x]$. On this interval $f'(z) = Rf(z) - g(z) < -R(\zeta - \eta)/3$. Recall that $f(x - 3/(R(\zeta - \eta))) < 1$. This implies that $f(x) < 0$, which is a contradiction.
iii) There is $x \ge x_0 + 2\zeta/(\varepsilon R)$ such that $g(z) > R(\zeta - \varepsilon/2)$ for all $z \in [x - \beta - 2\zeta/(\varepsilon R), x]$. If for some $y \in [x - \beta - 2\zeta/(\varepsilon R), x - 2\zeta/(\varepsilon R)]$ we had $f(y) \le \zeta - \varepsilon$, then $f'(y) = Rf(y) - g(y) < -R\varepsilon/2$ would be decreasing. Thus, $f(z) \le \zeta - \varepsilon$ for $z \in [y, x]$. This would imply that $f(x) < 0$. Therefore, $f(y) > \zeta - \varepsilon$ for all $y \in [x - \beta - 2\zeta/(\varepsilon R), x - 2\zeta/(\varepsilon R)]$. □

We can now repeat the proof of Theorem 4.11.

Theorem 4.26. *Suppose that $C_- > 0$. Then $\lim_{x\to\infty} \psi(x)e^{Rx} = \zeta$.* □

We also get that the strategy converges.

Corollary 4.27. *Suppose that $C_- > 0$. Then we have $\lim_{x\to\infty} f'(x) = \lim_{x\to\infty} f''(x) = 0$ and $\lim_{x\to\infty} A(x) = A^*$.*

Proof. From $f'(x) = Rf(x) - g(x)$ we see that $\liminf_{x\to\infty} f'(x) = 0$. Suppose that $\limsup_{x\to\infty} f'(x) = \varepsilon > 0$. Then $\liminf_{x\to\infty} g(x) = R\zeta - \varepsilon$. For x_0 we can find an $x \ge x_0$ such that $g(x) < R\zeta - 2\varepsilon/3$. Because $g'(z) < Rg(z) < R^2\zeta$ whenever $g(z) < R\zeta$, we obtain $g(z) < R\zeta - \varepsilon/3$ for all $z \in [x, x + \varepsilon/(3R^2\zeta)]$. This implies that $f'(z) > \varepsilon/3$. But this contradicts that $f(x)$ is converging.

Because $\lim_{x\to\infty} g(x) = R\zeta$, it follows from bounded convergence that $\int_0^x (g(x-y) - g(x))(1 - G(y))e^{Ry}\,dy$ converges to zero as $x \to \infty$. Thus,

$$\lim_{x\to\infty} \frac{g(x)}{Rg(x)/g'(x) - 1} = \lim_{x\to\infty} \frac{g'(x)g(x)}{Rg(x) - g'(x)} = 0\,.$$

This implies that $\lim_{x\to\infty} g'(x) = 0$. From $f''(x) = Rf'(x) - g'(x)$ we conclude that $f''(x)$ converges to zero.

We now can express the limit of the strategy:

$$A(x) = -\frac{m\psi'(x)}{\sigma^2\psi''(x)} = \frac{mg(x)}{\sigma^2(Rg(x) - g'(x))}\,.$$

The latter converges to $m/(\sigma^2 R) = A^*$. □

Remark 4.28. From $f'(x) \to 0$ and $f''(x) \to 0$, we see that $\psi'(x) \sim -\zeta R \mathrm{e}^{-Rx}$ and $\psi''(x) \sim \zeta R^2 \mathrm{e}^{-Rx}$. Thus, in the Cramér–Lundberg approximation the limit and derivatives can be interchanged. ■

4.2.4 Optimal Proportional Reinsurance and Investment

Consider again the case where investment and proportional reinsurance are possible. Let R be the maximal adjustment coefficient $R(A, b)$. We assume that R solves (4.6) for $b = b^*$. For any A, b, r where $M_Y(br) < \infty$ let

$$\theta(r; A, b) = \lambda(M_Y(br) - 1) - (c(b) + mA)r + \tfrac{1}{2}A^2\sigma^2 r^2\,.$$

Because $\theta(r; A, b)$ is a convex function in r, we can rewrite the definition of R as

$$\inf_{b\in[0,1], A\in\mathbb{R}} \{\lambda(M_Y(br) - 1) - (c(b) + mA)r + \tfrac{1}{2}A^2\sigma^2 r^2\} = 0\,.$$

The optimal value A^* is then $m/(\sigma^2 R)$, i.e., R solves

$$\inf_{b\in[0,1]} \Bigl\{\lambda(M_Y(br) - 1) - c(b)r - \frac{m^2}{2\sigma^2}\Bigr\} = 0\,.$$

We immediately get an upper bound, which is proved in the same way as Proposition 4.21.

Proposition 4.29. *The ruin probability is bounded by $\psi(x) < \mathrm{e}^{-Rx}$.* □

We now assume that $M_Y(R) < \infty$. We again let the strategy be $(A(x), b(x)) = (0, 1)$ whenever $x < 0$. We next construct the martingale

$$M_t = \exp\Bigl\{-R(X_t - x) - \int_0^t \theta(R; A(X_s), b(X_s))\,\mathrm{d}s\Bigr\}\,.$$

Similarly as with Lemma 4.22, one can prove the following

Lemma 4.30. *The process M is a martingale.* □

We again define the measure $\mathbb{P}^*$ as $\mathbb{P}^*[A] = \mathbb{E}[M_t; A]$ for $A \in \mathcal{F}_t$. Similarly as in Lemmata 4.8 and 4.23, one gets the following law of X under $\mathbb{P}^*$.

Lemma 4.31. *Under $\mathbb{P}^*$ the process X is a jump diffusion process. The claim intensity is $\lambda_x^* = \lambda M_Y(b(x)R)$, the claim size distribution is $G_x^*(y) = (M_Y(b(x)R))^{-1} \int_0^y e^{Rb(x)z} \, dG(z)$, the drift of the diffusion is $c_x^* = c + A(x)m - \sigma^2 A^2(x)R$, and the diffusion rate is $\sigma_x^* = \sigma A(x)$.* □

The following lower bound can be proved as in Proposition 4.9.

Proposition 4.32. *Let C_- be defined by (4.8). Then $\psi(x) \ge C_- e^{-Rx}$.* □

In the rest of this section we will assume that $C_- > 0$. From (2.23) we find that $\psi(x)$ solves

$$-\frac{m^2}{2\sigma^2} \frac{\psi'(x)^2}{\psi''(x)} + c(b(x))\psi'(x) + \lambda\Big[\int_0^{x/b(x)} \psi(x - b(x)y) \, dG(y) + 1 - G(x/b(x)) - \psi(x)\Big] = 0 \,.$$

Using integration by parts, this can be written as

$$-\frac{m^2}{2\sigma^2} \frac{\psi'(x)^2}{\psi''(x)} + c(b(x))\psi'(x) - \lambda \int_0^x \psi'(y)[1 - G((x - y)/b(x))] \, dy + \lambda\delta(0)[1 - G(x/b(x))] = 0 \,,$$

where again $\delta(0) = 1 - \psi(0)$. We let $f(x) = \psi(x)e^{Rx}$ and $g(x) = Rf(x) - f'(x) = -\psi'(x)e^{Rx}$. By the corresponding properties of $\psi(x)$, we get $f(x) \ge C_-$, $g(x) > 0$, and $g'(x) < Rg(x)$. Using the definition of $f(x)$, the Hamilton–Jacobi–Bellman equation reads

$$-\frac{m^2}{2\sigma^2} \frac{(Rf(x) - f'(x))^2}{R^2 f(x) - 2Rf'(x) + f''(x)} - c(b(x))(Rf(x) - f'(x)) + \lambda\Big[\int_0^{x/b(x)} f(x - b(x)y)e^{Rb(x)y} \, dG(y) + (1 - G(x/b(x)))e^{Rx} - f(x)\Big] = 0 \,.$$

The first two terms are negative, while the others are bounded. Thus, the first two terms are bounded, too. From $f'(x) < Rf(x)$ it follows that $f'(x)$ is bounded from above. Clearly, $f'(x)$ cannot be monotonically decreasing to minus infinity, because $f(x)$ is bounded. Looking at a point where $f''(x) = 0$, one concludes that $f'(x)$ is also bounded from below. In particular, $g(x)$ is a bounded function.

With the help of the function $g(x)$, the Hamilton–Jacobi–Bellman equation reads

$$-\frac{m^2}{2\sigma^2} \frac{g(x)^2}{Rg(x) - g'(x)} - c(b(x))g(x) + \lambda \int_0^x g(x - y)[1 - G(y/b(x))]e^{Ry} \, dy + \lambda\delta(0)[1 - G(x/b(x))]e^{Rx} = 0 \,. \tag{4.19}$$

Because $b(x)$ is the point where the infimum on the left-hand side is taken, by replacing $b(x)$ by b^*, we obtain

$$-\frac{m^2}{2\sigma^2}\frac{g(x)^2}{Rg(x)-g'(x)} - c(b^*)g(x) \tag{4.20}$$
$$+\lambda\int_0^x g(x-y)[1-G(y/b^*)]\mathrm{e}^{Ry}\,\mathrm{d}y + \lambda\delta(0)[1-G(x/b^*)]\mathrm{e}^{Rx} \ge 0\,.$$

We multiply (4.6) by $g(x)/R$ and subtract this from the last equation:

$$-\frac{m^2}{2\sigma^2 R}\frac{g(x)g'(x)}{Rg(x)-g'(x)} + \lambda\int_0^x (g(x-y)-g(x))[1-G(y/b^*)]\mathrm{e}^{Ry}\,\mathrm{d}y$$
$$+\lambda g(x)\int_x^\infty [1-G(y/b^*)]\mathrm{e}^{Ry}\,\mathrm{d}y + \lambda\delta(0)[1-G(x/b^*)]\mathrm{e}^{Rx} \ge 0\,.$$

We again let $\zeta = R^{-1}\limsup_{x\to\infty} g(x)$, and find that $\zeta > 0$. Using the inequality above, we can prove similarly as in Lemma 4.25 the following

Lemma 4.33. i) *For any* $\varepsilon, \beta, x_0 > 0$, *there is* $x \ge x_0$ *such that* $g(z) > R\zeta - \varepsilon$ *for* $z \in [x-\beta, x]$.
ii) *We have* $\limsup_{x\to\infty} f(x) = \zeta$.
iii) *For any* $\varepsilon, \beta, x_0 > 0$, *there is* $x \ge x_0$ *such that* $f(z) > \zeta - \varepsilon$ *for* $z \in [x-\beta, x]$. □

Repeating the proof of Theorem 4.11, we obtain the following

Theorem 4.34. *Suppose that* $C_- > 0$. *Then* $\lim_{x\to\infty}\psi(x)\mathrm{e}^{Rx} = \zeta$. □

Corollary 4.35. *Suppose that* $C_- > 0$. *Then we have* $\lim_{x\to\infty} f'(x) = \lim_{x\to\infty} f''(x) = 0$. *In particular,* $\lim_{x\to\infty} A(x) = A^*$. *If* b^* *is uniquely determined, then also* $\lim_{x\to\infty} b(x) = b^*$.

Proof. That $f'(x)$ converges to zero follows as in the proof of Corollary 4.27. Recall that $g(x)$ converges to $R\zeta$. Let $\{x_n\}$ be a sequence such that $g'(x_n)$ converges to some value γ. By taking a subsequence we can assume that $b(x_n)$ converges to some value b_0. Letting $n\to\infty$ in (4.20), we obtain

$$0 \le -\frac{m^2}{2\sigma^2}\frac{R^2\zeta^2}{R^2\zeta-\gamma} - c(b^*)R\zeta + \lambda\zeta(M_Y(b^*R)-1) = -\frac{m^2}{2\sigma^2}\frac{\gamma\zeta}{R^2\zeta-\gamma}\,.$$

In the last inequality we used (4.6). This implies that $\gamma \le 0$. Letting $n\to\infty$ in (4.19), we obtain

$$-\frac{m^2}{2\sigma^2}\frac{R^2\zeta^2}{R^2\zeta-\gamma} - c(b_0)R\zeta + \lambda\zeta(M_Y(b_0R)-1) = 0\,.$$

The infimum is taken at b^*; thus,

$$0 \ge -\frac{m^2}{2\sigma^2}\frac{R^2\zeta^2}{R^2\zeta-\gamma} - c(b^*)R\zeta + \lambda\zeta(M_Y(b^*R)-1) = -\frac{m^2}{2\sigma^2}\frac{\gamma\zeta}{R^2\zeta-\gamma}\;.$$

It follows that $\gamma \ge 0$, and therefore $\gamma = 0$. This shows that $\lim_{x\to\infty} g'(x) = 0$, and therefore also $\lim_{x\to\infty} f''(x) = 0$. We also have

$$-\frac{m^2\zeta}{2\sigma^2} - c(b_0)R\zeta + \lambda\zeta(M_Y(b_0R)-1) = 0\;.$$

If b^* is unique, then $b_0 = b^*$. That $\lim_{x\to\infty} A(x) = A^*$ follows as in the proof of Corollary 4.27. □

Remark 4.36. We have proved again that $\psi'(x) \sim -R\zeta\mathrm{e}^{-Rx}$ and $\psi''(x) \sim R^2\zeta\mathrm{e}^{-Rx}$. Thus, the derivatives and limit can be interchanged. ■

Bibliographical Remarks

The results in this section for proportional reinsurance (with and without investment) were found by Schmidli [162]. The case of excess of loss reinsurance is treated in Vogt [184]; see also Schmidli [164]. The asymptotic behaviour in the case with investment but without reinsurance is also published by Hipp and Schmidli [99], where, in particular, the case of exponentially distributed claim sizes is discussed. Some similar results were found by Gaier et al. [71] and Grandits [87].

4.3 The Heavy-Tailed Case

In this section we discuss the case of heavy-tailed claim size distributions. Because the behaviour of the ruin probability is different in the cases with/without investment and with/without reinsurance, we have to treat the three cases separately. The results will hold for different classes of heavy-tailed claim size distributions. Thus, we will have to consider different classes of heavy-tailed distribution functions in the different sections.

4.3.1 Proportional Reinsurance

Suppose that the claim sizes have a regularly varying tail with index $-\alpha$, i.e.,

$$\lim_{x\to\infty}\frac{1-G(xt)}{1-G(x)} = t^{-\alpha}\;.$$

From the asymptotic behaviour of $\psi^A(x)$ for a constant strategy $A_t = A$, we expect that $\psi(x)/\int_x^\infty(1-G(z))\,\mathrm{d}z$ converges to a constant. In "nice" cases one can even expect that $\psi'(x)/(1-G(x))$ converges to a constant. We therefore let $g(x) = -\psi'(x)/(1-G(x))$.

Consider the Hamilton–Jacobi–Bellman equation (2.14) divided by $1 - G(x)$. It can be written as

$$\inf_{b\in[\underline{b},1]}\Bigl\{\lambda\Bigl[\delta(0)\frac{1-G(x/b)}{1-G(x)}+\int_0^x g(z)\frac{(1-G(z))(1-G((x-z)/b))}{1-G(x)}\,\mathrm{d}z\Bigr] - c(b)g(x)\Bigr\} = 0\,.$$

Because the first term is bounded away from zero, we conclude that $g(x)$ is bounded away from zero. Because $\psi(x)/\int_x^\infty(1-G(z))\,\mathrm{d}z$ is bounded, we know also that $g_0 < \infty$.

Let x_n be a sequence tending to infinity such that $g(x_n)$ tends to $g_0 := \liminf_{x\to\infty} g(x) > 0$. By considering a subsequence we can assume that $b(x_n)$ converges to some $b_0 \in [\underline{b}, 1]$. Then $(1-G(x_n/b(x_n)))/(1-G(x_n))$ tends to b_0^α. By continuity $c(b(x_n))g(x_n)$ tends to $c(b_0)g_0$. Next we consider the integral. We divide the integration area into $(0, x_n/2]$ and $(x_n/2, x_n]$. The first integral can be estimated as

$$\int_0^{x_n/2} g(z)(1-G(z))\frac{1-G(x_n/\{2b(x_n)\})}{1-G(x_n)}\,\mathrm{d}z \le -C\int_0^{x_n/2}\psi'(z)\,\mathrm{d}z$$

for some constant C. By the dominated convergence theorem, we can interchange the limit and integration and obtain

$$\lim_{n\to\infty}\int_0^{x_n/2} g(z)(1-G(z))\frac{1-G((x_n-z)/b(x_n))}{1-G(x_n)}\,\mathrm{d}z = b_0^\alpha\psi(0)\,.$$

The second part of the integral can be written as

$$\int_0^{x_n/2} g(x_n-z)\frac{1-G(x_n-z)}{1-G(x_n)}(1-G(z/b(x_n)))\,\mathrm{d}z\,.$$

The limes inferior is bounded from below by

$$g_0\liminf_{n\to\infty}\int_0^{x_n/2}\frac{1-G(x_n-z)}{1-G(x_n)}(1-G(z/b(x_n)))\,\mathrm{d}z\,.$$

Again using bounded convergence gives the lower bound $g_0 b_0 \mu$. Thus,

$$\lambda b_0^\alpha + \lambda\mu b_0 g_0 - c(b_0)g_0 \le 0\,.$$

This shows that $c(b_0) - \lambda\mu b_0 > 0$. We therefore can conclude that

$$g_0 \ge \frac{\lambda b_0^\alpha}{c(b_0)-\lambda\mu b_0}\,.$$

We can now prove the following result.

Theorem 4.37. *Suppose that the tail of the claim size distribution is regularly varying with index $-\alpha$. Then*

$$\lim_{x\to\infty} \frac{\psi(x)}{\int_x^\infty 1-G(z)\,\mathrm{d}z} = \inf_{b\in(0,1]} \frac{\lambda b^\alpha}{(c(b)-\lambda\mu b)^+}\;,$$

where $1/0$ is interpreted as ∞. If there is a unique value b_0 for which the infimum is taken, we also have $\lim_{x\to\infty} b(x) = b_0$.

Proof. Note that by continuity there is a value b_0 at which the infimum is taken, and this value must fulfil $c(b_0) - \lambda\mu b_0 > 0$. Denote by $\psi^0(x)$ the ruin probability for the constant strategy $b_t = b_0$. Then by (D.7)

$$\lim_{x\to\infty} \frac{\psi^0(x)}{\int_x^\infty 1-G(z/b_0)\,\mathrm{d}z} = \frac{\lambda}{c(b_0)-\lambda\mu b_0}\;.$$

The integral behaves asymptotically like

$$\int_x^\infty 1-G(z/b_0)\,\mathrm{d}z = b_0 \int_{x/b_0}^\infty 1-G(z)\,\mathrm{d}z \sim b_0^\alpha \int_x^\infty 1-G(z)\,\mathrm{d}z$$

by [58, p. 281]. We conclude that

$$\limsup_{x\to\infty} \frac{\psi(x)}{\int_x^\infty 1-G(z)\,\mathrm{d}z} \le \frac{\lambda b_0^\alpha}{c(b_0)-\lambda\mu b_0}\;.$$

From the consideration above we also conclude that there is a b_1 such that

$$\liminf_{x\to\infty} \frac{-\psi'(x)}{1-G(x)} \ge \frac{\lambda b_1^\alpha}{c(b_1)-\lambda\mu b_1}\;.$$

This implies that

$$\liminf_{x\to\infty} \frac{\psi(x)}{\int_x^\infty 1-G(z)\,\mathrm{d}z} \ge \frac{\lambda b_1^\alpha}{c(b_1)-\lambda\mu b_1}\;.$$

Thus, we have proved the asymptotic behaviour of $\psi(u)$.

We next want to show that $g(x)$ converges. Suppose first that $g(x)$ is unbounded. Then there exists a sequence $\{x_n\}$ tending to infinity such that $g(x_n) = \sup_{0\le x\le x_n} g(x)$. From the Hamilton–Jacobi–Bellman equation we conclude that

$$\begin{aligned}
0 &\le \lambda\Bigl[\frac{\delta(0)}{g(x_n)}\frac{1-G(x_n/b_0)}{1-G(x_n)} \\
&\quad + \int_0^{x_n} \frac{g(z)}{g(x_n)}\frac{(1-G(z))(1-G((x_n-z)/b_0))}{1-G(x_n)}\,\mathrm{d}z\Bigr] - c(b_0) \\
&\le \lambda\Bigl[\frac{\delta(0)}{g(x_n)}\frac{1-G(x_n/b_0)}{1-G(x_n)} + \int_0^{x_n/2} \frac{g(z)}{g(x_n)}\frac{(1-G(z))(1-G((x_n-z)/b_0))}{1-G(x_n)}\,\mathrm{d}z \\
&\quad + \int_0^{x_n/2} \frac{(1-G(x_n-z))(1-G(z/b_0))}{1-G(x_n)}\,\mathrm{d}z\Bigr] - c(b_0)\;.
\end{aligned}$$

As above, we see that we can interchange the limit and integration. For the first integral, we obtain

$$\begin{aligned}\lim_{n\to\infty}\int_0^{x_n/2}&\frac{g(z)}{g(x_n)}\frac{(1-G(z))(1-G((x_n-z)/b_0))}{1-G(x_n)}\,\mathrm{d}z\\&=\lim_{n\to\infty}\frac{\int_0^{x_n/2}g(z)(1-G(z))b_0^\alpha\,\mathrm{d}z}{g(x_n)}=\lim_{n\to\infty}b_0^\alpha\frac{\psi(0)-\psi(x_n/2)}{g(x_n)}=0\,.\end{aligned}$$

Letting $n\to\infty$ in the inequality above shows that $\lambda\mu b_0-c(b_0)\ge 0$. But we have seen above that for b_0 the net profit condition is fulfilled. This is a contradiction. Thus, $g(x)$ is bounded.

Let $g_1=\limsup_{x\to\infty}g(x)$. Choose a sequence $\{x_n\}$ such that $g(x_n)$ tends to g_1. Then

$$\begin{aligned}\lambda\Bigl[\delta(0)\frac{1-G(x_n/b_0)}{1-G(x_n)}+\int_0^{x_n}g(z)\frac{(1-G(z))(1-G((x_n-z)/b_0))}{1-G(x_n)}\,\mathrm{d}z\Bigr]&\\-c(b_0)g(x_n)\ge 0\,.&\end{aligned}$$

In the same way as above,

$$\lambda b_0^\alpha+\limsup_{n\to\infty}\int_0^{x_n/2}g(x_n-z)\frac{1-G(x_n-z)}{1-G(x_n)}(1-G(z/b_0))\,\mathrm{d}z-c(b_0)g_1\ge 0\,.$$

Taking the limit inside the integral shows that

$$\lambda b_0^\alpha+(\lambda\mu b_0-c(b_0))g_1\ge 0\,.$$

We obtain $g_1\le g_0$. Thus, $\lim_{x\to\infty}g(x)=g_0$. Now let $\{x_n\}$ be a sequence such that $b(x_n)$ converges to a value b_2. Letting $n\to\infty$ in the Hamilton–Jacobi–Bellman equation divided by $(1-G(x_n))$ gives as above the limit

$$\lambda b_2^\alpha+(\lambda\mu b_2-c(b_2))g_0=0\,.$$

We conclude that the net profit condition is fulfilled. Hence, we can write

$$g_0=\frac{\lambda b_2^\alpha}{c(b_2)-\lambda\mu b_2}\,.$$

If the value b_0 is unique, we find that $b_2=b_0$. □

Remarks 4.38. i) The above theorem shows that asymptotically we cannot perform better than with a fixed strategy. This is in contrast to the small claims case where the constant before the exponential term could be made smaller.

ii) From the limit of $g(x)$, we see that the limit and derivative can be interchanged. ■

We do not get as explicit results if $1 - G(x)$ is of rapid variation.

Theorem 4.39. *Suppose that $G(x) \in \mathcal{S}^*$ is of rapid variation. Let $b_0 = \inf\{b : c(b) > \lambda\mu b\}$. Then for any $b > b_0$,*

$$\lim_{x\to\infty} \frac{\psi(x)}{\int_x^\infty 1 - G(z/b)\,\mathrm{d}z} = 0\;.$$

For the strategy we obtain that $\limsup_{x\to\infty} b(x) = b_0$.

Proof. Choose $b > b_0$. Let $g(x) = -\psi'(x)/(1 - G(x/b))$. There is $b_0 < b_1 < b$ such that $c(b_1) > \lambda\mu b_1$. Let us for the moment assume also that $c(b) > \lambda\mu b$. Dividing the Hamilton–Jacobi–Bellman equation by $1 - G(x/b)$ gives

$$\lambda\Bigl[\delta(0)\frac{1 - G(x/b_1)}{1 - G(x/b)} + \int_0^x g(z)\frac{(1 - G(z/b))(1 - G((x - z)/b_1))}{1 - G(x/b)}\,\mathrm{d}z\Bigr] - c(b_1)g(x) \ge 0\;.$$

The first term converges to zero. We first show that $g(x)$ is bounded. Suppose that this was not the case. Then there is a sequence $\{x_n\}$ tending to infinity such that $g(x_n) = \sup_{0\le x\le x_n} g(x)$. We divide the equation by $g(x_n)$. Consider the integral on $(0, x_n/2]$. By the rapid variation for n large enough we have $(1 - G(x/b_1))/(1 - G(x/b)) < \varepsilon$ for all $x \ge x_n/2$. Thus, we can estimate the integral

$$\begin{aligned}
&\int_0^{x_n/2} \frac{g(z)}{g(x_n)}\frac{(1 - G(z/b))(1 - G((x_n - z)/b_1))}{1 - G(x_n/b)}\,\mathrm{d}z\\
&\qquad\le \varepsilon\int_0^{x_n/2} \frac{(1 - G(z/b))(1 - G((x_n - z)/b))}{1 - G(x_n/b)}\,\mathrm{d}z\\
&\qquad= \varepsilon b\int_0^{x_n/(2b)} \frac{(1 - G(z))(1 - G(x_n/b - z))}{1 - G(x_n/b)}\,\mathrm{d}z\;.
\end{aligned}$$

The right-hand side converges to $\varepsilon b\mu$ as $n \to \infty$. This implies that

$$\lim_{n\to\infty}\int_0^{x_n/2} \frac{g(z)}{g(x_n)}\frac{(1 - G(z/b))(1 - G((x_n - z)/b_1))}{1 - G(x_n/b)}\,\mathrm{d}z = 0\;.$$

The second part of the integral can be estimated by

$$\int_0^{x_n/2} \frac{(1 - G((x_n - z)/b))(1 - G(z/b))}{1 - G(x_n/b)}\,\mathrm{d}z\;.$$

This converges to μb. Thus, letting $n \to \infty$ gives $\lambda\mu b - c(b) \ge 0$. This contradicts our assumption. Therefore, $g(x)$ is bounded. Let $g_0 = \limsup_{x\to\infty} g(x)$ and choose a sequence $\{x_n\}$ such that $g(x_n)$ tends to g_0. The considerations above show that in the limit we obtain $(\lambda\mu b - c(b))g_0 \ge 0$. Because

$\lambda\mu b - c(b) < 0$, we conclude that $g_0 = 0$. Integration over (x, ∞) proves the limit if $c(b) > \lambda\mu b$. If $c(b) \le \lambda\mu b$, then the result follows because

$$\lim_{x\to\infty} \frac{\int_x^\infty 1 - G(z/b_1)\,\mathrm{d}z}{\int_x^\infty 1 - G(z/b)\,\mathrm{d}z} = 0$$

for any $b_1 < b$, and we just proved that $\psi(x) = o(\int_x^\infty 1 - G(z/b_1)\,\mathrm{d}z)$.

Now let $\{x_n\}$ be a sequence such that $b(x_n)$ converges to some value b_2. Suppose that $b_2 > b_0$. Let $b_1 = (2b_0 + b_2)/3$ and $g(x) = -\psi'(x)/(1 - G(x/b_1))$. Then we just proved that $g(x)$ tends to zero. In particular, $c(b(x_n))g(x_n)$ tends to zero. For n large enough, $b(x_n) > b_3 = (b_0 + 2b_2)/3$. Thus,

$$\lim_{n\to\infty} \delta(0)\frac{1 - G(x_n/b(x_n))}{1 - G(x_n/b_1)} \ge \lim_{n\to\infty} \delta(0)\frac{1 - G(x_n/b_3)}{1 - G(x_n/b_1)} = \infty\,.$$

This means that the Hamilton–Jacobi–Bellman equation divided by $1 - G(x_n/b_1)$ would converge to infinity. This would be a contradiction. Therefore, $\limsup_{x\to\infty} b(x) \le b_0$. On the other hand, if $\limsup_{x\to\infty} b(x) < b_0$ held, then for x large enough the infinitesimal drift of the process would be negative. Thus, the process would return to some value x_0 again as long as ruin has not occurred. Recall that X_t has to tend to ∞ on $\{\tau = \infty\}$. Therefore, ruin would occur almost surely. This shows that $\limsup_{x\to\infty} b(x) = b_0$. □

The result shows that for large capital the insurer reduces the drift in order to make the "large" claims smaller. Comparing this with the regular variation case it seems paradoxical that the insurer would be more concerned about the smaller claims. The reason is that even with reinsurance the claim sizes are regularly varying with the same parameter α. Thus, reinsurance does not change the distribution tail of the claims but only the mean value. Hence, for a small b, ruin will occur anyway if one cannot get away from small x quickly enough. In the rapidly varying case, reinsurance makes the tail of the distribution smaller. Therefore, the insurer will try to get as much reinsurance as possible.

4.3.2 Excess of Loss Reinsurance

Here we use the notation of Section 4.2.2. Suppose that $M_Y(r) = \infty$ for all $r > 0$. This implies that $\theta(r; a) < \infty$ for some $r > 0$ if and only if $a \ne \infty$. In particular, R exists and $a^* < \infty$. Because $a(x) = \infty$ is possible for some x (this will be the case, in fact, for x small under quite mild conditions), the process M defined as in (4.10) is in general not a martingale because it may not be integrable. We therefore cannot apply the approach of Section 4.2.2 directly. But as we will see, a slight change yields the Cramér–Lundberg approximation.

Recall that $f(x) = \psi(x)\mathrm{e}^{Rx}$ and $g(x) = -\psi'(x)\mathrm{e}^{Rx}$ with R defined as the largest r solving $\inf_{a\ge 0}\theta(r; a) = 0$. We denote by a^* an argument for which $\theta(R, a^*) = 0$. We first observe that for $x \ge a^*$ it follows from (4.12) that

$$\int_0^{a^*} (g(x-z) - g(x)) e^{Rz} (1 - G(z)) \, dz \geq 0 \, .$$

In the proof of Lemma 4.18 it is not used that $C_- > 0$. That means that Lemma 4.18 remains true. Because $g(x)$ is bounded, $(1 - G(x))e^{Rx} \to \infty$ and $\delta(0) > 0$, (4.11) implies that $a(x) < \infty$ for x large enough. Recall that $a(x) \leq x$ whenever $a(x) < \infty$. Thus, for such an x

$$\lambda \int_0^{a(x)} g(x-z) e^{Rz} (1 - G(z)) \, dz - c(a(x)) g(x) = 0 \, . \tag{4.21}$$

Suppose now that $\liminf_{x\to\infty} g(x) = 0$. Then there exists a sequence x_n tending to ∞ such that $g(x_n) < g(x)$ for all $x < x_n$ and $g(x_n) \to 0$. We conclude that

$$\begin{aligned} 0 &= \lambda \int_0^{a(x_n)} g(x_n - z) e^{Rz} (1 - G(z)) \, dz - c(a(x_n)) g(x_n) \\ &> g(x_n) \Bigl[\lambda \int_0^{a(x_n)} e^{Rz} (1 - G(z)) \, dz - c(a(x_n)) \Bigr] \geq 0 \, , \end{aligned} \tag{4.22}$$

where we used the definition of the adjustment coefficient in the last inequality. This is a contradiction, and therefore $\liminf_{x\to\infty} g(x) > 0$. We next conclude that $\liminf_{x\to\infty} f(x) > 0$. Let $\eta = \inf_{x>0} g(x)$. Suppose that $f(x) < \eta/R$ for some x. Then $f'(x) = Rf(x) - g(x) < 0$. Thus, $f(x)$ is decreasing, and $f'(z) < Rf(x) - \eta$ for all $z > x$. But this contradicts that $f(x) > 0$ for all x. Therefore, $f(x) \geq \eta/R$ for all x. We are now ready to prove the Cramér–Lundberg approximation in the heavy-tailed case.

Theorem 4.40. *Suppose that $M_Y(r) = \infty$ for all $r > 0$. Then there exists $\zeta \in (0,1)$ such that $\lim_{x\to\infty} \psi(x) e^{Rx} = \zeta$. If, moreover, a^* is unique, then $\lim_{x\to\infty} a(x) = a^*$.*

Proof. From (4.21) and the fact that $\inf_{x>0} g(x) > 0$, we can conclude that $a(x)$ is bounded for large x. That is, $a(x) \wedge x$ is bounded by η, say. Consider now the following model $\tilde{X}$. We use the reinsurance strategy $a(x)$, but the claim sizes are $Y_i \wedge 2\eta$. On the interval $[0, \tau)$ the two processes coincide and therefore have the same ruin probability. In the proof of Theorem 4.19 we did not use that $a(x)$ is the optimal strategy. Because the conditions of Theorem 4.19 are fulfilled for X^*, we have proved the Cramér–Lundberg approximation. The proof of Corollary 4.20 applies and shows convergence of the strategy. □

Let us now again consider the small claims case. All we needed in the proof above was that $(1 - G(x))e^{Rx}$ tends to infinity. This shows that we have also proved the result in the case where $M_Y(R) = \infty$ and $(1 - G(x))e^{Rx} \to \infty$. Suppose now that $M_Y(R) = \infty$ and $\liminf_{x\to\infty} (1 - G(x)) e^{Rx} < \infty$. We then still have $a^* < \infty$. We still can conclude that $g(x)$ is bounded. We want to show that $g(x)$ is also bounded away from zero. Suppose that $\liminf_{x\to\infty} g(x) = 0$. Let $\{x_n\}$ be a sequence as in (4.22). Then we conclude from (4.22) that $a(x_n) = \infty$. Thus, (4.11) reads

$$\lambda\delta(0)(1 - G(x_n))\mathrm{e}^{Rx_n} + \lambda\int_0^{x_n} g(x_n - z)\mathrm{e}^{Rz}(1 - G(z))\,\mathrm{d}z - cg(x_n) = 0\,.$$

Because $g(x_n) > 0$, we can divide by $g(x_n)$:

$$\lambda\delta(0)\frac{(1 - G(x_n))\mathrm{e}^{Rx_n}}{g(x_n)} + \lambda\int_0^{x_n}\frac{g(x_n - z)}{g(x_n)}\mathrm{e}^{Rz}(1 - G(z))\,\mathrm{d}z - c = 0\,.$$

Because $-c$ is the only negative term and $g(x_n) < g(x_n - z)$, we conclude that

$$c \ge \int_0^{x_n}\frac{g(x_n - z)}{g(x_n)}\mathrm{e}^{Rz}(1 - G(z))\,\mathrm{d}z > \int_0^{x_n}\mathrm{e}^{Rz}(1 - G(z))\,\mathrm{d}z\,.$$

But the right-hand side tends to infinity as $n \to \infty$. This is a contradiction and $g(x)$ is bounded away from zero. As above, we can now conclude that $a(x) \wedge x$ is bounded. Hence, Theorem 4.40 holds true whenever $M_Y(R) = \infty$.

Note that from the proof of Corollary 4.20 we have $\lim_{x\to\infty} g(x) = R\zeta$. Therefore, $\psi'(x) \sim -R\zeta\mathrm{e}^{Rx}$, that is, the limit and derivative can be interchanged.

4.3.3 Optimal Investment

Because the claim sizes remain heavy-tailed if one invests, we cannot expect that the ruin probability is exponentially decreasing. In [64] it is shown that if a constant part of the portfolio is invested — even if the claim sizes are light-tailed — then the ruin probability decays like a power function. Thus, at least if the distribution tail of the claim size distribution is not regularly varying, one would expect that $A(x)/x$ tends to zero.

In a perturbed risk model when the claim sizes are subexponential, the asymptotics is given by (D.7). Considering a constant strategy $A_t = A$, then for x large enough, one has to choose A large in order to make the ruin probability small. Thus, we expect that $A(x)$ tends to infinity.

We start by showing that $\psi(x)$ cannot be bounded by an exponential function.

Lemma 4.41. *Suppose that $M_Y(r) = \infty$ for all $r > 0$. Then for any $r > 0$, one has $\limsup_{x\to\infty} \psi(x)\mathrm{e}^{rx} = \infty$.*

Proof. Suppose that $\psi(x) \le C\mathrm{e}^{-Rx}$ for some $R, C > 0$. Let $f(x) = \psi(x)\mathrm{e}^{Rx}$ and $g(x) = Rf(x) - f'(x) = -\psi'(x)\mathrm{e}^{Rx} > 0$. Then $f(x) \le C$ and we know that $g'(x) < Rg(x)$ because $\psi(x)$ is strictly convex. We first show that $g(x)$ is bounded. Suppose that for some $x > 1/(2R)$ we have, $g(x) \ge 6RC$. Because $g'(z) < Rg(z)$ we have $g'(z) < 6R^2C$ whenever $g(z) \le 6RC$. Therefore, $g(z) > 3RC$ for $z \in [x - 1/(2R), x]$. On this interval we have $f'(z) = Rf(z) - g(z) < -2RC$. Thus, $f(x) < f(x - 1/(2R)) - 2RC/(2R) \le 0$, which is a contradiction. Therefore, $g(x)$ is bounded.

Because $M_Y(R/2) = \infty$, we have $\lim_{x\to\infty}(1-G(x))e^{Rx} = \infty$. Thus, (4.17) implies that $\lim_{x\to\infty} Rg(x) - g'(x) = 0$. Choose $\varepsilon > 0$. For x large enough we have $g'(x) > Rg(x) - R\varepsilon$. Thus, $g(x) \le \varepsilon$ for x large enough; otherwise, $g(x)$ would ultimately be increasing and could not be bounded. We conclude that $g(x)$ converges to zero. From (4.17) we conclude that $g(x)^2/(Rg(x) - g'(x))$ converges to infinity. Then for $x \ge x_0$, we get $g(x)^2 > Rg(x) - g'(x)$ and $g(x) < R/2$. Because $g(x) > 0$, we have $(\log g(x))' > R - g(x) > R/2$. Thus, $\log g(x)$ tends to infinity, which is a contradiction. □

We next show that $A(x)$ is unbounded.

Proposition 4.42 (Hipp and Plum [97]). *Suppose that $M_Y(r) = \infty$ for all $r > 0$. Then $A(x)$ is unbounded, i.e., $\limsup_{x\to\infty} A(x) = \infty$.*

Proof. Suppose that $A(x) \le \zeta$ for some $\zeta > 0$. From the definition of $A(x)$, we get that

$$\frac{m}{\sigma^2}\frac{-\psi'(x)}{\psi''(x)} \le \zeta\,.$$

This can be written as

$$\frac{-\psi''(x)}{-\psi'(x)} \le -\frac{m}{\zeta\sigma^2}\,.$$

Integrating yields

$$\log(-\psi'(x)) - \log(-\psi'(0)) \le -\frac{m}{\zeta\sigma^2}x$$

or

$$-\psi'(x) \le -\psi'(0)e^{-mx/(\zeta\sigma^2)}\,.$$

Recall that $\psi'(0) = -\lambda\delta(0)/c > -\infty$ is well defined. Integration over (x_0, ∞) shows that

$$\psi(x_0) \le \frac{-\psi'(0)\zeta\sigma^2}{m}e^{-mx_0/(\zeta\sigma^2)}\,.$$

But this contradicts Lemma 4.41. □

We now make more explicit assumptions. Suppose that the hazard rate $\ell(y) = G'(y)/(1 - G(y))$ tends to zero and that $G(y) \in \mathcal{S}^*$. Recall that this means that $\mathbb{E}[Y] < \infty$ and

$$\lim_{x\to\infty}\int_0^x \frac{(1 - G(x-y))(1 - G(y))}{1 - G(x)}\,dy = 2\mathbb{E}[Y]\,.$$

These conditions are fulfilled for the most common heavy-tailed distributions like the log-normal distribution, the heavy-tailed Weibull distribution, or a distribution with a regularly varying distribution tail like the Pareto distribution.

Proposition 4.43. *Suppose that the assumptions above hold. Then we have convergence of the strategy:* $\lim_{x\to\infty} A(x) = \infty$.

Proof. Let $g(x) = -\psi'(x)/(1 - G(x))$. The Hamilton–Jacobi–Bellman equation (4.15) then reads

$$-\frac{m^2}{2\sigma^2}\frac{g(x)^2(1-G(x))^2}{g(x)G'(x) - g'(x)(1-G(x))} - cg(x)(1-G(x))$$
$$+ \lambda \int_0^x g(x-y)(1-G(x-y))(1-G(y))\,dy + \lambda(1-G(x))\delta(0) = 0\,.$$

Because $(1 - G(x)) > 0$, we can divide by $1 - G(x)$ and obtain

$$-\frac{m^2}{2\sigma^2}\frac{g(x)}{\ell(x) - \dfrac{g'(x)}{g(x)}} - cg(x)$$
$$+ \lambda \int_0^x g(x-y)\frac{(1-G(x-y))(1-G(y))}{1-G(x)}\,dy + \lambda\delta(0) = 0\,. \tag{4.23}$$

Because $\psi''(x) > 0$, we have $\ell(x) - g'(x)/g(x) > 0$. We first show that $g(x)$ is bounded. Assume that this is not the case. Let $\{x_n\}$ be a sequence tending to infinity such that $g(x_n) \ge g(z)$ for all $z \le x_n$ and $g(x_n) \to \infty$. Such a sequence exists if $g(x)$ is unbounded. Then $g'(x_n) \ge 0$. We can estimate

$$g(x_n)\Bigl(-\frac{m^2}{2\sigma^2\ell(x_n)} - c + \lambda \int_0^{x_n} \frac{(1-G(x_n-y))(1-G(y))}{1-G(x_n)}\,dy\Bigr) + \lambda\delta(0) \ge 0\,.$$

Because $G(x) \in \mathcal{S}^*$, the integral is bounded. Because the hazard rate $\ell(x)$ tends to zero, the left-hand side can be made arbitrarily small, in particular smaller than zero. This is a contradiction. Thus, $g(x)$ must be bounded. It follows that

$$\frac{g(x)}{\ell(x) - \dfrac{g'(x)}{g(x)}} \tag{4.24}$$

remains bounded. If $g(x)$ did not converge, then there must be a sequence $\{x_n\}$ tending to infinity such that $g(x_n) \to \limsup_{x\to\infty} g(x) > 0$ and $g'(x_n) = 0$. But then (4.24) would be unbounded. Thus, $g(x)$ converges. Suppose that the limit is not zero. Because (4.24) is bounded, this is only possible if $g'(x)/g(x) \le -\varepsilon$ for x large enough. Integration over $(x_0, z]$ yields $g(z) \le g(x_0)\mathrm{e}^{-\varepsilon(z-x_0)}$. Thus, $\lim_{x\to\infty} g(x) = 0$.

The Hamilton–Jacobi–Bellman equation (4.15) divided by $-\psi'(x)$ can be written as

$$-\tfrac{1}{2}mA(x) - c + \lambda \int_0^x \frac{\psi'(x-y)}{\psi'(x)}(1-G(y))\,\mathrm{d}y + \lambda\delta(0)/g(x) = 0\,.$$

The last term tends to infinity while the last term but one is positive. Thus, $A(x)$ also has to converge to infinity. □

Remark 4.44. Note that it follows from the proof that

$$\lim_{x\to\infty} \frac{\psi(x)}{\int_x^\infty (1-G(y))\,\mathrm{d}y} = 0\,.$$

Indeed, for x large enough, $-\psi'(x) < \varepsilon(1-G(x))$. Integration over (x,∞) shows that the limes superior is smaller than ε. Thus, $\psi(x)$ goes to zero at a faster rate than for any fixed investment strategy. The result could also be obtained from (D.7) by noting that the constant can be made arbitrarily small by choosing A large enough. ■

The integral in (4.23) can be written as

$$\int_0^{x/2} (g(y) + g(x-y))\frac{(1-G(x-y))(1-G(y))}{1-G(x)}\,\mathrm{d}y\,.$$

Choose $\varepsilon > 0$. There is x_0 such that $g(x) < \varepsilon$ for all $x \ge x_0$. Thus,

$$\begin{aligned}\int_0^{x/2} g(x-y)&\frac{(1-G(x-y))(1-G(y))}{1-G(x)}\,\mathrm{d}y \\ &< \varepsilon \int_0^{x/2} \frac{(1-G(x-y))(1-G(y))}{1-G(x)}\,\mathrm{d}y\end{aligned}$$

for $x > 2x_0$. Therefore,

$$\lim_{x\to\infty} \int_0^{x/2} g(x-y)\frac{(1-G(x-y))(1-G(y))}{1-G(x)}\,\mathrm{d}y = 0\,.$$

We also have

$$\begin{aligned}\int_{x_0}^{x/2} g(y)&\frac{(1-G(x-y))(1-G(y))}{1-G(x)}\,\mathrm{d}y \\ &< \varepsilon \int_{x_0}^{x/2} \frac{(1-G(x-y))(1-G(y))}{1-G(x)}\,\mathrm{d}y\,.\end{aligned}$$

Because by (D.6)

$$\lim_{x\to\infty} \int_0^{x_0} g(y)\frac{(1-G(x-y))(1-G(y))}{1-G(x)}\,\mathrm{d}y = \int_0^{x_0} g(y)(1-G(y))\,\mathrm{d}y\,,$$

it follows that

$$\lim_{x\to\infty} \int_0^x g(x-y) \frac{(1-G(x-y))(1-G(y))}{1-G(x)} \,dy = \int_0^\infty g(y)(1-G(y)) \,dy = \psi(0) .$$

From (4.23) we thus conclude that

$$\lim_{x\to\infty} \frac{\psi'(x)^2}{\psi''(x)(1-G(x))} = \frac{2\sigma^2\lambda}{m^2} =: \kappa . \tag{4.25}$$

From this limit we can find an asymptotic expression for $\psi(x)$.

Theorem 4.45. *Suppose that $G(y) \in \mathcal{S}^*$. Then*

$$\psi(x) \sim \kappa \int_x^\infty \frac{1}{\displaystyle\int_0^y \frac{1}{1-G(z)} \,dz} \,dy .$$

Proof. Let us first assume that $\ell(x) \to 0$. Define

$$\gamma(x) = \kappa \frac{\psi''(x)(1-G(x))}{\psi'(x)^2} .$$

By (4.25) it follows that $\gamma(x)$ converges to 1 as $x \to \infty$. Then

$$-\Bigl(\frac{1}{\psi'(x)}\Bigr)' = \frac{\gamma(x)}{\kappa(1-G(x))}$$

Integration yields

$$-\frac{1}{\psi'(x)} = -\frac{1}{\psi'(0)} + \frac{1}{\kappa} \int_0^x \frac{\gamma(z)}{1-G(z)} \,dz . \tag{4.26}$$

To simplify the notation, let $a = -\kappa/\psi'(0)$. Note that $\psi'(0) < 0$ because $\psi(x)$ is a convex function. Integrating once more yields

$$\psi(x) = \kappa \int_x^\infty \frac{1}{a + \displaystyle\int_0^y \frac{\gamma(z)}{1-G(z)} \,dz} \,dy .$$

Let $\bar\gamma = \sup_{x\ge 0} \gamma(z)$. This is finite because $\gamma(z)$ is continuous and converges to 1. Thus,

$$\psi(x) \ge \kappa \int_x^\infty \frac{1}{a + \displaystyle\int_0^y \frac{\bar\gamma}{1-G(z)} \,dz} \,dy .$$

We conclude that

$$\int_x^\infty \frac{1}{a/(2\bar\gamma) + \frac{1}{2}\int_0^y \frac{1}{1-G(z)}\,\mathrm{d}z}\,\mathrm{d}y$$

tends to zero. Because $\int_0^y 1/(1-G(z))\,\mathrm{d}z$ tends to infinity, we find that for x large enough, $\frac{1}{2}\int_0^y (1-G(z))^{-1}\,\mathrm{d}z > a/(2\bar\gamma)$ for $y \ge x$. Thus,

$$\int_x^\infty \frac{1}{\int_0^y \frac{1}{1-G(z)}\,\mathrm{d}z}\,\mathrm{d}y$$

tends to zero as $x \to \infty$. We therefore can use l'Hôpital's rule twice to obtain

$$\lim_{x\to\infty} \frac{\psi(x)}{\kappa \int_x^\infty \frac{1}{\int_0^y \frac{1}{1-G(z)}\,\mathrm{d}z}\,\mathrm{d}y} = \lim_{x\to\infty} \frac{\int_0^x \frac{1}{1-G(z)}\,\mathrm{d}z}{a + \int_0^x \frac{\gamma(z)}{1-G(z)}\,\mathrm{d}z} = \lim_{x\to\infty} \frac{1}{\gamma(x)} = 1\,.$$

This proves the result if $\ell(x)$ tends to zero.

Now let $G(y)$ be an arbitrary distribution from $\mathcal{S}^*$. There is a distribution function $\widetilde{G}(y)$ that is tail-equivalent, i.e., $(1-G(y))/(1-\widetilde{G}(y)) \to 1$, and $\tilde\ell(x)$ tends to zero; see [152, p. 57]. We can assume that $\tilde\ell(x)$ is bounded. Let $\varepsilon > 0$. Construct $G_1(x)$ in the following way. Let $x_0 = \sup\{x : (1+\varepsilon)(1-\widetilde{G}(x)) \le 1 - G(x)\} \vee \sup\{x : (1+\varepsilon)(1-\widetilde{G}(x)) \ge 1\}$. We let $1-G_1(x) = (1+\varepsilon)(1-\widetilde{G}(x))$ for $x \ge 2x_0$. There exists a continuously differentiable and decreasing extension to $1-G_1(x)$ such that $G_1(x) = 0$ for $x \le x_0$, $G_1(x) \le G(x)$, and $G_1'(2x_0) = (1+\varepsilon)\widetilde{G}'(2x_0)$. As a continuous function, $G_1'(x)$ is bounded on $[x_0, 2x_0]$. Thus, the hazard rate $\ell_1(x)$ is bounded and tends to zero as $x \to \infty$. In particular, $G_1(x)$ has a bounded density and the corresponding optimal ruin probability $\psi_1(x)$ fulfils the Hamilton–Jacobi–Bellman equation. Theorem 4.45 holds therefore for $\psi_1(x)$. From the construction it follows that $1 - G_1(x) \ge (1-G(x))$, that is, the claims with distribution function $G_1(x)$ are stochastically larger than the claims with distribution function $G(x)$. This yields $\psi_1(x) \ge \psi(x)$. We also have $1 - G_1(x) \sim (1+\varepsilon)(1-\widetilde{G}(x)) \sim (1+\varepsilon)(1-G(x))$. L'Hôpital's rule and Remark 4.46 below show that

$$\limsup_{x\to\infty} \frac{\psi(x)}{\kappa \displaystyle\int_x^\infty \frac{1}{\displaystyle\int_0^y \frac{1}{1-G(z)}\,\mathrm{d}z}\,\mathrm{d}y} \le \lim_{x\to\infty} \frac{\psi_1(x)}{\kappa \displaystyle\int_x^\infty \frac{1}{\displaystyle\int_0^y \frac{1}{1-G(z)}\,\mathrm{d}z}\,\mathrm{d}y}$$

$$= \lim_{x\to\infty} \frac{(1+\varepsilon)\psi_1(x)}{\kappa \displaystyle\int_x^\infty \frac{1}{\displaystyle\int_0^y \frac{1}{1-G_1(z)}\,\mathrm{d}z}\,\mathrm{d}y}$$

$$= 1+\varepsilon\,.$$

A similar argument yields

$$\liminf_{x\to\infty} \frac{\psi(x)}{\kappa \displaystyle\int_x^\infty \frac{1}{\displaystyle\int_0^y \frac{1}{1-G(z)}\,\mathrm{d}z}\,\mathrm{d}y} \ge 1-\varepsilon\,.$$

Because ε is arbitrary, the result follows. □

Remark 4.46. Note that

$$\int_x^\infty \frac{1}{\displaystyle\int_0^y \frac{1}{1-G(z)}\,\mathrm{d}z}\,\mathrm{d}y \sim \int_x^\infty \frac{1}{\displaystyle\int_{x_0}^y \frac{1}{1-G(z)}\,\mathrm{d}z}\,\mathrm{d}y$$

for any $x_0 \ge 0$. Thus, the asymptotic behaviour of $\psi(x)$ is dependent on the tail of the distribution only and not on the whole distribution function. ■

Let us now turn to distribution functions with a regularly varying tail with index $-\alpha$ and $\mu < \infty$. Thus, $\alpha \ge 1$. Then we find the following simpler asymptotic behaviour.

Corollary 4.47 (Gaier and Grandits [70]). *Suppose that* $1-G(x)$ *is regularly varying with index* $-\alpha$. *Then*

$$\lim_{x\to\infty} \frac{\psi(x)}{1-G(x)} = \frac{2\lambda\sigma^2(\alpha+1)}{m^2\alpha}\,.$$

In particular, $\psi(x)$ *is regularly varying with index* $-\alpha$.

Proof. Clearly, $1/(1-G(x))$ is regularly varying with index α. From Karamata's theorem [58, p. 281], we conclude that

$$\int_0^x \frac{1}{1-G(z)}\,\mathrm{d}z \sim \frac{x}{(\alpha+1)(1-G(x))}$$

is regularly varying with index $\alpha+1$. Thus, $1/\int_0^x \frac{1}{1-G(z)}\,\mathrm{d}z$ is regularly varying with index $-\alpha-1$. Again, Karamata's theorem gives

$$\psi(x) \sim \kappa \int_x^\infty \frac{1}{\displaystyle\int_0^y \frac{1}{1-G(z)}\,\mathrm{d}z}\,\mathrm{d}y \sim \frac{\kappa x}{\alpha \int_0^x \frac{1}{1-G(z)}\,\mathrm{d}z} \sim \frac{\kappa(\alpha+1)}{\alpha}(1-G(x))\,.$$

This is the assertion. □

For not regularly varying distribution tails, we need an additional condition that is fulfilled, for example, for the log-normal or the heavy-tailed Weibull distributions. As shown in [83], this additional condition is quite weak.

A distribution function $G(y)$ is said to be in the *maximal domain of attraction* $\mathrm{MDA}(F(y))$ of a distribution function $F(y)$ if there are numbers a_n and b_n such that $\lim_{n\to\infty}\{G(a_n + b_n x)\}^n = F(x)$. That is, the maximum of n iid random variables with distribution function $G(y)$ properly normed tends weakly to the distribution function $F(y)$. There are only three types of limiting functions $F(y)$ possible. If $G(x) < 1$ for all x, there are only two types left: the Fréchet distribution $\exp\{-x^{-\alpha}\}$ and the Gumbel distribution $\exp\{-\mathrm{e}^{-x}\}$. $\mathrm{MDA}(\exp\{-x^{-\alpha}\})$ consists of the distributions with a regularly varying tail with index $-\alpha$. Because we already treated this case in Corollary 4.47 we suppose that $G(y) \in \mathrm{MDA}(\exp\{-\mathrm{e}^{-x}\})$. According to [15], $G(y)$ has the representation

$$1 - G(x) = c(x)\exp\Bigl\{-\int_0^x a(z)\,\mathrm{d}z\Bigr\}\,, \tag{4.27}$$

where $c(x)$ converges to 1 and $a(x) > 0$ is absolutely continuous such that the density of $1/a(x)$ tends to zero. The function $a(x)$ is called an *auxiliary function*. Because we consider subexponential distribution functions, we also get that $a(x)$ tends to zero. Indeed, from

$$1 = \lim_{x\to\infty} \frac{1-G(x-y)}{1-G(x)} = \lim_{x\to\infty} \frac{c(x-y)}{c(x)} \exp\Bigl\{\int_{x-y}^x a(z)\,\mathrm{d}z\Bigr\},$$

we conclude that $\int_{x-y}^x a(z)\,\mathrm{d}z$ tends to zero. From $(1/a(x))'$ tending to zero, we conclude that $a(x)$ tends to zero.

Corollary 4.48. *Suppose that* $G(x) \in \mathcal{S}^* \cap \mathrm{MDA}(\exp\{-\mathrm{e}^{-x}\})$. *Then*

$$\lim_{x\to\infty} \frac{\psi(x)}{1-G(x)} = \frac{2\sigma^2\lambda}{m^2}\,.$$

Remark 4.49. Using the representation (4.27), we have by l'Hôpital's rule $1/(xa(x)) = (1/a(x))/x \sim (1/a)'(x) \to 0$. We then find for all $t > 1$ that

$$\begin{aligned}\int_x^{tx} a(z)\,\mathrm{d}z &= x\int_1^t a(xz)\,\mathrm{d}z = x\int_1^t \frac{1}{1/a(x) + \int_x^{xz}(1/a)'(y)\,\mathrm{d}y}\,\mathrm{d}z\\ &= x\int_1^t \frac{1}{1/a(x) + x\int_1^z (1/a)'(xy)\,\mathrm{d}y}\,\mathrm{d}z\\ &= \int_1^t \frac{1}{1/(xa(x)) + \int_1^z (1/a)'(xy)\,\mathrm{d}y}\,\mathrm{d}z\,.\end{aligned}$$

It follows that $\int_x^{tx} a(z)\,\mathrm{d}z$ tends to infinity, and therefore

$$\frac{1-G(tx)}{1-G(x)} = \frac{c(tx)}{c(x)} \exp\Big\{-\int_x^{tx} a(z)\,\mathrm{d}z\Big\}$$

tends to zero. That is, $G(x)$ is of rapid variation. Thus, Corollary 4.47 remains true with $\alpha = \infty$. ■

Proof. As in the proof of Theorem 4.45, it is enough to prove the result for some tail-equivalent distribution. We can therefore assume that $c(x) = 1$ and $a(x) = \ell(x) > 0$. We have $G'(x) = \ell(x)(1 - G(x))$ and, in the sense of absolutely continuous functions, $G''(x) = [\ell'(x) - \ell(x)^2](1 - G(x))$. We find that

$$\lim_{x\to\infty} -\frac{G''(x)(1-G(x))}{G'(x)^2} = \lim_{x\to\infty} \frac{\ell(x)^2 - \ell'(x)}{\ell(x)^2} = 1 + \lim_{x\to\infty}\Big(\frac{1}{\ell(x)}\Big)' = 1\,.$$

Note that $G'(x)$ tends to zero as $x \to \infty$. We therefore conclude from l'Hôpital's rule that

$$\begin{aligned}\lim_{x\to\infty} \frac{\psi(x)}{\kappa(1-G(x))} &= \lim_{x\to\infty} \frac{\displaystyle\int_x^\infty \frac{1}{\int_0^y 1/(1-G(z))\,\mathrm{d}z}\,\mathrm{d}y}{1-G(x)} \\ &= \lim_{x\to\infty} \frac{1/G'(x)}{\int_0^x 1/(1-G(y))\,\mathrm{d}y} \\ &= \lim_{x\to\infty} -\frac{G''(x)(1-G(x))}{G'(x)^2} = 1\,.\end{aligned}$$

This proves the result. □

Let us now assume that $\psi(x) \sim C(1-G(x))$ for some $C > 0$. But now we allow $G(y) \in \mathcal{S}$ to be just a subexponential distribution. That is, $G(x)$ fulfils

$$\lim_{x\to\infty} \int_0^x \frac{1-G(x-y)}{1-G(x)}\,\mathrm{d}G(y) = 1\,.$$

Because $-\psi'(x)$ is decreasing, we still get by (D.6)

$$\limsup_{x\to\infty} \frac{-\psi'(x)}{1-G(x)} \le \lim_{x\to\infty} \frac{\psi(x-1)-\psi(x)}{1-G(x)} = 0\,.$$

Let us consider the integral in (4.14) divided by $1 - G(x)$ and write $\psi(x) = f(x)C(1-G(x))$. Then $f(x)$ tends to 1. There is an x_0 such that for all $x > x_0$ one has $|f(x) - 1| < \varepsilon$. Then

$$\begin{aligned}&\limsup_{x\to\infty} \int_{x-x_0}^x f(x-y)\frac{1-G(x-y)}{1-G(x)}\,\mathrm{d}G(y) \\ &\qquad\le \sup_{0\le y\le x_0} f(y) \lim_{x\to\infty}\Big[\frac{1-G(x-x_0)}{1-G(x)} - 1\Big] = 0\,.\end{aligned}$$

Similarly,

$$\limsup_{x\to\infty} \int_{x-x_0}^{x} \frac{1-G(x-y)}{1-G(x)} \,\mathrm{d}G(y) = 0\,.$$

Moreover,

$$\begin{aligned}\limsup_{x\to\infty} \int_0^{x-x_0} &|f(x-y)-1| \frac{1-G(x-y)}{1-G(x)} \,\mathrm{d}G(y) \\ &\le \varepsilon \lim_{x\to\infty} \int_0^{x} \frac{1-G(x-y)}{1-G(x)} \,\mathrm{d}G(y) = \varepsilon\,;\end{aligned}$$

see (D.5). Because ε is arbitrary, we obtain

$$\lim_{x\to\infty} \int_0^{x} f(x-y) \frac{1-G(x-y)}{1-G(x)} \,\mathrm{d}G(y) = 1\,.$$

From (4.14) we now see that (4.25) holds. From the proof of Proposition 4.43, we conclude that $A(x)$ tends to infinity.

We next want to see which limits C are possible. Suppose for the moment that $C \neq \kappa$. Let $\eta = C/\kappa - 1$. Then $(\psi'(x)/\psi(x))' \sim \eta(\psi'(x)/\psi(x))^2$. If $\eta < 0$, then $\psi'(x)/\psi(x)$ would be decreasing for x large enough. Hence, $\psi'(x)/\psi(x)$ would converge to $-\infty$. In particular, $\psi'(x) < -\psi(x)$ for x large enough. But then $\psi(x)$ would be exponentially decreasing, which is not possible due to Lemma 4.41. We therefore have $\eta \ge 0$. If $\eta > 0$, then the function $\gamma(x) = (\psi''(x)\psi(x)/\psi'(x)^2 - 1)/\eta$ tends to 1 as $x \to \infty$. Integration yields

$$\frac{\psi(x)}{-\psi'(x)} = \frac{\psi(0)}{-\psi'(0)} + \eta \int_0^x \gamma(z) \,\mathrm{d}z\,.$$

Dividing the equation by x and taking the limit shows that

$$\lim_{x\to\infty} \frac{\psi(x)}{-x\psi'(x)} = \eta \lim_{x\to\infty} \frac{1}{x} \int_0^x \gamma(z) \,\mathrm{d}z = \eta\,.$$

From Karamata's theorem [58, p. 281], we conclude that $\psi(x)$ and therefore also that $1-G(x)$ is regularly varying with index $1/\eta$. Because we assume a finite mean, we are in the situation of Corollary 4.47, which we have already treated. Thus, either $1-G(x)$ is regularly varying or $C = \kappa$. If $C = \kappa$, we conclude from (4.25) that

$$\lim_{x\to\infty} \frac{\psi(x)\psi''(x)}{\psi'(x)^2} = 1\,.$$

Therefore, we have the representation

$$\psi(x) = \psi(0) \exp\Bigl\{-\int_0^x a(z) \,\mathrm{d}z\Bigr\}\,,$$

where $a(x) = [-\log(\psi(x))]'$. The derivative of $1/a(x)$ tends to zero. Because $\psi(x) \sim \kappa(1 - G(x))$, it follows that $G(y) \in \mathrm{MDA}(\exp\{-\mathrm{e}^{-x}\})$. We see that $\psi(x) \sim C(1 - G(x))$ is only possible under maximal domain of attraction conditions.

If $\psi(x) \sim C(1 - G(x))$ and $G(x) \in \mathcal{S}$, or if $G(y) \in \mathcal{S}^*$ and $\ell(x) \to 0$, then we have seen that (4.25) holds. Integration shows that

$$\psi'(x) \sim \kappa\Bigl(\int_0^x \frac{1}{1 - G(z)}\,\mathrm{d}z\Bigr)^{-1} .$$

Thus, we can interchange the limit and derivative. Moreover,

$$\psi''(x) \sim \frac{\psi'(x)^2}{\kappa(1 - G(x))} \sim \frac{\frac{\kappa}{1 - G(x)}}{(\int_0^x \frac{1}{1-G(z)}\,\mathrm{d}z)^2} .$$

Thus the limit and second derivative can also be interchanged.

We now consider the asymptotic behaviour of $A(x)$.

Proposition 4.50. *Suppose that $G(y) \in \mathcal{S}$ and $\psi(x) \sim C(1 - G(x))$ for some $C > 0$ or that $G(y) \in \mathcal{S}^*$ and $\ell(x) \to 0$. Then*

$$\lim_{x\to\infty} -\psi'(x) \int_0^x \frac{1}{1 - G(z)}\,\mathrm{d}z = \kappa$$

and

$$A(x) \sim \frac{m}{\sigma^2} \int_0^x \frac{1 - G(x)}{1 - G(z)}\,\mathrm{d}z .$$

Proof. We have proved the first assertion above. By the definition of $A(x)$,

$$A(x) = -\frac{m\psi'(x)}{\sigma^2\psi''(x)} = \frac{m}{\sigma^2}\frac{\psi'(x)^2}{\psi''(x)(1 - G(x))}\frac{1 - G(x)}{-\psi'(x)} \sim \frac{m}{\sigma^2}\int_0^x \frac{1 - G(x)}{1 - G(z)}\,\mathrm{d}z .$$

This proves the result. □

Remark 4.51. We have seen that we can interchange the asymptotics and the first two derivatives. It should be noted that this is not necessarily the case for the representation of the asymptotics $C(1 - G(x))$. ■

Let us discuss the behaviour of $A(x)$ in the regularly varying case and for $G(y) \in \mathrm{MDA}(\exp\{-\mathrm{e}^{-x}\})\}$. We start with the regularly varying case.

Corollary 4.52 (Gaier and Grandits [70]). *Suppose that $G(x)$ is regularly varying with index $-\alpha$. Then*

$$\lim_{x\to\infty} \frac{A(x)}{x} = \frac{m}{\sigma^2(\alpha + 1)} ,$$

that is, for large x, a fraction of the surplus is invested.

Proof. This follows directly from Proposition 4.50 and Karamata's theorem [58, p. 281]. □

For large x the fraction of the wealth invested is approximately $m/(\sigma^2(\alpha+1))$. It turns out that asymptotically the strategy $A_t = mX_t/(\sigma^2(\alpha+1))$ is as good as the optimal strategy. But for moderate capital it will not be a good idea to use the latter strategy; compare it with Figure 2.4 in the case of proportional reinsurance.

Proposition 4.53. *Suppose that $G(x)$ has a regularly varying tail with index $-\alpha$. Let $\psi(x)$ be the ruin probability for the strategy $A_t = mX_t/(\sigma^2(\alpha+1))$. Then*

$$\lim_{x\to\infty} \frac{\psi(x)}{1-G(x)} = \frac{2\lambda\sigma^2(\alpha+1)}{m^2\alpha} .$$

Proof. By tail-equivalence it is enough to consider a distribution function $G(x)$ whose hazard rate fulfils $\lim_{x\to\infty} x\ell(x) = \alpha$. This is because $\alpha\int_x^\infty y^{-1}(1-G(y))\,\mathrm{d}y \sim 1-G(x)$. Let $\xi = 2\lambda\sigma^2(\alpha+1)/(m^2\alpha)$. The constant fraction strategy is then $A_t = 2\lambda X_t/(m\xi\alpha)$. Standard techniques give that $\psi(x)$ solves the equation

$$\frac{2\lambda^2\sigma^2}{m^2\xi^2\alpha^2}x^2\psi''(x) + \Bigl(c+\frac{2\lambda}{\xi\alpha}x\Bigr)\psi'(x) + \lambda\Bigl[\delta(0)(1-G(x)) - \int_0^x \psi'(x-y)(1-G(y))\,\mathrm{d}y\Bigr] = 0 .$$

Let $g(x) = -\psi'(x)/(1-G(x))$. Then $\psi''(x) = (\ell(x)g(x) - g'(x))(1-G(x))$, and the equation can be written as

$$\frac{1}{\xi\alpha(1+\alpha)}x^2(\ell(x)g(x)-g'(x)) - \Bigl(\frac{c}{\lambda}+\frac{2}{\xi\alpha}x\Bigr)g(x) + \delta(0) + \int_0^x g(x-y)\frac{(1-G(x-y))(1-G(y))}{1-G(x)}\,\mathrm{d}y = 0 .$$

Similarly as in the proof of Proposition 4.43, we conclude that $g(x)\to 0$. In the same way as in the proof of (4.25), it follows that

$$\lim_{x\to\infty}\int_0^x g(x-y)\frac{(1-G(x-y))(1-G(y))}{1-G(x)}\,\mathrm{d}y = \psi(0) .$$

Thus, we conclude that

$$\lim_{x\to\infty}\frac{1}{\xi\alpha(1+\alpha)}x^2(\ell(x)g(x)-g'(x)) - \frac{2}{\xi\alpha}xg(x) = -1 .$$

This is equivalent to

$$\lim_{x\to\infty} x^2g'(x) + (2+\alpha)xg(x) = \xi\alpha(1+\alpha) .$$

Let $\gamma(x) = x^2 g'(x) + (2+\alpha) x g(x)$. Then

$$g(x) = x^{-\alpha-2} \int_0^x \gamma(y) y^\alpha \, \mathrm{d}y \, .$$

This is, $g(x)$ is regularly varying with index -1. Thus, $-\psi'(x)$ is regularly varying with index $-\alpha - 1$. Hence,

$$\lim_{x\to\infty} \frac{\psi(x)}{1 - G(x)} = \lim_{x\to\infty} \frac{\psi(x)}{-x\psi'(x)} x g(x) = \frac{1}{\alpha} \frac{\xi\alpha(1+\alpha)}{\alpha+1} = \xi \, .$$

□

Corollary 4.54. *Suppose that $G(y) \in \mathcal{S}^* \cap \mathrm{MDA}(\exp\{-\mathrm{e}^{-x}\})$. Then*

$$\lim_{x\to\infty} A(x) a(x) = \frac{m}{\sigma^2} \, ,$$

where $a(x)$ is the auxiliary function from the representation (4.27). *In particular, $A(x)/x$ tends to zero.*

Proof. We need to find the asymptotic behaviour of

$$\int_0^x \frac{c(x)}{c(y)} \exp\Bigl\{ - \int_y^x a(z) \, \mathrm{d}z \Bigr\} \, \mathrm{d}y \, .$$

Because

$$\int_0^{x_0} \frac{c(x)}{c(y)} \exp\Bigl\{ - \int_y^x a(z) \, \mathrm{d}z \Bigr\} \, \mathrm{d}y$$

tends to zero as $x \to \infty$ for each fixed x_0 and $c(x)$ converges to 1 as $x \to \infty$, it is enough to consider the case $c(x) = 1$. Thus, by l'Hôpital's rule, we find that

$$\lim_{x\to\infty} \frac{\int_0^x \exp\{\int_0^y a(z) \, \mathrm{d}z\} \, \mathrm{d}y}{a(x)^{-1} \exp\{\int_0^x a(z) \, \mathrm{d}z\}} = \lim_{x\to\infty} \frac{1}{1 + (1/a(x))'} = 1$$

because $(1/a(x))'$ tends to zero. We now get

$$\lim_{x\to\infty} \frac{A(x)}{x} = \frac{m}{\sigma^2} \lim_{x\to\infty} \frac{1}{x} \int_0^x (1/a(z))' \, \mathrm{d}z = 0 \, .$$

This proves the result. □

Remarks 4.55. i) That $A(x)/x$ is converging to zero is not surprising. From [64] one knows that if a constant fraction of the portfolio is invested, then the ruin probability decays like a power function. Because $\psi(x) \sim \kappa(1 - G(x))$ decays faster than a power function, one expects indeed that $A(x)/x$ tends to zero.

ii) Similarly as in the proof of Proposition 4.53, one can show that the strategy $A_t = m/(\sigma^2 a(X_t))$ has the same asymptotic behaviour $\psi(x) \sim \kappa(1 - G(x))$ as the optimal strategy. ■

The Weibull distribution $G(y) = 1 - \exp\{-cx^\tau\}$ has the hazard rate $\ell(x) = c\tau x^{\tau-1}$. The derivative of $1/\ell(x)$ is $(1-\tau)x^{-\tau}/(c\tau)$, which tends to zero. Therefore, $a(x) = \ell(x)$ and $A(x) \sim mx^{1-\tau}/(c\tau\sigma^2)$. Thus, $A(x)/x$ quickly goes to zero if τ is close to 1 (close to the exponential distribution) and slowly goes to zero if τ is close to zero (close to the Pareto distribution).

For the log-normal distribution, the calculations are more complicated. Let $\Phi(x)$ denote the distribution function of the standard normal distribution. Let $G(y) = \Phi((\log y - a)/b)$. One readily sees that $(1/\ell(y))'$ tends to zero exactly if $G''(y)(1 - G(y))/G'(y)^2$ tends to -1. We find that

$$\frac{G''(y)(1-G(y))}{G'(y)^2} = \frac{[\Phi''((\log y - a)/b) - b\Phi'(\log y - a)/b)]\Phi(-(\log y - a)/b)}{[\Phi'((\log y - a)/b)]^2}\;,$$

where we multiplied the numerator and denominator by b^2y^2. From the well-known relation $x\Phi(-x) \sim \Phi'(x)$, we see that $\Phi(-x)/\Phi'(x)$ tends to zero. From $-\Phi''(x) \sim x\Phi'(x)$, we find that $(1/\ell(x))'$ tends to zero. Thus, $a(x) = \ell(x)$ and

$$A(x) \sim \frac{m}{\sigma^2}\frac{bx\Phi(-(\log x - a)/b)}{\Phi'((\log x - a)/b)} \sim \frac{bmx}{\sigma^2(\log x - a)/b} \sim \frac{b^2m}{\sigma^2}\frac{x}{\log x}\;.$$

We see that $A(x)/x$ goes to zero very slowly.

4.3.4 Optimal Proportional Reinsurance and Investment

If reinsurance and investment are possible, we always get an exponential decay of the ruin probability. Indeed, one could reinsure the whole portfolio and then get a positive drift of the process just by investing. For simplicity we will only treat the case of proportional reinsurance in this section. We consider heavy-tailed claim size distributions in the sense that $M_Y(r) = \infty$ for all $r > 0$. We use the notation of Section 4.2.4.

Since for a constant strategy $b > 0$ we cannot get an exponential decay of the ruin probability, we must have $b^* = 0$. From (4.6) we find that

$$R = -\frac{m^2}{2\sigma^2 c(0)}\;, \qquad A^* = \frac{m}{\sigma^2 R} = -\frac{2c(0)}{m}\;.$$

We obtain the upper bound.

Proposition 4.56. *The ruin probability is bounded by* $\psi(x) \le \psi(0)\mathrm{e}^{-Rx}$, *where the strict inequality holds if* $x \neq 0$. *Moreover,* R *is the Lundberg coefficient in the sense that* $\lim_{x\to\infty} \psi(x)\mathrm{e}^{rx} = \infty$ *for all* $r > R$.

Proof. Choosing the constant strategy $(A_t, b_t) = (A^*, 0)$ until the point zero is reached for the first time and thereafter the optimal strategy yields the ruin probability $\psi(0)e^{-Rx}$; see (A.2). Because this is not the optimal strategy whenever $x \neq 0$ [we know that $\lim_{x\downarrow 0} A(x) = 0$], we get the strict inequality.

Let $\varepsilon > 0$ and choose $b > 0$ such that $c(b) < 0$ and

$$-\frac{m^2}{2\sigma^2 c(b)} < -\frac{m^2}{2\sigma^2 c(0)} + \frac{\varepsilon}{2} = R + \frac{\varepsilon}{2}\,.$$

From Corollary 4.59 ahead we find that $b(x)$ converges to zero. Thus, there is an x_0 such that for $x \geq x_0$, $b(x) \leq b$. Consider now the following process X^*. If $X_t^* < x_0$, then the optimal strategy is used. If $X_t^* \geq x_0$, then the full reinsurance $b_t = 0$ is chosen, but only the premium $c - c(b)$ is paid, i.e., the premium left to the insurer is $c(b)$. The amount invested is $-2c(b)/m$, i.e., the optimal amount of full reinsurance is chosen. Clearly, the corresponding ruin probability fulfils $\psi^*(x) \leq \psi(x)$. Above the level x_0 the process $\{X_t^*\}$ is a diffusion process. Since ruin can only occur by passing the level x_0 whenever $x > x_0$, we find, using (A.2), that

$$\psi^*(x) = \psi^*(x_0) \exp\Bigl\{\frac{m^2}{2\sigma^2 c(b)}(x - x_0)\Bigr\}\,.$$

Therefore, we have that $\psi^*(x)\exp\{(R+\varepsilon)x\}$ tends to infinity as $x \to \infty$. Thus, the same property holds for $\psi(x)$. □

As in Section 4.2.4, we define the functions $f(x) = \psi(x)e^{Rx}$ and $g(x) = Rf(x) - f'(x) = -\psi'(x)e^{Rx}$. The function $g(x)$ then fulfils (4.19).

Theorem 4.57. *Suppose that $M_Y(r) = \infty$ for all $r > 0$. Then there exists $\zeta \in [0, \psi(0))$ such that $\lim_{x\to\infty} \psi(x)e^{Rx} = \zeta$. Moreover, the functions $f(x)$ and $g(x)$ are decreasing.*

Remark 4.58. In general it is hard to show that $\zeta > 0$. However, according to Proposition 4.56, R is the best possible exponent. ■

Proof. Replacing $b(x)$ by $b^* = 0$ in (4.19), we find that

$$-\frac{m^2}{2\sigma^2}\frac{g(x)^2}{Rg(x) - g'(x)} - c(0)g(x) \geq 0\,.$$

Because $g(x) > 0$, we can divide by $g(x)/R$ and obtain

$$0 \leq -\frac{m^2}{2\sigma^2}\frac{Rg(x)}{Rg(x) - g'(x)} - Rc(0) = -\frac{m^2}{2\sigma^2}\frac{g'(x)}{Rg(x) - g'(x)}\,,$$

where we used the definition of R in the last equation. Thus, $g'(x) \leq 0$, and $g(x)$ is decreasing. Suppose that $f'(x) > 0$ for some x. Then $f'(x) = Rf(x) -$

$g(x)$ is increasing at this point. In particular, $f'(z) \ge f'(x)$ for all $z \ge x$. But then $f(x)$ would tend to infinity. Thus, $f'(x) \le 0$. Because $f(x)$ is decreasing and positive, it must converge to $\zeta \in [0, f(0)]$. That $\zeta < f(0) = \psi(0)$ follows from Proposition 4.56. □

We next prove convergence of the strategy $b(x)$.

Corollary 4.59. *Suppose that the conditions of Theorem 4.57 hold. Then the strategy $b(x)$ converges to zero.*

Proof. The value $b(x)$ minimises

$$\lambda \int_0^x g(x-y)[1-G(y/b(x))]e^{Ry}\, dy + \lambda\delta(0)[1-G(x/b(x))]e^{Rx} - c(b(x))g(x)\,.$$

In particular, this expression is bounded from above by $-c(0)g(x)$. Fix $b > 0$. Suppose that $b(x) \ge b$. Then because $g(x)$ is decreasing,

$$\lambda \int_0^x g(x-y)[1-G(y/b(x))]e^{Ry}\, dy \ge \lambda \int_0^x g(x-y)[1-G(y/b)]e^{Ry}\, dy \ge \lambda \int_0^x [1-G(y/b)]e^{Ry}\, dy\, g(x)\,.$$

There is an x_0 such that $\lambda \int_0^x [1-G(y/b)]\, e^{Ry}\, dy > c - c(0)$ for all $x > x_0$. We find that

$$\lambda \int_0^x g(x-y)[1-G(y/b(x))]e^{Ry}\, dy + \lambda\delta(0)[1-G(x/b(x))]e^{Rx} - c(b(x))g(x) \ge [c - c(0) - c(b(x))]g(x) \ge -c(0)g(x)\,.$$

Thus, $b(x) \le b$ for all $x \ge x_0$. Because b was arbitrary, this proves that $b(x)$ tends to zero. □

The strategy $A(x)$ also converges.

Corollary 4.60. *Suppose that the conditions of Theorem 4.57 hold. Then the strategy $A(x)$ converges to A^*.*

Proof. Taking the difference of (4.19) and

$$-\Bigl(c(0) + \frac{m^2}{2\sigma^2 R}\Bigr)g(x) = 0$$

yields

$$-\frac{m^2}{2\sigma^2 R}\frac{g'(x)g(x)}{Rg(x)-g'(x)} - (c(b(x)) - c(0))g(x) \tag{4.28}$$
$$+\lambda\int_0^x g(x-y)[1-G(y/b(x))]\,\mathrm{e}^{Ry}\,\mathrm{d}y + \lambda\delta(0)[1-G(x/b(x))]\,\mathrm{e}^{Rx} = 0\,.$$

Because $g'(x) \le 0$, $(c(b(x)) - c(0))g(x)$ is the only strictly negative term [provided that $b(x) \neq 0$]. Dividing this term by $g(x)$, it still converges to zero because $b(x)$ tends to zero. Thus, the positive terms also have to converge to zero when divided by $g(x)$; in particular,

$$\lim_{x\to\infty} -\frac{g'(x)}{Rg(x)-g'(x)} \le \lim_{x\to\infty} c(b(x)) - c(0) = 0\,.$$

Thus, $g'(x)/g(x)$ tends to zero, and

$$\lim_{x\to\infty} A(x) = \lim_{x\to\infty} \frac{m}{\sigma^2}\frac{g(x)}{Rg(x)-g'(x)} = A^*\,.$$

Note that $A(x) = A^*$ if $g'(x) = 0$, that is, if $b(x) = 0$. □

If $\zeta > 0$, then the limit of $g(x)$ must be $R\zeta$. Thus, $\psi'(x) \sim -R\zeta\mathrm{e}^{-Rx}$. We just have proved that $g'(x)$ tends to zero. Thus, $f''(x)$ tends to zero, too. We find that $\psi''(x) \sim \zeta R^2\mathrm{e}^{-Rx}$. Therefore, in this case we can interchange the limit and the first two derivatives.

An interesting question is to find conditions under which $\zeta > 0$. The following result shows that this is the case if the tail of the claim size distribution is "heavy enough."

Proposition 4.61. *Suppose that there exists $K > 0$ such that $c(b) - c(0) \le Kb$. Suppose, moreover, that there are constants $\alpha > 0$ and $0 < \gamma < \frac{1}{2}$ such that*

$$1 - G(y) \ge \alpha\exp\{-x^\gamma\}\,.$$

Then $\zeta = \lim_{x\to\infty}\psi(x)\mathrm{e}^{Rx} > 0$.

Remark 4.62. If the premium is obtained via a change of measure, as is usual in mathematical finance, one would have

$$c(b) - c(0) = (c - c(0)) - (c - c(b)) = \mathbb{E}^*\Bigl[\sum_{i=1}^{N_1} Y_i - \sum_{i=1}^{N_1}(1-b)Y_i\Bigr]\,,$$

where $\mathbb{P}^*$ is a measure equivalent to $\mathbb{P}$. Thus, $c(b) - c(0) = b\mathbb{E}^*[\sum_{i=1}^{N_1} Y_i]$, and the condition $c(b) - c(0) \le Kb$ is a natural condition on $c(b)$. In our case of a compound Poisson distribution it seems reasonable that the claims process remains a compound Poisson process under the measure $\mathbb{P}^*$; see also [45]. This would lead to $c(b) - c(0) = b\lambda^*\mu^*$, where λ^* and μ^* are the claim intensity and the mean claim size, respectively. ■

Proof. Let $h(x) = -g'(x)/g(x)$. Recall from the proof of Corollary 4.60 that $h(x) \to 0$ as $x \to \infty$. Then we can write

$$g(x) = g(0) \exp\Bigl\{-\int_0^x h(v)\,\mathrm{d}v\Bigr\}\,.$$

$\zeta = 0$ is then equivalent to $\int_0^\infty h(x)\,\mathrm{d}x = \infty$. For $h(x) \le R$, i.e., for x large enough, we find that

$$\begin{aligned}\frac{m^2}{4\sigma^2 R^2} h(x) &\le \frac{m^2}{2\sigma^2 R} \frac{h(x)}{R + h(x)} = -\frac{m^2}{2\sigma^2 R} \frac{g'(x)}{Rg(x) - g'(x)} \le c(b(x)) - c(0) \\ &\le K b(x)\,.\end{aligned}$$

The last inequality but one follows from (4.28). This shows that $\int_0^\infty b(x)\,\mathrm{d}x < \infty$ implies that $\zeta > 0$. From (4.19), the fact that $g(x)$ converges, and the definition of R, we conclude that $[1 - G(x/b(x))]\mathrm{e}^{Rx}$ converges to zero. Thus,

$$\alpha \exp\{Rx - (x/b(x))^\gamma\} \le [1 - G(x/b(x))]\mathrm{e}^{Rx} \le \alpha$$

for x large enough. This is equivalent to $b(x) \le R^{-1/\gamma} x^{1-1/\gamma}$. If $\gamma < \frac{1}{2}$ we find that $b(x)$ is integrable. □

We have seen that $b(x)$ tends to zero. A natural question is therefore whether $b(x) = 0$ for some finite x.

Proposition 4.63. i) *If* $\limsup_{b \downarrow 0} b^{-1}(c(b) - c(0)) > \lambda \mathbb{E}[Y]$, *then* $b(x) > 0$ *for all* x.

ii) *If* $\limsup_{b \downarrow 0} b^{-1}(c(b) - c(0)) < \lambda \mathbb{E}[Y]$, *then* $b(x) = 0$ *for all* x *large enough.*

Remark 4.64. Using utility theory for pricing shows that

$$c(b) - c(0) > \mathbb{E}\Bigl[\sum_{i=1}^{N_1} Y_i - \sum_{i=1}^{N_1} (1 - b) Y_i\Bigr] = b\lambda \mathbb{E}[Y]\,.$$

If there is a price for risk, one should have $\liminf_{b \downarrow 0} b^{-1}(c(b) - c(0)) \ge \lambda \mathbb{E}[Y]$. The assumption in part i) is therefore only slightly stronger. Note that the argument in Remark 4.62 shows that the strict inequality is natural. The condition in part ii) would mean that to change from almost full reinsurance to full reinsurance is too cheap. Therefore, it is not surprising that full reinsurance is chosen from a certain point on. ■

Proof. i) Take the difference of the right-hand side of (4.19) with $b(x)$ replaced by b and $b(x)$ replaced by zero:

$$[c(0) - c(b)]g(x) + \lambda \int_0^x g(x - y)[1 - G(y/b)]\mathrm{e}^{Ry}\,\mathrm{d}y + \lambda\delta(0)[1 - G(x/b)]\mathrm{e}^{Rx}\,.$$

We divide the expression by $\lambda b g(x)$:

$$\frac{c(0)-c(b)}{\lambda b}+\int_0^x \frac{g(x-y)\mathrm{e}^{Ry}}{g(x)}[1-G(y/b)]\frac{1}{b}\,\mathrm{d}y+\delta(0)\frac{x}{b}[1-G(x/b)]\frac{\mathrm{e}^{Rx}}{xg(x)}\;. \tag{4.29}$$

As $b \downarrow 0$, the last term tends to zero by Lemma A.11. The integral can be written as

$$\int_0^{x/b} \frac{g(x-by)\mathrm{e}^{Rby}}{g(x)}[1-G(y)]\,\mathrm{d}y\;.$$

By bounded convergence, this tends to $\mathbb{E}[Y]$ as $b \downarrow 0$. Consider now a sequence $\{b_n\}$ tending to zero such that

$$\lim_{n\to\infty} b_n^{-1}(c(b_n)-c(0))=\limsup_{b\downarrow 0} b^{-1}(c(b)-c(0))\;.$$

Thus, the limit of (4.29) (with $b=b_n$ as $n\to\infty$) is bounded from above by $-\lambda^{-1}\limsup_{b\downarrow 0} b^{-1}(c(b)-c(0))+\mathbb{E}[Y]<0$. That is, through the sequence $\{b_n\}$ the expression is decreasing in b close to zero. Therefore, the infimum cannot be taken in zero. Because $b(x)$ is the value at which the infimum is taken, we have $b(x)>0$.

ii) Suppose that $x>1$. Then because $g(x)$ is decreasing,

$$\int_0^x \frac{g(x-y)\mathrm{e}^{Ry}}{g(x)}[1-G(y/b)]\frac{1}{b}\,\mathrm{d}y > \int_0^1 [1-G(y/b)]\frac{1}{b}\,\mathrm{d}y=\int_0^{1/b}[1-G(y)]\,\mathrm{d}y\;.$$

The right-hand side converges to $\mathbb{E}[Y]$ uniformly in x. Thus, the limit of (4.29) as $b\downarrow 0$ is bounded from below by $-\lambda^{-1}\limsup_{b\downarrow 0} b^{-1}(c(b)-c(0))+\mathbb{E}[Y]>0$ uniformly in $x>1$. Thus, there is a $b_0>0$ such that $b(x)\notin(0,b_0]$. Because $b(x)$ tends to zero, this means that $b(x)=0$ for x large enough. □

Bibliographical Remarks

The results of Section 4.3.1 are new. Section 4.3.2 follows Schmidli [164]. Lemma 4.41 and Proposition 4.42 are from Hipp and Plum [97]. The rest of the section follows Schmidli [165]. Corollaries 4.47 and 4.52 were first obtained by Gaier and Grandits [70]. Section 4.3.4 follows Schmidli [163]. A model where deterministic interest is earned on the capital not invested was considered in Gaier and Grandits [70]. The asymptotic behaviour in the subexponential case in the latter model was obtained by Grandits [88].

We divide the expression by $\Lambda b\psi(x)$

$$\frac{c(0)-c(b)}{\Lambda b} + \int_0^x \frac{\psi(x-y)e^{\delta y}}{\psi(x)}[1-G(y/b)]\frac{1}{b}\,dy + \delta(0)\frac{x}{b}[1-G(x/b)]\frac{e^{\delta x}}{x\psi(x)}\,. \tag{4.29}$$

As $b\downarrow 0$, the last term tends to zero by Lemma A.12b. The integral can be written as

$$\int_0^{x/b} \frac{\psi(x-by)e^{\delta by}}{\psi(x)}[1-G(y)]\,dy\,.$$

By bounded convergence, this tends to $\mathbb{E}[Y_i]$ as $b\downarrow 0$. Consider now a sequence $\{b_n\}$ tending to zero such that

$$\lim_{n\to\infty} b_n^{-1}(c(b_n)-c(0)) = \limsup_{b\downarrow 0} b^{-1}(c(b)-c(0))\,.$$

Thus, the limit of (4.29) (with $b=b_n$ as $n\to\infty$) is bounded from above by $-\Lambda^{-1}\limsup_{b\downarrow 0} b^{-1}(c(b)-c(0)) + \mathbb{E}[Y_i] < 0$. That is, through the sequence $\{b_n\}$ the expression is decreasing in b close to zero. Therefore, the infimum cannot be taken in zero. Because $b(x)$ is the value at which the infimum is taken, we have $b(x) > 0$.

ii) Suppose that $x > 1$. Then because $\psi(x)$ is decreasing,

$$\int_0^x \frac{\psi(x-y)e^{\delta y}}{\psi(x)}[1-G(y/b)]\frac{1}{b}\,dy > \int_0^1 [1-G(y/b)]\frac{1}{b}\,dy = \int_0^{1/b}[1-G(y)]\,dy\,.$$

The right-hand side converges to $\mathbb{E}[Y_i]$ uniformly in x. Thus, the limit of (4.29) as $b\downarrow 0$ is bounded from below by $-\Lambda^{-1}\limsup_{b\downarrow 0} b^{-1}(c(b)-c(0)) + \mathbb{E}[Y_i] > 0$ uniformly in $x > 1$. Thus, there is a $b_0 > 0$ such that $b(x)\notin(0,b_0]$. Because $b(x)$ tends to zero, this means that $b(x)=0$ for x large enough. □

Bibliographical Remarks

The results of Section 4.3.1 are new. Section 4.3.2 follows Schmidli [164]. Lemma 4.41 and Proposition 4.42 are from Hipp and Plum [97]. The rest of the section follows Schmidli [164]. Corollaries 4.47 and 4.52 were first obtained by Gaier and Grandits [76]. Section 4.3.4 follows Schmidli [165]. A model where deterministic interest is earned on the capital not invested was considered in Gaier and Grandits [77]. The asymptotic behaviour in the subexponential case in the latter model was obtained by Grandits [88].

A

Stochastic Processes and Martingales

A.1 Stochastic Processes

Let I be either $\mathbb{N}$ or $\mathbb{R}_+$. A *stochastic process* on I with state space E is a family of E-valued random variables $X = \{X_t : t \in I\}$. We only consider examples where E is a Polish space. Suppose for the moment that $I = \mathbb{R}_+$. A stochastic process is called *cadlag* if its paths $t \mapsto X_t$ are right-continuous (a.s.) and its left limits exist at all points. In this book we assume that every stochastic process is cadlag. We say a process is *continuous* if its paths are continuous. The above conditions are meant to hold with probability 1 and not to hold pathwise.

A.2 Filtration and Stopping Times

The information available at time t is expressed by a σ-subalgebra $\mathcal{F}_t \subset \mathcal{F}$. An increasing family of σ-algebras $\{\mathcal{F}_t : t \in I\}$ is called a *filtration*. If $I = \mathbb{R}_+$, we call a filtration *right-continuous* if $\mathcal{F}_{t+} := \bigcap_{s>t} \mathcal{F}_s = \mathcal{F}_t$. If not stated otherwise, we assume that all filtrations in this book are right-continuous. In many books it is also assumed that the filtration is *complete*, i.e., $\mathcal{F}_0$ contains all $\mathbb{P}$-null sets. We do not assume this here because we want to be able to change the measure in Chapter 4. Because the changed measure and $\mathbb{P}$ will be singular, it would not be possible to extend the new measure to the whole σ-algebra $\mathcal{F}$.

A stochastic process X is called $\mathcal{F}_t$*-adapted* if X_t is $\mathcal{F}_t$-measurable for all t. If it is clear which filtration is used, we just call the process *adapted*. The *natural filtration* $\{\mathcal{F}_t^X\}$ is the smallest right-continuous filtration such that X is adapted. If we consider a process X, we always work with the natural filtration unless stated otherwise.

Suppose that $I = \mathbb{N}$. A stochastic process X is called *predictable* if $\{X_{t+1}\}$ is $\mathcal{F}_t$-adapted and X_0 is not stochastic. Suppose that $I = \mathbb{R}_+$. Let $\mathcal{F}_{t-} =$

$\bigvee_{s<t}\mathcal{F}_s$ be the smallest σ-algebra containing all $\mathcal{F}_s$ for $s < t$. We say that a stochastic process X is *predictable* if X is adapted to $\{\mathcal{F}_{t-}\}$. Note that if X is adapted, then $\{X_{t-}\}$ (with $X_{0-} = X_0$ deterministic) is predictable. Indeed,

$$\{X_{t-} \geq a\} = \bigcap_{n\in\mathbb{N}^*} \bigcup_{s\in\mathbb{Q}\cap[0,t)} \bigcap_{v\in\mathbb{Q}\cap(s,t)} \{X_v > a - n^{-1}\}\,.$$

Because $\{X_v > a - n^{-1}\} \in \mathcal{F}_v \subset \mathcal{F}_{t-}$, we have $\{X_{t-} \geq a\} \in \mathcal{F}_{t-}$.

A random variable $T \in I \cup \{\infty\}$ is called $\mathcal{F}_t$*-stopping time* if $\{T \leq t\} \in \mathcal{F}_t$ for all $t \in I$. Note that deterministic times are stopping times because $\{s \leq t\} \in \{\emptyset, \Omega\}$. If X is an adapted stochastic process and T is a stopping time, then the stopped process $\{X_{T\wedge t}\}$ is also adapted. Let $A \subset E$ be a Borel set. Define the first entrance times

$$T_A = \inf\{t \in I : X_t \in A\}\,, \qquad T_A^* = \inf\{t \in I : X_t \in A \text{ or } X_{t-} \in A\}\,.$$

The first entrance times are important examples of stopping times. If A is open, then T_A is a stopping time. If A is closed, then T_A^* is a stopping time. If A is compact or X is (pathwise) continuous, then T_A is a stopping time.

In many examples one needs the information up to a stopping time. For a stopping time T we define the σ-algebra

$$\mathcal{F}_T = \{A \in \mathcal{F} : A \cap \{T \leq t\} \in \mathcal{F}_t \text{ for all } t \in I\}\,.$$

The σ-algebra $\mathcal{F}_T$ contains all events observable until the stopping time T.

A.3 Martingales

An important concept for stochastic processes is the notion of a fair game, called a *martingale*. Now let $E = \mathbb{R}$. An adapted stochastic process M is called an $\mathcal{F}_t$*-submartingale* if $\mathbb{E}[|M_t|] < \infty$ for all $t \in I$ and

$$\mathbb{E}[M_{t+s} \mid \mathcal{F}_t] \geq M_t \qquad \text{for all } t, s \in I. \tag{A.1}$$

If the inequality (A.1) is reversed, the process M is an $\mathcal{F}_t$*-supermartingale*. A stochastic process that is both a sub- and a supermartingale is called a *martingale*.

If diffusion processes are involved, it often turns out that the notion of a martingale is too strong. For example, stochastic integrals with respect to Brownian motion are martingales if one stops them when reaching a finite level. This is very close to a martingale. We therefore say that a stochastic process M is an $\mathcal{F}_t$*-local martingale* if there exists a sequence of $\mathcal{F}_t$-stopping times $\{T_n : n \in \mathbb{N}\}$ with $\lim_{n\to\infty} T_n = \infty$ such that the stopped processes $\{M_{T_n\wedge t}\}$ are $\mathcal{F}_t$-martingales. Such a sequence of stopping times is called a *localisation sequence*. A localisation sequence can always be chosen to be increasing and such that $\{M_{T_n \wedge t}\}$ is a uniformly integrable martingale for each n.

The following results make martingales an important tool. The proofs can be found in [57] or [152].

Proposition A.1 (Convergence theorem). *Let M be a submartingale, and suppose that*

$$\sup_{t \geq 0} \mathbb{E}[(M_t)^+] < \infty .$$

Then there exists a random variable M_∞ such that $\lim_{t \to \infty} M_t = M_\infty$ and $\mathbb{E}[|M_\infty|] < \infty$. □

It should be noted that one cannot necessarily interchange the limit and integration. There are many examples where $M_0 = \lim_{t \to \infty} \mathbb{E}[M_t] \neq \mathbb{E}[M_\infty]$.

Proposition A.2 (Stopping theorem). *Let M be a submartingale, $t \geq 0$, and T_1, T_2 be stopping times. Then*

$$\mathbb{E}[M_{T_1 \wedge t} \mid \mathcal{F}_{T_2}] \geq M_{T_1 \wedge T_2 \wedge t} .$$

In particular, the stopped process $\{M_{T_1 \wedge t}\}$ is a submartingale. □

When dealing with local martingales, the following result is often helpful.

Lemma A.3. *Let M be a positive local martingale. Then M is a supermartingale.*

Proof. Let $\{T_n\}$ be a localisation sequence. By Fatou's lemma,

$$\begin{aligned} \mathbb{E}[M_{t+s} \mid \mathcal{F}_t] &= \mathbb{E}[\lim_{n \to \infty} M_{T_n \wedge (t+s)} \mid \mathcal{F}_t] \leq \lim_{n \to \infty} \mathbb{E}[M_{T_n \wedge (t+s)} \mid \mathcal{F}_t] \\ &= \lim_{n \to \infty} M_{T_n \wedge t} = M_t . \end{aligned}$$

Thus, M is a supermartingale. □

A.4 Poisson Processes

An increasing stochastic process with values in $\mathbb{N}$ and $N_0 = 0$ is called a *point process*. If the jumps are only one unit, i.e., $N_t - N_{t-} \in \{0, 1\}$ for all t, then we call a point process *simple*.

A special case of a simple point process is the homogeneous *Poisson process*. This process has some special properties that make it easy to handle. It can be defined via one of the following equivalent conditions. A proof can be found in [152].

Proposition A.4. *Let N be a point process with occurrence times $0 = T_0 < T_1 < \cdots$ and $\lambda > 0$. The following conditions are equivalent:*

i) *N has stationary and independent increments such that*

$$\mathbb{P}[N_h = 0] = 1 - \lambda h + o(h)\;, \qquad \mathbb{P}[N_h = 1] = \lambda h + o(h)\;, \qquad \text{as } h \downarrow 0.$$

ii) *N has independent increments and N_t is Poisson-distributed with parameter λt for each $t > 0$.*

iii) *The interarrival times $\{T_k - T_{k-1} : k \geq 1\}$ are independent and exponentially distributed with parameter λ.*

iv) *For each $t > 0$, N_t is Poisson-distributed with parameter λt and given $\{N_t = n\}$ the occurrence points $T_1, \ldots, T_n$ have the same distribution as the order statistics of n independent uniformly on $(0, t)$ distributed random variables.*

v) *N has independent increments, $\mathbb{E}[N_1] = \lambda$, and given $\{N_t = n\}$ the occurrence points $T_1, \ldots, T_n$ have the same distribution as the order statistics of n independent uniformly on $(0, t)$ distributed random variables.*

vi) *N has stationary and independent increments such that $\mathbb{E}[N_1] = \lambda$ and $\mathbb{P}[N_h \geq 2] = o(h)$ as $h \downarrow 0$.* □

We call λ the *rate* of the Poisson process.

Let $\{Y_i\}$ be a sequence of iid random variables independent of N. The process $\{\sum_{i=1}^{N_t} Y_i\}$ is called a *compound Poisson process.* The Poisson process is just a special case with $Y_i = 1$. As the Poisson process, the compound Poisson process has independent and stationary increments. This leads to the following martingales.

If $\mathbb{E}[|Y_i|] < \infty$, then

$$\Bigl\{\sum_{i=1}^{N_t} Y_i - \lambda\mathbb{E}[Y_i]t\Bigr\}$$

is a martingale. If $\mathbb{E}[Y_i^2] < \infty$, then

$$\Bigl\{\Bigl(\sum_{i=1}^{N_t} Y_i - \lambda\mathbb{E}[Y_i]t\Bigr)^2 - \lambda\mathbb{E}[Y_i^2]t\Bigr\}$$

is a martingale. If the moment-generating function $\mathbb{E}[\exp\{rY_i\}]$ exists, then the process

$$\Bigl\{\exp\Bigl\{r\sum_{i=1}^{N_t} Y_i - \lambda\mathbb{E}[\exp\{rY_i\} - 1]t\Bigr\}\Bigr\}$$

is a martingale. The martingale property can easily be verified from the independent and stationary increments property.

A.5 Brownian Motion

A (cadlag) stochastic process W is called a *standard Brownian motion* if it has independent increments, $W_0 = 0$, and W_t is normally distributed with mean value 0 and variance t. Wiener [186] proved that the Brownian motion exists. A process $\{X_t = mt + \sigma W_t\}$ is called an (m, σ^2) Brownian motion. It turns out that a Brownian motion also has stationary increments and that its paths are continuous (a.s.). It is easy to verify that a standard Brownian motion is a martingale.

A Brownian motion fluctuates rapidly. If we let $T = \inf\{t : X_t < 0\}$, then $T = 0$. The Brownian motion therefore reaches strictly positive and strictly negative values immediately. The reason is that the Brownian motion has unbounded variation; see the definition ahead. One can also show that the sample path of a Brownian motion is nowhere differentiable.

Brownian motion has bounded *quadratic variation*. Namely, for $t > 0$,

$$\lim_{n\to\infty} \sum_{i=1}^n (W_{t_i} - W_{t_{i-1}})^2 = t \,,$$

where $0 = t_0 < t_1 < \cdots < t_n = t$ is a partition and $\sup_k t_k - t_{k-1}$ has to converge to zero with n. Indeed,

$$\mathbb{E}\Bigl[\sum_{i=1}^n (W_{t_i} - W_{t_{i-1}})^2\Bigr] = t$$

and

$$\operatorname{Var}\Bigl[\sum_{i=1}^n (W_{t_i} - W_{t_{i-1}})^2\Bigr] = \sum_{i=1}^n \operatorname{Var}[(W_{t_i} - W_{t_{i-1}})^2] = \sum_{i=1}^n 2(t_i - t_{i-1})^2 \,.$$

Thus, the variance converges to zero with n. Take a subsequence n_k such that the sum $\sum_{k=1}^\infty \sum_{i=1}^{n_k} (t_i^{(n_k)} - t_{i-1}^{(n_k)})^2$ is finite. Then for any $\varepsilon > 0$, by Chebyshev's inequality, we have that

$$\sum_{k=1}^\infty \mathbb{P}\Bigl[\Bigl|\sum_{i=1}^{n_k} (W_{t_i^{(n_k)}} - W_{t_{i-1}^{(n_k)}})^2 - t\Bigr| > \varepsilon\Bigr] \le \varepsilon^{-2} \sum_{k=1}^\infty \sum_{i=1}^{n_k} (t_i^{(n_k)} - t_{i-1}^{(n_k)})^2 \,.$$

By the Borel–Cantelli lemma, we get that

$$\lim_{k\to\infty} \sum_{i=1}^{n_k} (W_{t_i^{(n_k)}} - W_{t_{i-1}^{(n_k)}})^2 = t \,.$$

Because the subsequence was arbitrary, the limit is independent of the chosen subsequence.

For a Brownian motion first passage times can easily be obtained. Consider an (m, σ^2) Brownian motion X with $m > 0$. Then one easily verifies from the stationary and independent increments property that M defined as

$$M_t = \exp\Bigl\{-\frac{2m}{\sigma^2} X_t\Bigr\}$$

is a martingale. Let $\tau = \inf\{t : X_t = -x\}$ for some $x > 0$ and $\psi(x) = \mathbb{P}[\tau < \infty]$. By the stopping theorem $\{M_{\tau\wedge t}\}$ is a bounded martingale. Thus, $M_{\tau\wedge t}$ converges to M_τ. Because $X_{t+1} - X_t$ is normally distributed with mean m and variance σ^2, it is not possible that $X_\tau \in (-x, \infty)$. Thus, $X_\tau \in \{-x, \infty\}$. Therefore, by bounded convergence, $1 = M_0 = \mathbb{E}[M_\tau]$ and

$$\psi(x) = \mathbb{P}\Bigl[M_\tau = \exp\Bigl\{\frac{2mx}{\sigma^2}\Bigr\}\Bigr] = \mathbb{E}[M_\tau] \exp\Bigl\{-\frac{2mx}{\sigma^2}\Bigr\} = \exp\Bigl\{-\frac{2mx}{\sigma^2}\Bigr\}. \tag{A.2}$$

For ruin in finite time, one has the formula

$$\mathbb{P}[\tau \le t] = \Phi\Bigl(-\frac{mt + x}{\sigma\sqrt{t}}\Bigr) + \mathrm{e}^{-2mx/\sigma^2} \Phi\Bigl(\frac{mt - x}{\sigma\sqrt{t}}\Bigr), \tag{A.3}$$

where $\Phi(x)$ is the distribution function of the standard normal distribution; see [152, p. 428]. Formula (A.3) also holds for $m \le 0$.

A.6 Stochastic Integrals and Itô's Formula

For a stochastic process X, we define the *variation* over $(0, t]$ by

$$V_X(t) = \sup\Bigl\{\sum_{i=1}^n |X_{t_i} - X_{t_{i-1}}|\Bigr\},$$

where the supremum is taken over all $n \in \mathbb{N}^*$ and all partitions $0 = t_0 < t_1 < \cdots < t_n = t$. If $V_X(t) < \infty$ for all t, we call the process X to be of *bounded variation.* There exist increasing processes $\overline{X}$ and $\underline{X}$ such that $X = X_0 + \overline{X} - \underline{X}$ and $V_X = \overline{X} + \underline{X}$. If there are two increasing processes A and B such that $X = X_0 + A - B$, then $\overline{X}_t \le A_t$ and $\underline{X}_t \le B_t$. A typical example of a process of bounded variation are the integrals $X_t = \int_0^t Y_s \,\mathrm{d}s$, where Y is a stochastic process. In this case $\overline{X}_t = \int_0^t (Y_s)^+ \,\mathrm{d}s$ and $\underline{X}_t = \int_0^t (Y_s)^- \,\mathrm{d}s$.

Let X and Y be stochastic processes. If X is of bounded variation, one can define the process $\{\int_0^t Y_{s-} \,\mathrm{d}X_s\}$, which itself is a process of bounded variation. If X is of unbounded variation, the stochastic integral cannot be defined in the classical way. Here wc will here motivate the definition of a stochastic integral if X is a standard Brownian motion.

Suppose first that Y is a pure jump process with N_t jumps in $(0, t]$ and $\mathbb{E}[N_t] < \infty$. Denote the jump times by T_i. Then, letting $T_0 = 0$,

$$M_t = \int_0^t Y_s \, dW_s = \sum_{i=0}^{N_t - 1} Y_{T_i}(W_{T_{i+1}} - W_{T_i}) + Y_t(W_t - W_{T_{N_t}}) \, .$$

The process M is continuous. Let $\tau_n = \inf\{t : |M_t| > n\}$. Then $\{M_{T_n \wedge t}\}$ is a martingale. Thus, M is a local martingale. For any (possibly not cadlag) process Y such that $\int_0^t Y_s^2 \, ds < \infty$ for all t, it is possible to approximate Y by pure jump processes Y^n and $\int_0^t Y_s^n \, dW_s$ converges almost surely to some process M. The details can be found in [57] or [152]. The process M is a local martingale.

It would be beyond the scope of this book to give a construction of the stochastic integral with respect to Brownian motion. In order for the reader to be able to work with stochastic integrals, we review some properties of stochastic integrals.

Proposition A.5. *Suppose that X and Y are some (not necessarily cadlag) processes such that $\int_0^t (Y_s^2 + X_s^2) \, ds < \infty$, and let W be a standard Brownian motion. Then the processes*

$$\int_0^t X_s \, dW_s \tag{A.4}$$

and

$$\int_0^t X_s \, dW_s \int_0^t Y_s \, dW_s - \int_0^t X_s Y_s \, ds \tag{A.5}$$

are local martingales. If $\int_0^t \mathbb{E}[Y_s^2 + X_s^2] \, ds < \infty$, then (A.4) *and* (A.5) *are true martingales.* □

In particular, if Y is a bounded process, then the stochastic integral is a martingale.

Let $Z_t = \int_0^t X_s \, dW_s + V_t$, where $\int_0^t X_s^2 \, ds < \infty$ and V is a process of bounded variation. If Y is a process, such that $\int_0^t Y_s^2 X_s^2 \, ds < \infty$ and $\mathbb{E}[|Y_{t-} \Delta V_t|] < \infty$, then we can define the stochastic integral $\int_0^t Y_{s-} \, dZ_t = \int_0^t Y_s X_s \, dW_s + \int_0^t Y_{s-} \, dV_s$. Recall that $\Delta V_t = V_t - V_{t-}$.

If M is a local martingale of bounded variation, then the stochastic integral $\{\int_0^t X_{s-} \, dM_s\}$ also becomes a local martingale. It is important that the integrand is X_{s-} and not X_s, whereas in the integral with respect to Brownian motion it does not matter. The reason is that if jumps can occur at the same time for X and M, the martingale property is lost otherwise. Take, for example, a Poisson jump process N with rate 1, and consider the martingale $\{M_t = N_t - t\}$. Then

$$M_t' = \int_0^t N_{s-} \, dM_s = \sum_{i=1}^{N_t} (i-1) - \int_0^t N_s \, ds = \tfrac{1}{2} N_t(N_t - 1) - \int_0^t N_s \, ds \, ,$$

which can be verified to be a martingale. If we instead calculate

$$\int_0^t N_s \, \mathrm{d}M_s = M'_t + N_t \,,$$

we do not end up with a martingale.

From the construction via pure jump integrands it is not surprising that the stochastic integral is linear, i.e., for $a, b \in \mathbb{R}$,

$$\int_0^t (aX_s + bY_s) \, \mathrm{d}W_s = a \int_0^t X_s \, \mathrm{d}W_s + b \int_0^t Y_s \, \mathrm{d}W_s \,.$$

Let $\{W^i : 1 \le i \le n\}$ be independent Brownian motions and V be a process of bounded variation. A process X of the form

$$X_0 + \sum_{i=1}^n \int_0^t Y_s^i \, \mathrm{d}W_s^i + \int_0^t Y_{s-}^0 \, \mathrm{d}V_s$$

is called an *Itô process* if it is well defined.

Remark A.6. In the definition of an Itô process we could just let $n = 1$. We have chosen this slightly more complicated definition in order to be able to deal with dependent Brownian motions in Proposition A.7 below in a simple way. ■

A very useful result is the following proposition, whose proof can be found in [57] or [152].

Proposition A.7 (Itô's formula). *Let $n, d \in \mathbb{N}^*$, let $\{W^i : 1 \le i \le n\}$ be independent Brownian motions, and consider the Itô processes $\{X^i : 1 \le i \le d\}$ with*

$$X_t^i = X_0^i + \sum_{j=1}^n \int_0^t Y_s^{ij} \, \mathrm{d}W_s^j + \int_0^t Y_{s-}^{i0} \, \mathrm{d}V_s^i \,.$$

Let $g : \mathbb{R}^{d+1} \to \mathbb{R}$, $(t, x_1, \ldots, x_d) \mapsto g(t, x_1, \ldots, x_d)$ be a function that is twice continuously differentiable with respect to $(x_1, \ldots, x_d)$ and continuously differentiable with respect to t. Then the stochastic process $\{g(t, X_t^1, \ldots, X_t^d)\}$ is an Itô process and

$$\begin{aligned} g(t, X_t^1, \ldots, X_t^d) &= g(0, X_0^1, \ldots, X_0^d) + \int_0^t g_t(s, X_s^1, \ldots, X_s^d) \, \mathrm{d}s \\ &\quad + \sum_{i=1}^d \int_0^t g_{x_i}(s, X_{s-}^1, \ldots, X_{s-}^d) Y_{s-}^{i0} \, \mathrm{d}V_s^i \\ &\quad + \sum_{i=1}^d \sum_{j=1}^n \int_0^t g_{x_i}(s, X_s^1, \ldots, X_s^d) Y_s^{ij} \, \mathrm{d}W_s^j \\ &\quad + \tfrac{1}{2} \sum_{i=1}^d \sum_{j=1}^d \sum_{k=1}^n \int_0^t g_{x_i x_j}(s, X_s^1, \ldots, X_s^d) Y_s^{ik} Y_s^{jk} \, \mathrm{d}s \,, \end{aligned}$$

where g_{x_j} and g_t denote the partial derivative of g with respect to x_j and t, respectively, and $g_{x_i x_j}$ is the second-order partial derivative. □

The last term comes from the quadratic variation of the Brownian motion and is not present in the ordinary differential calculus.

For completeness we give the following definition. A process X is called a *semi-martingale* if it is of the form $X = M + V$ where M is a local martingale and V is a process of bounded variation. All processes considered in this book are semi-martingales. A famous example of a process that is not a semi-martingale is fractional Brownian motion. And of course, in most problems considered here we could choose a control process that is not a semi-martingale. However, the optimal controls found are semi-martingales. This follows from the Markov property of the processes considered.

To change the measure, exponents of Itô processes are important. These exponents should be martingales. The following condition is known as *Novikov's condition*.

Proposition A.8. *Let Y be a stochastic process such that*

$$\mathbb{E}\Big[\exp\Big\{\tfrac{1}{2}\int_0^t Y_s^2 \,\mathrm{d}s\Big\}\Big] < \infty \tag{A.6}$$

for all $t > 0$. Then the process

$$\Big\{\exp\Big\{\int_0^t Y_s \,\mathrm{d}W_s - \tfrac{1}{2}\int_0^t Y_s^2 \,\mathrm{d}s\Big\}\Big\}$$

is a martingale. □

Often the following result is helpful, whose proof can be found, for instance, in [152, p. 566].

Lemma A.9. *Let M be a local martingale of bounded variation. If M is continuous, then M is constant.* □

A continuous martingale can be expressed as a time-changed Brownian motion. Let $M_t = \int_0^t X_s \,\mathrm{d}W_s$ be well defined, and consider $Q_t = \int_0^t X_s^2 \,\mathrm{d}s$. We consider the inverse function $D_s := \inf\{t : Q_t > s\}$ for $s \ge 0$. We define the filtration $\mathcal{G}_t = \mathcal{F}_{D_t}$.

Lemma A.10. *The process $\{M_{D_t}\}$ is a $\mathcal{G}_t$-standard Brownian motion.* □

A.7 Some Tail Asymptotics

Lemma A.11. *Suppose that $\mathbb{E}[|Y|^n] < \infty$. Then $\lim_{x\to\infty} x^n(1 - G(x)) = 0$ and $\lim_{x\to-\infty} x^n G(x) = 0$.*

Proof. We can assume that $Y \geq 0$ because $\mathbb{E}[(Y^+)^n] < \infty$. Suppose that $\limsup_{x\to\infty} x^n(1-G(x)) > 0$. Then there are points $x_1 < 2x_1 \leq x_2 < 2x_2 \leq x_3 < \cdots$ such that $x_k^n(1-G(x_k)) > 2^n\varepsilon$ for some $\varepsilon > 0$. For $y \in [x_k/2, x_k]$ we conclude that

$$y^n(1-G(y)) \geq 2^{-n}x_k^n(1-G(x_k)) > \varepsilon\,.$$

This yields

$$\int_{x_k/2}^{x_k} y^{n-1}(1-G(y))\,\mathrm{d}y > \int_{x_k/2}^{x_k} \frac{\varepsilon}{y}\,\mathrm{d}y = \varepsilon(\log x_k - \log(x_k/2)) = \varepsilon \log 2\,.$$

This implies that

$$\mathbb{E}[Y^n] = \int_0^\infty ny^{n-1}(1-G(y))\,\mathrm{d}y \geq n\sum_{k=1}^\infty \varepsilon\log 2 = \infty\,.$$

The second assertion follows from considering $-Y$. □

Lemma A.12. *Suppose that $M_Y(r) < \infty$ for some $r > 0$. Then*

$$\lim_{x\to\infty}(1-G(x))\,\mathrm{e}^{rx} = 0\,.$$

Proof. Suppose that $\limsup_{x\to\infty}(1-G(x))\,\mathrm{e}^{rx} > 0$. Then there are points $x_1 < x_1+1 \leq x_2 < x_2+1 \leq x_3 \cdots$ such that $(1-G(x_k))\,\mathrm{e}^{rx_k} > \varepsilon\mathrm{e}^r$. Then

$$\int_{x_k-1}^{x_k} (1-G(y))\,\mathrm{e}^{ry}\,\mathrm{d}y \geq (1-G(x_k))\,\mathrm{e}^{r(x_k-1)} > \varepsilon\,.$$

Thus,

$$\mathbb{E}[\mathrm{e}^{rY}] = \int_{-\infty}^\infty r\mathrm{e}^{ry}(1-G(y))\,\mathrm{d}y > r\sum_{k=1}^\infty \varepsilon = \infty\,.$$

□

Bibliographical Remarks

Introductions to the theory of stochastic processes can be found in many books, such as for Dellacherie and Meyer [44], Doob [48], Ethier and Kurtz [57], Karatzas and Shreve [117], Revuz and Yor [149], Rogers and Williams [151], Rolski et al. [152], or Shreve [167]. Introductions to stochastic calculus can be found in the books mentioned above, Øksendal [143], or Protter [148]. An alternative to Novikov's condition (A.6) is Portenko's condition, which is often easier to use; see Portenko [147].

B

Markov Processes and Generators

In this appendix we review some results on Markov processes. A general introduction to Markov processes can be found in [57].

B.1 Definition of Markov Processes

Let X be a stochastic process. X is called an *$\{\mathcal{F}_t\}$-Markov process* if it is adapted to $\{\mathcal{F}_t\}$ and

$$\mathbb{P}[X_{t+s} \in B \mid \mathcal{F}_t] = \mathbb{P}[X_{t+s} \in B \mid X_t] \,.$$

We just say that X is a Markov process if it is an $\{\mathcal{F}_t^X\}$-Markov process. This means that the future of the process depends on the present state of the process only. We define the *transition function*

$$P_t(s, x, B) = \mathbb{P}[X_{t+s} \in B \mid X_t = x] \,.$$

If the transition function does not depend on t, we call the Markov process *homogeneous*, and we will omit the index t.

Let T be a stopping time and X be a homogeneous Markov process. If

$$\mathbb{P}[X_{T+s} \in B \mid \mathcal{F}_T] = P(s, X_T, B) \,,$$

then we call X *strong Markov* at T. If X is strong Markov at all stopping times, we call X a *strong Markov process*.

B.2 The Generator

Let us denote by $B(E)$ the set of all measurable bounded real functions on E, and we endow $B(E)$ with the supremum-norm $\|f\| = \sup_{x \in E} |f(x)|$. We say that $f \in \mathcal{D}(\mathfrak{A})$ if the limit

$$\mathfrak{A}f(x) = \lim_{t\downarrow 0} \frac{1}{t}\mathbb{E}[f(X_t) - f(x) \mid X_0 = x]$$

exists [in the sense of uniform convergence on $B(E)$]. We call $\mathfrak{A}$ the *infinitesimal generator* and $\mathcal{D}(\mathfrak{A})$ the *domain of the generator* $\mathfrak{A}$.

The following result links the infinitesimal generator to martingales.

Theorem B.1 (Dynkin's theorem). *Let X be a (homogeneous) Markov process and $f \in \mathcal{D}(\mathfrak{A})$ be a function in the domain of the generator. The process*

$$\Bigl\{f(X_t) - f(X_0) - \int_0^t \mathcal{A}f(X_s)\,\mathrm{d}s\Bigr\}$$

is a martingale. □

One often wants to construct certain martingales by solving the equation $\mathfrak{A}f(x) = h(x)$. The desired solutions to this equation are often unbounded. One therefore defines a version of the generator that also acts on unbounded functions. We say that a measurable function f is in the *domain of the full generator* if there exists a measurable function g such that the process

$$\Bigl\{f(X_t) - f(X_0) - \int_0^t g(X_s)\,\mathrm{d}s\Bigr\} \tag{B.1}$$

is a martingale. Note that g is not uniquely defined. We identify all functions g such that (B.1) is a martingale and write $\mathfrak{A}f$ instead of g. $\mathfrak{A}$ is called the *full generator*. The domain is also denoted by $\mathcal{D}(\mathfrak{A})$. Note that the domain of the infinitesimal generator is contained in the domain of the full generator.

Sometimes, in particular for diffusion processes, it is easy to see that a process of the form (B.1) is a local martingale for some function g. One therefore defines the *extended generator* in the same way, but it is only required that (B.1) is a local martingale.

Suppose that X is a process of bounded variation. Let $f \in \mathcal{D}(\mathfrak{A})$ and h be a differentiable function. Then one could ask what the compensator is in the Dynkin martingale of the process $\{f(X_t)h(t)\}$. Instead of the Markov process X, here we consider the homogeneous Markov process $\{(X_t, t)\}$. The process

$$\int_0^t h(s)\,\mathrm{d}\Bigl(f(X_s) - \int_0^s \mathfrak{A}f(X_v)\,\mathrm{d}v\Bigr) = \int_0^t h(s)\,\mathrm{d}(f(X_s)) - \int_0^t h(s)\mathfrak{A}f(X_s)\,\mathrm{d}s$$

is then a (local) martingale. Using integration by parts yields

$$\int_0^t h(s)\,\mathrm{d}(f(X_s)) = h(t)f(X_t) - h(0)f(X_0) - \int_0^t h'(s)f(X_s)\,\mathrm{d}s\,.$$

Thus, the process

$$\left\{f(X_t)h(t) - \int_0^t (h(s)\mathfrak{A}f(X_s) + h'(s)f(X_s))\,\mathrm{d}s\right\}$$

is a (local) martingale. That is, the generator applied to $\{f(x)h(t)\}$ is $h(t)\mathfrak{A}f(x) + h'(t)f(x)$. If f is in the domain of the full generator and both $h(s)$ and $h'(s)$ are bounded on $(0,t]$, then we get that the above process is a martingale. We use this fact in Chapter 2 for the function $h(s) = \mathrm{e}^{-\delta s}$.

We now consider some examples. Let X be an (m, σ^2)-Brownian motion, i.e., $X_t = mt + \sigma W_t$ for some standard Brownian motion W. Suppose that $f(x)$ is a twice continuously differentiable function. By Ito's formula,

$$f(X_t) = f(X_0) + \int_0^t (mf'(X_s) + \tfrac{1}{2}\sigma^2 f''(X_s))\,\mathrm{d}s + \int_0^t \sigma f'(X_s)\,\mathrm{d}W_s\;,$$

or equivalently,

$$f(X_t) - f(X_0) - \int_0^t (mf'(X_s) + \tfrac{1}{2}\sigma^2 f''(X_s))\,\mathrm{d}s = \int_0^t \sigma f'(X_s)\,\mathrm{d}W_s\;.$$

The right-hand side is a local martingale; and therefore,

$$\mathfrak{A}f(x) = mf'(x) + \tfrac{1}{2}\sigma^2 f''(x)\;.$$

Let $X_t = ct - \sum_{i=1}^{N_t} Y_i$ be a compound Poisson process with drift, i.e., N is a Poisson process and $\{Y_i\}$ are iid and independent of N with distribution function $G(y)$. In order to analyse the process $\{f(X_t)\}$, we consider the process between jumps and compensate for the jumps. Then

$$f(X_t) = f(X_0) + \sum_{i=1}^{N_t}\Bigl[\int_{T_{i-1}}^{T_i} cf'(X_s)\,\mathrm{d}s + f(X_{T_i}) - f(X_{T_i-})\Bigr] + \int_{T_{N_t}}^t cf'(X_s)\,\mathrm{d}s\;,$$

where T_i are the jump times. Let us consider the jump part $\sum_{i=1}^{N_t} f(X_{T_i}) - f(X_{T_i-})$. We need to find a function g such that

$$\left\{\sum_{i=1}^{N_t} f(X_{T_i}) - f(X_{T_i-}) - \int_0^t g(X_s)\,\mathrm{d}s\right\}$$

becomes a martingale. Writing the latter as

$$\sum_{i=1}^{N_t} f(X_{T_i}) - f(X_{T_i-}) - \int_{T_{i-1}}^{T_i} g(X_s)\,\mathrm{d}s - \int_{T_{N_t}}^t g(X_s)\,\mathrm{d}s\;,$$

it turns out that it is enough to consider

$$(f(X_{T_1}) - f(X_{T_1-}))\mathbb{1}_{T_1 \le t} - \int_0^{T_1 \wedge t} g(X_s)\,\mathrm{d}s\;.$$

By the lack of memory property of the exponential distribution, we can just consider the expected value, and we do not have to take the conditional expectation with respect to $\mathcal{F}_s$. Thus, we need a function g such that

$$\mathbb{E}\Big[(f(X_{T_1}) - f(X_{T_1-}))\mathbb{1}_{T_1 \le t} - \int_0^{T_1 \wedge t} g(X_s)\,\mathrm{d}s\Big] = 0\,.$$

The expected value of the first term is

$$\int_0^t \int_{-\infty}^{\infty} (f(x + cs - y) - f(x + cs))\,\mathrm{d}G(y)\,\lambda \mathrm{e}^{-\lambda s}\,\mathrm{d}s\,.$$

The expected value of the second term is

$$\int_0^t \int_0^s g(x + cv)\,\mathrm{d}v \lambda \mathrm{e}^{-\lambda s}\,\mathrm{d}s + \mathrm{e}^{-\lambda t} \int_0^t g(x + cs)\,\mathrm{d}s\,.$$

Integration by parts yields

$$\int_0^t g(x + cv)\mathrm{e}^{-\lambda v}\,\mathrm{d}v\,.$$

We therefore have to choose

$$g(x) = \lambda \int_{-\infty}^{\infty} (f(x - y) - f(x))\,\mathrm{d}G(y)\,.$$

We find that

$$\mathfrak{A}f(x) = cf'(x) + \lambda \int_{-\infty}^{\infty} (f(x - y) - f(x))\,\mathrm{d}G(y)\,.$$

The extended generator is discussed in detail in [39] and [40]. A sufficient condition for f to be in the domain of the full generator is given in [38] and, with a proof, in [152].

Bibliographical Remarks

The theory of Markov processes is treated in many books, such as Davis [40], Dynkin [53, 54], Ethier and Kurtz [57], Jacobsen [112], Revuz and Yor [149], or Rogers and Williams [151]. For simplicity here we considered the semigroup $T_t f(x) = \mathbb{E}[f(X_t) \mid X_0 = x]$ operating on all bounded measurable functions $B(E)$. Working with Markov processes, in particular Feller processes, it is often better to work with a closed subspace $L \subset B(E)$ and a semigroup $T_t : L \to L$. For example, L could be the continuous functions $f(x)$ with the property that $\lim_{|x|\to\infty} f(x) = 0$. The details can be found in the references above.

C

Change of Measure Techniques

C.1 Introduction

Let M be a strictly positive martingale with expected value 1. On the σ-algebra $\mathcal{F}_t$ we can then define the equivalent measure $\mathbb{P}^*[A] = \mathbb{E}[M_t; A] := \mathbb{E}[M_t \mathbb{1}_A]$. At first sight it seems as if the measure would depend on t. Let $s < t$, and suppose that $A \in \mathcal{F}_s$. Then

$$\mathbb{E}[M_t; A] = \mathbb{E}[\mathbb{E}[M_t \mathbb{1}_A \mid \mathcal{F}_s]] = \mathbb{E}[\mathbb{E}[M_t \mid \mathcal{F}_s] \mathbb{1}_A] = \mathbb{E}[M_s; A]$$

by the martingale property. Thus, the measure is independent of t.

In many cases it is possible to extend the measure $\mathbb{P}^*$ to the whole σ-algebra $\mathcal{F}$. But usually the measures $\mathbb{P}^*$ and $\mathbb{P}$ are singular. For an example, see Section D.1.2. Let T be a stopping time and $A \in \mathcal{F}_T$, the σ-algebra generated by the process up to T. If $A \subset \{T < \infty\}$, then

$$\mathbb{P}^*[A] = \mathbb{E}[M_T; A] .$$

Indeed, because $A \cap \{T \le t\} \in \mathcal{F}_t$, we have $\mathbb{P}^*[A \cap \{T \le t\}] = \mathbb{E}[M_t; A \cap \{T \le t\}]$. By the martingale stopping theorem this can be written as $\mathbb{P}^*[A \cap \{T \le t\}] = \mathbb{E}[M_T; A \cap \{T \le t\}]$. The monotone limit theorem then yields the desired formula.

One often needs conditional probabilities. The following formula holds.

Lemma C.1. *The* conditional probabilities *can be obtained from*

$$\mathbb{P}^*[A \mid \mathcal{F}_t] = M_t^{-1} \mathbb{E}[M_s \mathbb{1}_A \mid \mathcal{F}_t]$$

for any $A \in \mathcal{F}_s$ *and* $s \ge t$.

Proof. First we note that $M_t^{-1} \mathbb{E}[M_s \mathbb{1}_A \mid \mathcal{F}_t]$ is $\mathcal{F}_t$-measurable. Let Z be a bounded $\mathcal{F}_t$-measurable variable. Then

$$\begin{aligned}\mathbb{E}^*[\mathbb{1}_A Z] &= \mathbb{E}[M_s Z \mathbb{1}_A] = \mathbb{E}[\mathbb{E}[M_s Z \mathbb{1}_A \mid \mathcal{F}_t]] = \mathbb{E}[\mathbb{E}[M_s \mathbb{1}_A \mid \mathcal{F}_t] Z] \\ &= \mathbb{E}^*[M_t^{-1} \mathbb{E}[M_s \mathbb{1}_A \mid \mathcal{F}_t] Z] .\end{aligned}$$

This shows the formula by the definition of the conditional expectation. □

C.2 The Brownian Motion

Consider now a Brownian motion with drift $X_t = x + mt + \sigma W_t$, where W is a standard Brownian motion. Let $\theta(r) = \frac{1}{2}\sigma^2 r^2 - mrt$. Then $M = \{\exp(-r(X_t - x) - \theta(r)t)\}$ is a martingale. The next result is known as *Girsanov's theorem.*

Proposition C.2. *Under the measure $\mathbb{P}^*$ the process W^* with $W^*_t = W_t + r\sigma t$ is a standard Brownian motion. In particular, under $\mathbb{P}^*$ the process X is an $(m - r\sigma^2, \sigma^2)$-Brownian motion.*

Proof. It is no loss of generality to assume that $m = 0$. Clearly, W^* has continuous paths and $W^*_0 = 0$ because the measures are equivalent on $\mathcal{F}_t$. For the increment we obtain by Lemma C.1 for $s < t$

$$\begin{aligned}
\mathbb{E}^*[\mathrm{e}^{\beta(W^*_t - W^*_s)} \mid \mathcal{F}_s] &= \mathrm{e}^{r\sigma W_s + \theta(r)s}\mathbb{E}[\mathrm{e}^{-r\sigma W_t - \theta(r)t}\mathrm{e}^{\beta(W^*_t - W^*_s)} \mid \mathcal{F}_s] \\
&= \mathbb{E}[\mathrm{e}^{-r\sigma(W_t - W_s)}\mathrm{e}^{\beta(W_t - W_s + r\sigma(t-s))} \mid \mathcal{F}_s]\mathrm{e}^{-\theta(r)(t-s)} \\
&= \exp\{\tfrac{1}{2}[(\beta - r\sigma)^2 + 2\beta r\sigma - r^2\sigma^2](t-s)\} \\
&= \exp\{\tfrac{1}{2}\beta^2(t-s)\}\,.
\end{aligned}$$

This is the moment-generating function of a normal distribution with mean zero and variance $t - s$. Thus, W^* has independent normally distributed increments. This shows that W^* is a standard Brownian motion. The process X can then be expressed as $X_t = x + (m - r\sigma^2) + \sigma W^*_t$. □

A consequence of Proposition C.2 is that it is possible to extend the measure $\mathbb{P}^*$ to $\mathcal{F}$ because the $(m - r\sigma^2, \sigma^2)$-Brownian motion exists. Note, however, that the filtration $\{\mathcal{F}_t\}$ cannot be completed. Extension is only possible because of our convention that we work with the smallest right-continuous filtration such that the processes are adapted. Note also that $\mathbb{P}$ and $\mathbb{P}^*$ are singular unless $r = 0$ because $\mathbb{P}^*[\lim_{t\to\infty} X_t/t = m] = 0 \neq 1 = \mathbb{P}[\lim_{t\to\infty} X_t/t = m]$.

The change of measure formula can be used to find the first passage probabilities. Suppose that $m > 0$. Then $\theta(r) = 0$ has the solutions $r = 0$ and $r = R := 2m/\sigma^2$. The drift under $\mathbb{P}^*$ is $-m$, and ruin thus occurs almost surely. The ruin probability can be expressed as

$$\psi(x) = \mathbb{E}^*[\mathrm{e}^{R(X_\tau - x)}; \tau < \infty] = \mathbb{E}^*[\mathrm{e}^{RX_\tau}]\,\mathrm{e}^{-Rx} = \mathrm{e}^{-2mx/\sigma^2}\,,$$

which is (A.2). If we now want to find the ruin probability in finite time, we calculate the Laplace transform of τ, $\mathbb{E}[\mathrm{e}^{-\beta\tau}1\!\!1_{\tau<\infty}]$. If we look for the positive solution to $\theta(r) = \beta$, we find that $r = \sigma^{-2}(m + \sqrt{m^2 + 2\beta\sigma^2})$. The drift of the measure-changed process becomes $-\sqrt{m^2 + 2\beta\sigma^2}$, and also here ruin probability occurs almost surely. From the change of measure formula we obtain

$$\begin{aligned}
\mathbb{E}[\mathrm{e}^{-\beta\tau}; \tau < \infty] &= \mathbb{E}^*[\mathrm{e}^{r(X_t - x) + \theta(r)\tau - \beta\tau}] = \mathbb{E}^*[\mathrm{e}^{rX_\tau}]\mathrm{e}^{-rx} \\
&= \mathrm{e}^{-\left(m + \sqrt{m^2 + 2\beta\sigma^2}\right)x/\sigma^2}\,.
\end{aligned}$$

This is the Laplace transform of the inverse Gaussian distribution (A.3).

C.3 The Classical Risk Model

Consider a classical risk model $\{X_t = x+ct-\sum_{i=1}^{N_t} Y_i\}$. This model is discussed in Section D.1. Here N is a Poisson process with rate λ, the $\{Y_i\}$ are iid, positive, and independent of N. We assume the net profit condition $c > \lambda\mu$. We denote by $M_Y(r) = \mathbb{E}[e^{rY}]$ the moment-generating function. We now assume that $M_Y(r_\infty) = \infty$, where $r_\infty = \sup\{r : M_Y(r) < \infty\}$. That is, $\lim_{r\uparrow r_\infty} M_Y(r) = \infty$. Let $\theta(r) = \lambda(M_Y(r)-1) - cr$. Then $M = \{\exp(-rX_t - \theta(r)t)\}$ is a martingale.

Under $\mathbb{P}^*$ we get the same sort of process.

Proposition C.3. *Under $\mathbb{P}^*$ the process X is again a classical risk model. The claim intensity is $\lambda M_Y(r)$, the claim sizes follow the distribution $G^*(x) = M_Y(r)^{-1}\int_0^x e^{ry}\,dG(y)$, and the premium rate is c.*

Proof. Because the slope of the process between claims is deterministic, the premium rate cannot change. We therefore have to find the joint distribution of N_t, $T_1,\dots,T_{N_t}$, and $Y_1,\dots,Y_{N_t}$. Let $n \in \mathbb{N}$, B_k be Borel sets, and $A \in \sigma(T_1,\dots,T_n)$. Then

$$
\begin{aligned}
&\mathbb{P}^*[N_t = n, (T_1,\dots,T_n) \in A, Y_k \in B_k, 1 \le k \le n] \\
&\quad = \mathbb{E}\Bigl[\exp\Bigl\{r\sum_{i=1}^{N_t} Y_i\Bigr\}; N_t = n, (T_1,\dots,T_n) \in A, Y_k \in B_k\Bigr] e^{-\lambda(M_Y(r)-1)t} \\
&\quad = \frac{(\lambda t)^n}{n!} e^{-\lambda M_Y(r)t}\,\mathbb{P}[(T_1,\dots,T_n) \in A \mid N_t = n] \prod_{k=1}^n \mathbb{E}[e^{rY_k}; Y_k \in B_k] \\
&\quad = \frac{(\lambda M_Y(r)t)^n}{n!} e^{-\lambda M_Y(r)t}\,\mathbb{P}[(T_1,\dots,T_n) \in A \mid N_t = n] \\
&\qquad \times \prod_{k=1}^n M_Y(r)^{-1}\mathbb{E}[e^{rY_k}; Y_k \in B_k]\,.
\end{aligned}
$$

This shows that $\{Y_i\}$ is independent of N and has the right distribution. N_t has a Poisson distribution with parameter $\lambda M_Y(r)t$. Given $N_t = n$, the occurrence times have the same conditional distribution as under the original measure. By Proposition A.4 this is a Poisson process with rate $\lambda M_Y(r)$. □

The above proposition shows that it is possible to extend the measure $\mathbb{P}^*$ to the whole σ-algebra $\mathcal{F}$. Let us calculate the drift of the process,

$$
c - \lambda M_Y(r)\frac{M_Y'(r)}{M_Y(r)} = c - \lambda M_Y'(r) = -\theta'(r)\,.
$$

Because the function $\theta(r)$ is strictly convex, we get $\theta'(r) \ne \theta'(\tilde r)$ for $r \ne \tilde r$. In particular, the measures $\mathbb{P}^*$ and $\mathbb{P}$ are singular. Denote by r_0 the (unique) solution to $\theta'(r) = 0$. Then the net profit condition is fulfilled for $r < r_0$, and ruin occurs almost surely for $r \geq r_0$.

As in Section C.2, the change of measure formula can be used to calculate ruin probability in finite and infinite time. However, because $X_\tau < 0$, the formulae are not as explicit as for the Brownian motion; see Section D.1.2.

Bibliographical Remarks

Early versions of the change of measure technique can be found, for example, in Asmussen [6] or Siegmund [172]. An introduction to the method presented in this chapter is found in Asmussen [7] or Rolski et al. [152]. Change of measure is an important technique for simulation; see Asmussen and Rubinstein [10], Glynn and Iglehart [82], Lehtonen and Nyrhinen [125], or Siegmund [171]. In the theory of large deviations it turns out that conditioned on crossing a large level the process follows the law obtained by changing the measure with the Lundberg exponent R; see, for instance, Asmussen [4].

D

Risk Theory

In this appendix we review the main results on the classical Cramér–Lundberg model. The results can also be found in [152]. For further reading also see the references therein. At the end of the chapter we will review the most important reinsurance treaties.

The reader should be aware that the model has to be considered as a *technical tool* only. It is used to "measure" the effect of a certain decision of the actuary on the risk. In this model the present environment of the insurer is fixed and cannot be changed in the future. Of course, in reality the environment does change. For example, changes in car construction do influence the claim size distribution for motor insurance. Legal changes may change the portfolio or the solvency level. And last but not least, if an insurer's capital went to infinity, both equity holders and the insured would claim to get part of it. Moreover, for tax reasons it is not advisable to have too large a surplus.

The time t in the model has to be considered as *operational time*. On the one hand, the insured's exposure to risk is not constant over time. On the other hand, the number of persons insured is not constant over time either. One therefore has to consider time as the integrated risk volume the company has taken over.

The ruin probabilities defined here are therefore not the probability that the company is ruined, even though for some claim size distributions this could be the case. Ruin means that the capital set aside for the risk considered was not enough. The ruin probability is then a measure for the risk. Ruin theory gives the actuary a tool to measure the risk in a simple way. The goal is therefore not to have "the realistic model" but a simple model that is able to characterise the risk connected to the business.

D.1 The Classical Risk Model

D.1.1 Introduction

A sound mathematical basis for the stochastic modelling of insurance risk goes back to the pioneering work by Filip Lundberg [127, 128] and Harald Cramér [36, 37]. Their *collective risk model* was obtained as a limit of a sum of *individual risk models* for an increasing number of individual contracts. It turns out that many of the basic constructions like adjustment coefficient, expense loading, premium structure, etc. needed in more general models are already present in this early model. Despite the obvious lack of reality of many in the assumptions made, one uses the *Cramér–Lundberg model* as a skeleton for many recently developed "more realistic" generalisations.

In a classical risk model the surplus of a collective contract or a large portfolio is modelled as

$$X_t = x + ct - \sum_{i=1}^{N_t} Y_i \;,$$

where N is a Poisson process with rate λ, the $\{Y_i\}$ are iid, (strictly) positive, and independent of N, $c > 0$ is the premium rate, and x is the initial capital. We denote the claim occurrence times by $T_1 < T_2 < \cdots$, and for convenience we let $T_0 = 0$. Because the process has stationary and independent increments, the process is always in its stationary state. It does not matter whether or not there was a claim at time zero. We denote the distribution of Y_i by $G(y)$, and its moments by $\mu_n = \mathbb{E}[Y_i^n]$. For simplicity we let $\mu = \mu_1$.

The main object of interest in risk theory is the *ruin probability*. Let $\tau = \inf\{t : X_t < 0\}$ be the *time of ruin*. As usual, we let $\inf \emptyset = \infty$. The *probability of ultimate ruin* $\psi(x) = \mathbb{P}[\tau < \infty]$ is the probability that ruin occurs in finite time. From the theory of random walks one knows that $\tau < \infty$ (a.s.) if and only if $\mathbb{E}[c(T_i - T_{i-1}) - Y_i] \le 0$. That is, $\psi(x) = 1$ for all x if $c \le \lambda\mu$. One therefore usually assumes the *net profit condition* $c > \lambda\mu$. Note that $\mathbb{E}[X_t - x] = (c - \lambda\mu)t$, which explains the name "net profit condition."

The ruin probability is absolutely continuous and differentiable at all points y where $G(y)$ is continuous. The density $\psi'(x)$ fulfils

$$c\psi'(x) + \lambda\Bigl[\int_0^x \psi(x-y)\,\mathrm{d}G(y) + 1 - G(x) - \psi(x)\Bigr] = 0\;. \tag{D.1}$$

Note that for the Markov process $\{X_{\tau\wedge t}\}$ this is $\mathfrak{A}\psi(x) = 0$, where $\mathfrak{A}$ is the infinitesimal generator. Conversely, if there is a solution (in the sense of absolutely continuous functions) on $[0, \infty)$ to (D.1) with $\psi(x) \to 0$ as $x \to \infty$, then this is the ruin probability.

Integration and reformulation of the equation yields

$$c(\psi(0) - \psi(x)) = \lambda\Big[\int_0^x (1 - G(y))\,dy - \int_0^x \psi(x-y)(1 - G(y))\,dy\Big]\,.$$

Letting $x \to \infty$ and noting that $\psi(x) \to 0$ yields

$$\psi(0) = \frac{\lambda\mu}{c}\,.$$

Then

$$\psi(x) = \int_0^x \psi(x-y)\frac{\lambda}{c}(1 - G(y))\,dy + \int_x^\infty \frac{\lambda}{c}(1 - G(y))\,dy\,. \tag{D.2}$$

Let $B(x) = \mu^{-1}\int_0^x (1 - G(y))\,dy$. If $x = 0$ and $\tau < \infty$, then $-X_\tau$ has the distribution $B(x)$. We can therefore see that $B(x)$ is the strictly descending ladder height distribution, given that there is a ladder epoch. $\lambda\mu B(y)/c$ is the improper ladder height distribution.

(D.2) has the following interpretation. Let $\tau_1 = \inf\{t : X_t < x\}$ be the first ladder epoch and $L_1 = x - X_{\tau_1}$ be the first ladder height. Then L_1 has the defective distribution $\lambda\mu B(y)/c$. If $y = L_1 \le x$, then the capital is $x - y$ and the ruin probability after time τ_1 is $\psi(x-y)$. If $y > x$, then ruin occurs at time τ_1.

D.1.2 Small Claims

Let $M_Y(r) = \mathbb{E}[e^{rY_i}]$ denote the moment-generating function of the claim sizes. Let $\theta(r) = \lambda(M_Y(r) - 1) - cr$. Clearly, $\theta(0) = 0$. The second derivative is $\theta''(r) = \lambda\mathbb{E}[Y_i^2 e^{rY_i}] > 0$. Thus, $\theta(r)$ is a strictly convex function. Because $\theta'(r) = \lambda\mathbb{E}[Y_i e^{rY_i}] - c$, the derivative in zero is $\theta'(0) = -(c - \lambda\mu)$. This is strictly negative under the net profit condition. Under mild conditions one could therefore have that there is a strictly positive R such that $\theta(R) = 0$. For example, if $r_\infty = \sup\{r : M_Y(r) < \infty\}$ is the right endpoint where the moment-generating function is finite, and $\lim_{r\uparrow r_\infty} M_Y(r) = \infty$, then R exists. This is always the case if $r_\infty = \infty$.

Suppose that there is an $R \neq 0$ such that $\theta(R) = 0$. Then $R > 0$ and the process $\{M_t = e^{-R(X_t - x)}\}$ becomes a martingale. On the σ-algebra $\mathcal{F}_t$ we can define the new measure $\mathbb{P}^*[A] = \mathbb{E}[M_t; A] := \mathbb{E}[M_t \mathbb{I}_A]$. The measure is independent of t. Moreover, it is possible to extend the measure to the whole σ-algebra $\mathcal{F}$. However, the measure is equivalent to $\mathbb{P}$ on $\mathcal{F}_t$ but singular on $\mathcal{F}$. The process X remains a classical risk model with premium rate c, with claim intensity $\lambda M_Y(R)$, and with claim size distribution $\int_0^y e^{Rz}\,dG(z)/M_Y(R)$; see Proposition C.3. The net profit condition is not fulfilled, and thus $\mathbb{P}^*[\tau < \infty] = 1$. The ruin probability can then be expressed as

$$\psi(x) = \mathbb{E}^*[(M_\tau)^{-1}; \tau < \infty] = \mathbb{E}^*[e^{RX_\tau}; \tau < \infty]\,e^{-Rx} = \mathbb{E}^*[e^{RX_\tau}]\,e^{-Rx}\,.$$

Because $X_\tau < 0$, we obtain *Lundberg's inequality*:

$$\psi(x) < \mathrm{e}^{-Rx} . \tag{D.3}$$

By conditioning on $X_{\tau-}$ we can also obtain a lower bound. We have

$$\begin{aligned}
\mathbb{E}^*[\mathrm{e}^{RX_\tau} \mid X_{\tau-} = z] &= \mathbb{E}^*[\mathrm{e}^{-R(Y-z)} \mid Y > z] = \frac{\mathrm{e}^{Rz}(1-G(z))}{\int_z^\infty \mathrm{e}^{Ry} \,\mathrm{d}G(y)} \\
&= \frac{1}{\mathbb{E}[\mathrm{e}^{R(Y-z)} \mid Y > z]} .
\end{aligned}$$

Defining

$$C_- = \inf_z \frac{1}{\mathbb{E}[\mathrm{e}^{R(Y-z)} \mid Y > z]} , \tag{D.4}$$

we get the lower Lundberg bound $\psi(x) \ge C_- \mathrm{e}^{-Rx}$. Here we take the infimum over all z such that $\mathbb{P}[Y > z] > 0$.

The lower Lundberg bound only makes sense if $C_- > 0$. This is clearly fulfilled if Y_i has bounded support. If Y_i has unbounded support, we have, for example, the following sufficient condition. This condition is only slightly stronger than the condition that $r_\infty > R$.

Lemma D.1. *Suppose that $G(y)$ is absolutely continuous, and denote by $\ell(y) = G'(y)/(1-G(y))$ its hazard rate. If $\liminf_{y\to\infty} \ell(y) > R$, then $C_- > 0$.*

Proof. We need to show that $\mathbb{E}[\mathrm{e}^{R(Y-z)} \mid Y > z]$ is bounded. We start by rewriting this expression:

$$\begin{aligned}
\mathbb{E}[\mathrm{e}^{R(Y-z)} \mid Y > z] &= \frac{\int_z^\infty \mathrm{e}^{R(y-z)} \,\mathrm{d}G(y)}{1-G(z)} \\
&= 1 + \frac{\int_z^\infty \int_0^{y-z} R\mathrm{e}^{Rv} \,\mathrm{d}v \,\mathrm{d}G(y)}{1-G(z)} \\
&= 1 + R\frac{\int_z^\infty \int_z^y \mathrm{e}^{R(v-z)} \,\mathrm{d}v \,\mathrm{d}G(y)}{1-G(z)} \\
&= 1 + R\frac{\int_z^\infty (1-G(v))\mathrm{e}^{R(v-z)} \,\mathrm{d}v}{1-G(z)} \\
&= 1 + R\frac{\int_z^\infty \exp\{-\int_0^v \ell(y) \,\mathrm{d}y\}\mathrm{e}^{R(v-z)} \,\mathrm{d}v}{\exp\{-\int_0^z \ell(y) \,\mathrm{d}y\}} \\
&= 1 + R\int_z^\infty \exp\Bigl\{\int_z^v (R-\ell(y)) \,\mathrm{d}y\Bigr\} \,\mathrm{d}v .
\end{aligned}$$

Let $r = \liminf_{y\to\infty}(\ell(y) - R)/2$. There is a z_0 such that $\ell(y) > R + r$ for $z \ge z_0$. Then for $z \ge z_0$ we have $\mathbb{E}[\mathrm{e}^{R(Y-z)} \mid Y > z] < 1 + R/r$. Because $\mathbb{E}[\mathrm{e}^{R(Y-z)} \mid Y > z]$ is bounded for $z \in [0, z_0]$, the result follows. □

The function $f(x) = \mathbb{E}^*[e^{RX_\tau} \mid X_0 = x]$ fulfils a renewal equation. From the key renewal theorem one obtains

$$\lim_{x\to\infty} \psi(x)e^{Rx} = \frac{c - \lambda\mu}{\lambda M_Y'(R) - c} ,$$

which is called the *Cramér–Lundberg approximation*. It turns out that the approximation

$$\psi(x) \sim \frac{c - \lambda\mu}{\lambda M_Y'(R) - c} e^{-Rx}$$

works quite well also for quite small initial capital x. For an alternative approach, see [152].

D.1.3 Large Claims

A distribution function $F(x)$ is called *subexponential* if

$$\lim_{x\to\infty} \frac{1 - F^{*n}(x)}{1 - F(x)} = n$$

for some $n \ge 2$ and therefore all $n \ge 2$, where $F^{*n}(x)$ denotes the n-fold convolution. Because for iid $\{X_i\}$ with distribution $F(x)$

$$\lim_{x\to\infty} \frac{\mathbb{P}[\max\{X_1, \dots, X_n\} > x]}{1 - F(x)} = n ,$$

the interpretation is the following. If $F(x)$ is subexponential, then the sum of n random variables exceeds a large level x typically if the largest of the n variables exceeds the level x. Something similar can be observed for many insurance portfolios. A small number of claims is responsible for almost all the aggregate claim. We denote the class of subexponential distributions by $\mathcal{S}$.

Using $F^{*n}(x) = \int_0^x F^{*(n-1)}(x - y) \, dF(y)$, the definition of subexponentiality can be written as

$$\lim_{x\to\infty} \int_0^x \frac{1 - F^{*n}(x - y)}{1 - F(x)} \, dF(y) = n \tag{D.5}$$

for some and hence all $n \ge 1$.

A slightly smaller class of distribution functions are the distribution functions with finite mean such that

$$\lim_{x\to\infty} \int_0^x \frac{(1 - F(x - y))(1 - F(y))}{1 - F(x)} \, dy = 2\mu_F ,$$

where $\mu_F := \int_0^\infty (1 - F(z)) \, dz$. We denote this class by $\mathcal{S}^*$. The motivation for this definition is that one often needs the distribution function $F^s(x) :=$

$\mu_F^{-1} \int_0^x (1-G(y))\, dy$ to be subexponential. One possible criterion is l'Hôpital's rule

$$\lim_{x\to\infty} \frac{\int_x^\infty \int_0^y (1-F(y-z))(1-F(z))\, dz\, dy}{\mu_F \int_x^\infty (1-F(y))\, dy}$$

$$= \lim_{x\to\infty} \int_0^x \frac{(1-F(x-z))(1-F(z))}{\mu_F(1-F(x))}\, dz\,.$$

Thus, $F(x) \in \mathcal{S}^*$ implies that $F^{\mathrm{s}}(x) \in \mathcal{S}$. One can show, moreover, that $F(x) \in \mathcal{S}^*$ also implies that $F(x) \in \mathcal{S}$.

By symmetry the definition of $\mathcal{S}^*$ can be written as

$$\lim_{x\to\infty} \int_0^{x/2} \frac{(1-F(x-y))(1-F(y))}{1-F(x)}\, dy = \mu_F\,.$$

Note that from (D.6) below we see that the limit and integration can be interchanged in this alternative definition but not in the first definition.

A subexponential distribution has the property that

$$\lim_{x\to\infty} \frac{1-F(x-y)}{1-F(x)} = 1\,. \tag{D.6}$$

Note that for any distribution function $F(x)$,

$$\liminf_{x\to\infty} \int_0^{x/2} \frac{(1-F(x-y))(1-F(y))}{1-F(x)}\, dy$$

$$\ge \int_0^\infty \liminf_{x\to\infty} \frac{(1-F(x-y))(1-F(y))}{1-F(x)} \mathbb{1}_{y\le x/2}\, dy = \mu_F\,.$$

A distribution function $F(x)$ has a *regularly varying tail* with index $-\alpha$ if

$$\lim_{x\to\infty} \frac{1-F(xy)}{1-F(x)} = y^{-\alpha}$$

for all $y > 0$. One can show that the convergence is uniform for $y \in [a,b]$ for $0 < a < b < \infty$. It is clear that $(1-F(x-y))/(1-F(x)) \to 1$ as $x \to \infty$. If the mean value is finite, then $\alpha \ge 1$. We find in this case that

$$\int_0^{x/2} \frac{(1-F(x-y))(1-F(y))}{1-F(x)}\, dy \le \int_0^{x/2} \frac{(1-F(x/2))}{1-F(x)}(1-F(y))\, dy\,.$$

Because $(1-F(x/2))/(1-F(x))$ converges to 2^α, the limit and integration can be interchanged. Thus, all distributions with regularly varying tail and finite mean are in $\mathcal{S}^*$.

If $\alpha = \infty$, i.e., $(1-F(xy))/(1-F(x)) \to 0$ for all $y > 1$, we call the distribution tail of *rapid variation.* Distributions like the log-normal distribution, the Weibull distribution, or all the light-tailed distributions belong to this class.

Another useful criterion for subexponentiality is the following. Suppose that the hazard rate $\ell_F(x) = F'(x)/(1-F(x))$ exists and is eventually decreasing to zero. Then $F(x) \in \mathcal{S}^*$ if and only if

$$\lim_{x\to\infty} \int_0^x \exp\{y\ell_F(x)\}(1-F(y)) = \mu_F < \infty\,.$$

This implies that the heavy-tailed Weibull distribution $F(x) = 1-\exp\{-cx^a\}$ for $0 < a < 1$ and the log-normal distribution belong to $\mathcal{S}^*$.

The property $(1-F(x-y))/(1-F(x)) \to 1$ as $x \to \infty$ implies in particular that $M_Y(r) = \infty$ for all $r > 0$. Therefore, no exponential moments exist, and the adjustment coefficient does not exist. Thus, the theory of Section D.1.2 does not apply.

Suppose that $B(x) = \mu^{-1}\int_0^x (1-G(y))\,dy$ is subexponential. Then

$$\lim_{x\to\infty} \frac{\psi(x)}{1-B(x)} = \frac{\lambda\mu}{c-\lambda\mu}\,. \tag{D.7}$$

The limit is the expected number of ladder heights. Thus, we can interpret the result that ruin occurs for a large initial capital x typically if the largest ladder height exceeds the level x.

D.2 Perturbed Risk Models

In order to model some uncertainties, Gerber [73] added a Brownian motion to a classical risk model:

$$X_t = x + ct - \sum_{i=1}^{N_t} Y_i + \sigma W_t\,.$$

We define ruin as before. The difference is now that ruin can also occur because the perturbation $\{\sigma W_t\}$ leads to ruin. In this case $X_\tau = 0$.

If $G(y)$ has a bounded density, then $\psi(x)$ is twice continuously differentiable. In this case $\psi(x)$ fulfils

$$\frac{\sigma^2}{2}\psi''(x) + c\psi'(x) + \lambda\Bigl[\int_0^x \psi(x-y)\,\mathrm{d}G(y) + 1 - G(x) - \psi(x)\Bigr] = 0\,. \tag{D.8}$$

This is again $\mathfrak{A}\psi(x) = 0$. Conversely, any twice continuously differentiable solution to (D.8) on $[0,\infty)$ with $\psi(x) \to 0$ as $x \to \infty$ is the ruin probability.

For small claims one defines the Lundberg coefficient $R \neq 0$ as the solution to

$$\lambda(M_Y(R) - 1) - cR + \frac{\sigma^2}{2}R^2 = 0\,. \tag{D.9}$$

As for the classical model the process $\{M_t = e^{-R(X_t - x)}\}$ is a martingale. Define the new measure on $\mathcal{F}_t$, $\mathbb{P}^*[A] = \mathbb{E}[M_t; A]$. The measure can be extended on $\mathcal{F}$. Under $\mathbb{P}^*$ the process $\{X_t\}$ is again a perturbed classical risk model. The premium rate is $c - \sigma^2 R$, the claim intensity is $\lambda M_Y(R)$, and the claim size distribution is $\int_0^y e^{Rz}\, dG(z)/M_Y(R)$. The volatility σ does not change. We have $\mathbb{P}^*[\tau < \infty] = 1$. The ruin probability is then

$$\psi(x) = \mathbb{E}^*[M_\tau^{-1}; \tau < \infty] = \mathbb{E}^*[e^{RX_\tau}]e^{-Rx} .$$

Thus, *Lundberg's inequality*

$$\psi(x) < e^{-Rx} \tag{D.10}$$

again holds. Through a renewal approach we also get the *Cramér–Lundberg approximation*

$$\lim_{x\to\infty} \psi(x)e^{Rx} = \frac{c - \lambda\mu}{\lambda M_Y'(R) - c + \sigma^2 R} .$$

The large claim case works similarly as for the classical model. If $B(y)$ is a subexponential distribution, then (D.7) holds.

D.3 Diffusion Approximations

It is often difficult to calculate characteristics of the classical risk model, such as ruin probabilities. One therefore looks for approximations. A simple idea, and quite successful in queueing theory, is diffusion approximations. The idea is to consider a sequence of classical risk models, such that the models converge weakly to a diffusion process.

Because a classical risk process has stationary and independent increments, a diffusion approximation only makes sense if the diffusion process has also stationary and independent increments. Thus, the limiting process should be a Brownian motion. Diffusion approximations to the ruin probability make sense because of the following result for ruin probabilities proved in [126] for finite time and [155] for infinite time.

Proposition D.2. *Let $X^{(n)}$ be a sequence of classical risk models with $X_0^{(n)} = x$, and let W be a standard Brownian motion. Suppose that $X^{(n)}$ converges weakly to X with $X_t = x + mt + \sigma W_t$. Denote by $\tau^{(n)}$ the ruin time of $X^{(n)}$, by τ the ruin time of X. Then $\tau^{(n)}$ converges weakly to τ. If, moreover, $\limsup_{n\to\infty} \mathrm{Var}[X_1^{(n)}] < \infty$, then also $\lim_{n\to\infty} \mathbb{P}[\tau^{(n)} < \infty] = \mathbb{P}[\tau < \infty]$.* □

If one now wants to approximate the ruin probability by the ruin probability of the diffusion approximation, the infinite ruin probability is given by (A.2). The ruin probability in finite time is given by (A.3).

The simplest diffusion approximation is obtained in the following way. In order for the limit to be a diffusion, one needs to increase the number of claims and to make them smaller. If one likes to keep the first two moments of the risk process constant, one should have $c^{(n)} - \lambda^{(n)}\mu^{(n)}$ and $\lambda^{(n)}\mu_2^{(n)}$ remaining constant. In order for $\lambda^{(n)} \to \infty$, we let $\lambda^{(n)} = \lambda n$. Then $\mu_2^{(n)} = \mu_2/n$. We therefore choose $G^{(n)}(y) = G(y\sqrt{n})$. It follows then that $c^{(n)} = c+(\sqrt{n}-1)\lambda\mu$. Note that this implies that the safety loading

$$\rho^{(n)} = \frac{c^{(n)} - \lambda^{(n)}\mu^{(n)}}{\lambda^{(n)}\mu^{(n)}} = \frac{c - \lambda\mu}{\lambda\mu\sqrt{n}} = \frac{\rho}{\sqrt{n}}$$

converges to zero. This should in fact be the case for any meaningful diffusion approximation. It turns out that diffusion approximations work well as long as the safety loading is small. Numerical examples can be found in [85].

If enough exponential moments exist, it is possible to consider *corrected diffusion approximations*. The basic idea is to consider an exponential class of risk models and to change the parameter determining which class is considered in an appropriate way. This type of approximation is described in [5], [156], or [172].

As we see in Chapter 2, an optimal control is usually of the feedback form $U_t^* = u(X_{t-}^*)$, where $u(x)$ is some function and $\{X_t^*\}$ is the process under the optimal control U^*. The following result from [155] motivates us to approximate a control problem by a diffusion approximation.

Proposition D.3. *Let $m : \mathbb{R} \to \mathbb{R}$ be a Lipschitz continuous function, $\{X^{(n)} : n \in \mathbb{N}\}$ be semi-martingales such that $X_0^{(n)} = 0$, and X be a semi-martingale such that $X_0 = 0$. Suppose that $Z^{(n)}$ fulfils the stochastic integral equation*

$$Z_t^{(n)} = x + X_t^{(n)} + \int_0^t m(Z_s^{(n)}) \,\mathrm{d}s$$

and Z fulfils

$$Z_t = x + X_t + \int_0^t m(Z_s) \,\mathrm{d}s \,.$$

Then $Z^{(n)}$ converges weakly to Z if and only if $X^{(n)}$ converges weakly to X. □

Note that the result does not imply that the control process converges weakly. Often it would be enough if the sequence of optimal control processes is relatively compact. However, one needs to verify some condition that the limit and expectation can be interchanged. This is often not a trivial task to show.

D.4 Premium Calculation Principles

For completeness we review here some of the most popular premium calculation principles. In the following we denote by S a risk; for example, the (annual) loss of some insurance contract. Any reasonable premium p should consist of the net premium $\mathbb{E}[S]$ and some security loading, i.e., $p > \mathbb{E}[S]$.

Expected value principle

The premium is calculated by $p = (1+\theta)\mathbb{E}[S]$ for some safety loading $\theta > 0$.

Variance principle

The premium is $p = \mathbb{E}[S] + \alpha \operatorname{Var}[S]$ for some $\alpha > 0$.

Modified variance principle

The premium is $p = \mathbb{E}[S] + \alpha \operatorname{Var}[S]/\mathbb{E}[S]$ for some $\alpha > 0$.

Standard deviation principle

The premium is $p = \mathbb{E}[S] + \alpha\sqrt{\operatorname{Var}[S]}$ for some $\alpha > 0$.

Exponential principle

The premium is $p = \alpha^{-1} \log \mathbb{E}[\exp\{\alpha S\}]$ for some $\alpha > 0$.

Zero utility principle

Let $u(x)$ be some strictly increasing, strictly concave function. The zero utility premium is the unique solution to the equation $u(w) = \mathbb{E}[u(w+p-S)]$. Here w is considered as the insurer's initial wealth. One then compares the utility of the initial wealth (no risk is taken over) with the expected utility of the wealth after the risk is taken over. The exponential premium principle is a special case with $u(x) = -\mathrm{e}^{-\alpha x}$.

Adjusted risk principle

Denote by $F(x)$ the distribution function of the risk S and we assume that $S \geq 0$. The premium is calculated as $p = \int_0^\infty (1-F(x))^\theta \,\mathrm{d}x$ for some $\theta \in (0,1)$. Note that $\theta = 1$ would give the net premium.

D.5 Reinsurance

A portfolio of a first insurer is often too small to make the risk small. In this context a first insurer is usually called a *cedent.* The supervising authority will then ask for large investments from the shareholders, i.e., a large initial capital. If several cedents constituted a pool, the portfolio would become larger. And with it also the capital requirements would become smaller for each of the participating companies.

In practise, a reinsurance company will take over part of the risk from different cedents. Because the reinsurer collects risk from several cedents, some sort of pool is created. In such a way, the requirement for initial capital is reduced for the cedent. On the other hand, the reinsurer needs some initial capital. However, the sum of the required initial capitals is usually smaller than it would be without reinsurance. General rules used as a guideline for legislation are formulated in Solvency II, the equivalent to Basle II for insurers.

There are three different types of reinsurance treaties: reinsurance acting on individual claims, reinsurance acting on the aggregate claim over a certain period, or reinsurance acting on the k largest claims occurring during a certain period. In this book we only consider reinsurance acting on individual claims. We will review here some basic reinsurance forms, for completeness from all three sorts of claims. Note that in practise combinations of the reinsurance treaties mentioned here apply; for example, proportional reinsurance in a certain layer.

Reinsurance acting on individual claims:

For each claim Y the part of the claim left to the insurer is $0 \le s(Y) \le Y$. The reinsurer pays $Y - s(Y)$. The function $s(y)$ is called the *self-insurance function.*

- **Full reinsurance:** The self-insurance function is $s(Y) = 0$, i.e., the reinsurer pays all the claim. This form is not used in practise. But it is a popular form of a contract between cedent and policyholder.
- **Proportional reinsurance:** The self-insurance function is $s(Y) = bY$ for a retention level $b \in (0, 1)$. The reinsurer pays $(1 - b)Y$.
- **Excess of loss reinsurance:** The self-insurance function is $s(Y) = \min\{Y, b\}$ for some deductible $b \in (0, \infty)$. The reinsurer pays $(Y - b)^+$.
- **First risk deductible:** The reinsurer pays $\min\{Y, b\}$ for some deductible $b \in (0, \infty)$. Thus, the self-insurance function is $s(Y) = (Y - b)^+$.
- **Proportional reinsurance in a layer:** The self-insurance function is $s(Y) = \min\{Y, a\} + (Y - a - \gamma)^+ + b \min\{(Y - a)^+, \gamma\}$ for some $a, \gamma > 0$ and $b \in (0, 1)$.

Reinsurance acting on the aggregate claim:

For the aggregate sum of claims $S = \sum_{i=1}^{N} Y_i$, the insurer pays the amount $s(S)$ with $0 \le s(S) \le S$. The reinsurer pays $S - s(S)$.

- **Proportional reinsurance:** The self-insurance function is $s(S) = bS$ for a retention level $b \in (0, 1)$. This is the same as *proportional reinsurance* acting on individual claims.
- **Stop-loss reinsurance:** The self-insurance function is $s(S) = \min\{S, b\}$ for some deductible $b \in (0, \infty)$.
- **First risk deductible:** The self-insurance function is $s(S) = (S - b)^+$ for some deductible $b \in (0, \infty)$.

Reinsurance acting on the k largest claims:

Let $Y_{1:N} \geq Y_{2:N} \geq \cdots \geq Y_{N:N}$ denote the order statistics of the N claims occurring in the period. If $k \geq N$, the reinsurance acts on all claims; if $N > k$, the reinsurance acts on $Y_{1:N}, \dots, Y_{k:N}$. Let $S_k = \sum_{i=1}^{N \wedge k} Y_{i:N}$ denote the sum of the k largest claims.

- **Full reinsurance on the k largest claims:** The self-insurance function is $s = S - S_k$.
- **Proportional reinsurance on the k largest claims:** The self-insurance function is $s = bS_k + (S - S_k)$, with retention level $b \in [0, 1)$.
- **Stop-loss reinsurance on the k largest claims:** The self-insurance function is $s = S - (S_k - b)^+$ with $b \in (0, \infty)$.
- **Excess of loss reinsurance on the k largest claims:** The self-insurance function is $s = S - \sum_{i=1}^{N \wedge k} (Y_{i:N} - b)^+$ with $b \in (0, \infty)$.
- **ECOMOR:** The self-insurance function is

$$s = \sum_{i=1}^{N} \min\{Y_i, Y_{k+1:N}\} ,$$

where $Y_{i:n} = 0$ if $i > n$. Here the deductible is random and equal to the largest claim that is not covered.

Bibliographical Remarks

The classical risk model was first considered by Lundberg [127, 128] and Cramér [36, 37], where the results of Sections D.1.1 and D.1.2 can be found. The lower bound (D.4) was found by Taylor [179]. Subexponential distributions were introduced by Chistyakov [34]. The distributions were also considered in Chover et al. [35]. The class $\mathcal{S}^*$ was introduced by Klüppelberg [119]. Result (D.7) goes back to Embrechts and Veraverbeke [56]. More literature on regular variation can be found in Bingham et al. [22] or Feller [58]. Further literature on classical risk processes are Asmussen [7], Dickson [46], Embrechts et al. [55], Gerber [76], Grandell [86], and Rolski et al. [152]. Perturbed risk processes were introduced by Gerber [73]. The light-tailed case is also considered by Dufresne and Gerber [52], Furrer and Schmidli [66], and Schmidli [157]. The large claim case was solved by Veraverbeke [183]. Generalisations were considered in Furrer [65] and Schmidli [159]. A review is found in Schmidli [158]. Diffusion approximations in insurance were first considered by Hadwiger [89]. Based on weak convergence, diffusion approximations were treated by Asmussen [5], Grandell [85], Iglehart [108], Schmidli [155, 156], and Siegmund [172]. For an introduction to the weak convergence of stochastic processes, see, for instance, Billingsley [21] or Ethier and Kurtz [57]. For more literature on reinsurance, see, for example, Goovaerts et al. [84] or Sundt [178].

E

The Black–Scholes Model

One of the most important and successful models used in financial mathematics is the Black–Scholes model. Most of the models used for pricing a risky asset are generalisations of this model. The reason for its popularity is the simple way financial derivatives can be priced; see, for instance, the Black–Scholes formula below.

Denote the price at time t of a risky asset, or a financial portfolio, by Z_t. We model the process Z as

$$Z_t = Z_0 \exp\{\sigma W_t + (m - \tfrac{1}{2}\sigma^2)t\}\;, \tag{E.1}$$

where W is a standard Brownian motion and $m, \sigma > 0$. In differential form the process is determined by the stochastic differential equation

$$\mathrm{d}Z_t = \sigma Z_t \,\mathrm{d}W_t + m Z_t \,\mathrm{d}t\;; \tag{E.2}$$

see Proposition A.7. It follows from the definition that $Z_t > 0$ for all t. In addition there is a riskless bond

$$Z_t^{\mathrm{r}} = \exp\{\delta t\}\;, \qquad \mathrm{d}Z_t^{\mathrm{r}} = \delta Z_t^{\mathrm{r}}\,\mathrm{d}t\;. \tag{E.3}$$

A trader has now the possibility to invest into the risky asset and the riskless bond. At time t the trader holds θ_t units of the asset and θ_t^{r} units of the riskless bond. The value at time t of the portfolio is then

$$V_t = \theta_t Z_t + \theta_t^{\mathrm{r}} Z_t^{\mathrm{r}}\;.$$

We often will need a portfolio such that no money has to be added but no money is consumed either.

Definition E.1. *An investment strategy* $\{(\theta_t, \theta_t^{\mathrm{r}})\}$ *is called* self-financing *if*

$$V_t = V_0 + \int_0^t \theta_s \,\mathrm{d}Z_s + \int_0^t \theta_s^{\mathrm{r}}\,\mathrm{d}Z_s^{\mathrm{r}}$$

for all $t \geq 0$.

The process $M = \{\exp\{-\nu W_t - \nu^2 t/2\}$ is a martingale. Choose $\nu = (m-\delta)/\sigma$ and consider the measure $\mathbb{P}^*[A] = \mathbb{E}[M_T; A]$ on $\mathcal{F}_T$. Then $W_t^* = W_t + \nu t$ is a Brownian motion under $\mathbb{P}^*$; see Section C.2. Then

$$\mathrm{d}Z_t = \sigma Z_t \,\mathrm{d}W_t^* + \delta Z_t \,\mathrm{d}t \,.$$

In particular, $\{Z_t^0 = Z_t/Z_t^{\mathrm{r}}\}$ is a local martingale under $\mathbb{P}^*$. It follows then that $V_t^0 = V_t/Z_t^{\mathrm{r}}$ solves

$$V_t^0 = V_0 + \int_0^t \theta_s \,\mathrm{d}Z_t^0$$

if $\{(\theta_t, \theta_t^{\mathrm{r}})\}$ is self-financing. Thus, $\{V_t^0\}$ is a local martingale under $\mathbb{P}^*$. If we now fix $\{\theta_t\}$, it follows that $\{(\theta_t, \theta_t^{\mathrm{r}})\}$ is self-financing if $\theta_t^{\mathrm{r}} = V_t^0 - \theta_t Z_t^0$. Thus, for any $\{\theta_t\}$, we can choose $\{\theta_t^{\mathrm{r}}\}$ such that we obtain a self-financing investment strategy.

The following result is fundamental.

Theorem E.2. *Let X be $\mathcal{F}_T^W$-measurable and $\mathbb{E}[X^2] < \infty$. Then there is a stochastic process θ such that the corresponding self-financing trading strategy $\{(\theta_t, \theta_t^{\mathrm{r}})\}$ reproduces X, i.e.,*

$$X = V_0 + \int_0^T \theta_s \,\mathrm{d}Z_s + \int_0^T \theta_s^{\mathrm{r}} \,\mathrm{d}Z_s^{\mathrm{r}} \,.$$

□

Because $\{V_t^0\}$ is a local martingale, it follows that $V_0 = \mathbb{E}^*[X/Z_T^{\mathrm{r}}]$. This is the price at time zero of the contingent claim X.

The price of a contingent claim is thus determined under the measure $\mathbb{P}^*$ instead of the physical measure $\mathbb{P}$. That means no loading for the risk is added. We therefore call the measure $\mathbb{P}^*$ the *risk-neutral probability measure.*

If $X = (Z_T - K)^+$ (European call option), the price at time zero is the famous *Black–Scholes formula*

$$V_t = Z_t \Phi(d_1(t)) - K\mathrm{e}^{-\delta(T-t)}\Phi(d_2(t)) \,,$$

where

$$d_{1/2}(t) = \frac{\log Z_t/K + (\delta \pm \frac{1}{2}\sigma^2)(T-t)}{\sigma\sqrt{T-t}} \,,$$

and $\Phi(x) = (2\pi)^{-1/2} \int_{-\infty}^x \mathrm{e}^{-y^2/2} \,\mathrm{d}y$ is the distribution function of the standard normal distribution.

We see that the price does not depend on the parameter m because under $\mathbb{P}^*$, m is replaced by δ. This is one reason why Black–Scholes theory is so popular. The agent's belief in μ does not influence the price. The problem is, however, that σ has to be estimated. And the Black–Scholes formula is quite sensible to the choice of σ because σ typically is small.

An important property of the Black–Scholes model is the absence of arbitrage. We call a self-financing trading strategy $\{(\theta_t, \theta_t^{\mathrm{r}})\}$ an *arbitrage* if $V_0 = 0$, $V_t \geq -c$ for all $t \in [0, T]$ and some $c > 0$, $V_T \geq 0$, and $\mathbb{P}[V_T > 0] > 0$.

Proposition E.3. *There is no arbitrage in the Black–Scholes model.*

Proof. Let $\{(\theta_t, \theta_t^r)\}$ be a self-financing trading strategy with $V_0 = 0$, $V_t \ge -c$, and $V_T \ge 0$. Then $\{V_t^0\}$ is a local martingale under $\mathbb{P}^*$ that is bounded from below. Hence, it is a supermartingale by Lemma A.3. From $0 \le \mathbb{E}^*[V_T] \le \mathbb{E}^*[V_0] = 0$, it follows that $\mathbb{P}^*[V_T = 0] = 1$. □

Bibliographical Remarks

There are many textbooks on mathematical finance. A short list of references is Dothan [50], Duffie [51], Karatzas and Shreve [117, 118], Lamberton and Lapeyre [122], and Shreve [167]. A shorter introduction can also be found in Rolski et al. [152]. Brownian motion as a driving force for option prices has been introduced by Bachelier [13]. The geometric Brownian motion (E.1) as a model was introduced by Samuelson [153]. The Black–Scholes formula was established by Black and Scholes [23]. The connection to martingales was found by Harrison and Kreps [91] and Harrison and Pliska [92, 93].

F

Life Insurance

F.1 Classical Life Insurance

In a classical life insurance contract there are two types of payments: payments at death or payments for survival. Denote by T_x the time of death of a person with age x at time zero. Then we denote by ${}_tp_x = \mathbb{P}[T_x > t]$ the survival probability of the insured. If the discount rate is constant [see (E.3)], the value of one unit to be paid at time T provided that the insured is still alive is $e^{-\delta T}\,{}_tp_x$.

Suppose that ${}_tp_x$ is absolutely continuous. Then the hazard rate

$$\mu_{x+t} = -\frac{d}{dt} \log {}_tp_x = -\frac{\frac{d}{dt}{}_tp_x}{{}_tp_x}$$

is called the *force of mortality*. The density of the time of death is therefore $\mu_{x+t}\,{}_tp_x$. The survival probability can then be written as

$${}_tp_x = \exp\Bigl\{-\int_0^t \mu_{x+s}\, ds\Bigr\}\,.$$

The value of one unit paid upon death, if death occurs before time T, is then

$$\int_0^T e^{-\delta t}\mu_{x+t}\,{}_tp_x\, dt = \int_0^T \mu_{x+t} \exp\Bigl\{-\int_0^t (\delta + \mu_{x+v})\, dv\Bigr\}\, dt\,.$$

Modelling the state of the contract, we would consider the process $\mathbb{1}_{T_x>t}$. This is a simple example of an inhomogeneous Markov chain. The state zero is absorbing, and μ_{x+t} is the intensity of changing from state one to state zero.

A life insurance contract can also contain a pension to be paid after time T, or a pension to be paid whenever the insured gets disabled. Also, lump sum payments could be possible, for example, at the time the insured becomes

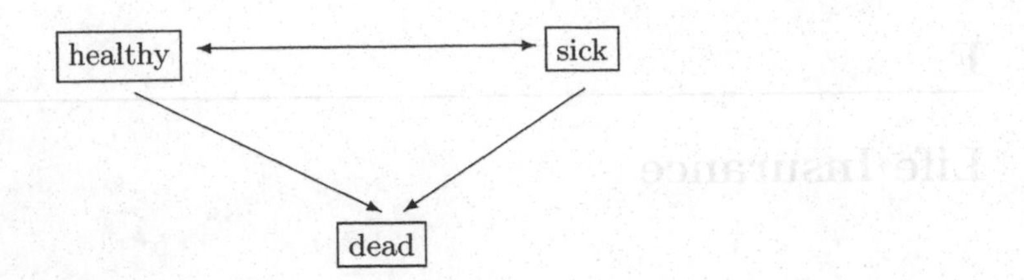

Fig. F.1. State of the insured for a simple insurance contract.

disabled. Motivated by the considerations above, we suppose that J_t describes the state of the contract at time t. We assume that $\{J_t\}$ is an inhomogeneous Markov chain with state space $\mathcal{J} = \{0, 1, \ldots, J\}$. A simple example with three states is shown in Figure F.1. Let $\boldsymbol{\mu}(t) = (\mu_{ij}(t))_{ij}$ denote the intensity matrix, which we assume to exist. That is, if $J_t = i$, $\tau_t = \inf\{s > t : J_t \neq i\}$, we have

$$\mathbb{P}[\tau_t \le s, J_{\tau_t} = j \mid J_t = i] = \int_t^s \mu_{ij}(v) \exp\Bigl\{\int_t^v \mu_{ii}(w)\, \mathrm{d}w\Bigr\}\, \mathrm{d}v\;, \qquad j \neq i,$$

$$\mathbb{P}[\tau_t > s \mid J_t = i] = \exp\Bigl\{\int_t^s \mu_{ii}(v)\, \mathrm{d}v\Bigr\}\;.$$

Note that $\mu_{ii}(t) = -\sum_{j \neq i} \mu_{ij}(t)$.

For simplicity we assume that $J_0 = 1$ and that the state 0 (dead) is absorbing. We now introduce the counting processes

$$N_t^{ij} = \sum_{s \le t} \mathbb{1}_{J_s = j, J_{s-} = i}\;, \qquad i \neq j\;,$$

the number of times the process jumps from state i to state j. $N_t^j = \sum_{i \neq j} N_t^{ij}$ is the number of times the process visits state j. For notational convenience we also define $N_t^{ii} = 0$. The dividend process is modelled as

$$D_t = \sum_{i=1}^J \int_0^t \mathbb{1}_{J_s = i} d_s^i\, \mathrm{d}s + \sum_{i=1}^J \sum_{j=0}^J d_s^{ij}\, \mathrm{d}N_s^{ij} + \sum_{i=1}^J \Delta D_T^i \mathbb{1}_{t \ge T}\;. \tag{F.1}$$

d_t^i is the dividend paid continuously when J is in state i, and d_t^{ij} is a lump sum payment due at the transition from state i to j. The payment ΔD_T^i is a payment at the expiration of the contract (pure endowment insurance). Note that even though the contract expires (usually) at time T, it is possible that dividends have to be paid until the state zero is reached. An example would be a disability pension.

Let the riskless interest rate be constant as in (E.3). Then the value of the insurance contract is

$$V(t,i) = \mathbb{E}\Big[\int_t^T \mathrm{e}^{-\delta(s-t)}\,\mathrm{d}D_s \,\Big|\, J_t = i\Big]\,. \tag{F.2}$$

Calculating the premium for the contract is usually done on a *technical basis*, or *first-order basis*. The insurer assumes a constant interest rate that should be smaller than the interest rate of a riskless bond. For the intensity matrix the values $\mu_{ij}(t)$ are chosen such that the unfavourable states are more likely than under the "real-world" intensities. This is the classical approach in actuarial mathematics and different from the approach in financial mathematics, where prices are calculated under an equivalent measure; see Appendix E.

The insurer invests the premium and gets a return that hopefully is larger than the technical interest rate. In reality, the outgo for a portfolio of contracts would also be smaller than expected because the technical intensities are not the real intensities. For simplicity in this book we will assume that the real intensities are used in the calculations.

Suppose that the insurer invests the reserves into a risky portfolio. Assuming a Black–Scholes model (E.1), the surplus of the contract is determined by the stochastic differential equation

$$\mathrm{d}X_t = mX_t\,\mathrm{d}t + \sigma X_t\,\mathrm{d}W_t - \mathrm{d}D_t\,, \qquad X_0 = x\,. \tag{F.3}$$

The value x will usually just be the single premium of the contract.

Remark F.1. The constants m and σ will, in general, not be the market constants. Because there is some guarantee for the insured, which is covered by the capital resources of the company, some part of the return from investment will be paid to the shareholders and not credited to the insurance contract. That means that σ and m are chosen such that the part of the surplus remaining in the portfolio follows (F.3). ■

F.2 Bonus Schemes

Calculation of prices from the technical basis is unfair to the customer. The insurer could make a (market value) profit because the calculations used for the valuation of the contract uses larger risks than the risk that is taken over. The guaranteed interest rate is usually chosen so small that despite the interest rate guarantee, its market price would be much smaller. The reason is that the insurer wants to be on the safe side. The guarantees are given over 30 years or more. Financial products for hedging the interest rate guarantee for such a long time horizon had not been on the market until recently and are quite expensive. Even if one buys an interest rate guarantee for such a long period, one still faces the counterparty's credit risk. Because interest rate risk influences all contracts in the same way, the default risk for a life insurer would be too large. Examples for defaults due to interest rate guarantees are

Nissan Mutual Life (Japan) or Equitable Life (UK). The insurer thus has to find a way to limit the interest rate risk in such a way that mainly the (almost hedgeable) mortality risk remains in the contract.

The solutions life insurance companies have found are *bonus schemes.* The dividend is not determined deterministically. If the surplus of the contract is large enough, part of the surplus is credited to a bonus account of the insured. More specifically, additional "contracts" are credited to the insured's account in such a way that the price of the insurance contract becomes a fair value of the "real" dividends paid to the insured.

Let g_t^i, g_t^{ij}, and ΔG_T^i be the guaranteed payments and b_t^i, b_t^{ij}, and ΔB_T^i be the dividends of the bonus contract. The number of bonus contracts credited to the insured at time t is denoted by k_t. The process $\{k_t\}$ should then be an increasing process with $k_0 = 0$. The actual dividend payments are then $d_t^i = g_t^i + k_t b_t^i$, $g_t^{ij} = g_t^{ij} + k_t b_t^{ij}$, and $\Delta D_T^i = \Delta G_T^i + k_t \Delta B_T^i$. The process $\{k_t\}$ is chosen in such a way that the initial premium is the market value of the dividend process.

F.3 Unit-Linked Insurance Contracts

A problem with bonus schemes is that it is difficult for the insured to control whether the payments are correct. This is because the insurer's balance sheet determines the bonus. Thus, the insured is punished for the insurer's bad investment. Another problem is that the life insurer has to use a careful investment strategy in order to meet the liabilities. Many customers would prefer a more risky investment strategy with a higher return.

The solution the insurance industry found is the *unit-linked* insurance contracts. Let Z_t be the price of some reference portfolio or of some index. The dividend payoff is then $d_t^i(Z_t)$, $d_t^{ij}(Z_t)$, and $\Delta S_T^i(Z_T)$. In this way the bonus is fixed at the beginning of the contract and the payoff is not dependent on the insurer's real investment portfolio.

In order to determine the value of the contract, one usually takes methods from financial mathematics to determine the value of the financial risk and actuarial methods to determine the mortality risk. The use of the actuarial method is justified by saying that with a huge number of contracts mortality risk is hedged away by the strong law of large numbers.

Consider, for example, a term insurance contract, that is, a contract with a payment in the case of death only. Let the reference portfolio be described by the Black–Scholes model (E.1) and the riskless bond be described by (E.3). The payoff in the case of death is a function $H(t, Z_t)$. The value at time zero of this payoff is then $\mathbb{E}^*[H(t, S_t)]/Z_t^r$, where $\mathbb{P}^*$ is the risk-neutral probability measure introduced in Appendix E. Under the assumption that mortality and financial risk are independent, we obtain the single premium

$$\int_0^T \mathbb{E}^*[H(t,S_t)]/Z_t^{\mathrm{r}}\,\mathrm{d}\mathbb{P}[\tau_x \le t]$$
$$= \int_0^T \mathbb{E}^*[H(t,S_t)]\mu_{x+t}\exp\Big\{-\int_0^t (\delta+\mu_{x+v})\,\mathrm{d}v\Big\}\,\mathrm{d}t\,.$$

Similarly, for a pure endowment insurance with payoff $H(S_T)$ at time T, provided that the insured is still alive at time T, the value at time zero becomes

$$\mathbb{E}^*[H(S_T)]\,\mathrm{e}^{-\delta T}\,{}_Tp_x\,.$$

Bibliographical Remarks

Introductions to life insurance mathematics can be found in the textbooks by Gerber [78], Koller [120], and Møller and Steffensen [135]. The multi-state Markov model can be found, for instance, in Hoem [102] or Møller and Norberg [133]. A theory on bonus schemes and their reserves is found in Norberg [141, 142], Steffensen [173, 175], or Møller and Steffensen [135]. Literature on unit-linked insurance contracts is Aase and Persson [2], Bacinello and Ortu [14], Delbaen [43], or Nielsen and Sandmann [136]. Møller [134] (see also Møller and Steffensen [135]) applied variance-minimising techniques and found that the measure used in Section F.3 is the variance-minimising martingale measure, which is in some sense the fair distribution of the risk between the insurer and the insured. Valuation of life insurance contracts is also treated in Chen and Xiang [33].

References

1. Aase, K.K. (1984). Optimum portfolio diversification in a general continuous time model. *Stochastic Process. Appl.* **18**, 81–98.
2. Aase, K.K. and Persson, S.-A. (1994). Pricing of unit-linked life insurance policies. *Scand. Actuarial J.*, 26–52.
3. Albrecher, H. and Tichy, R.F. (2000). Zur Konvergenz eines Lösungsverfahrens für ein Risikomodell mit gammaverteilten Schäden. *Schweiz. Verein. Versicherungsmath. Mitt.*, 115–127.
4. Asmussen, S. (1982). Conditioned limit theorems relating a random walk to its associate, with applications to risk reserve process and the GI/G/1 queue. *Adv. in Appl. Prob.* **14**, 143–170.
5. Asmussen, S. (1984). Approximations for the probability of ruin within finite time. *Scand. Actuarial J.*, 31–57.
6. Asmussen, S. (1989). Risk theory in a Markovian environment. *Scand. Actuarial J.*, 66–100.
7. Asmussen, S. (2000). *Ruin Probabilities*. World Scientific, Singapore.
8. Asmussen, S., Barndorff-Nielsen, O.E., and Schmidli, H. (eds). The interplay between insurance, finance and control. *Special Issue of Insurance Math. Econ.* **22** (1), 1–122.
9. Asmussen, S., Højgaard, B., and Taksar, M. (2000). Optimal risk control and dividend distribution policies. Example of excess-of-loss resinsurance for an insurance corporation. *Finance Stoch.* **4**, 299–324.
10. Asmussen, S. and Rubinstein, R.Y. (1995). Steady-state rare events simulation in queueing models and its complexity properties. In: Dshalalov, J.H. (ed.) *Advances in Queueing*. CRC Press, Boca Raton, FL.
11. Asmussen, S. and Taksar, M. (1997). Controlled diffusion models for optimal dividend pay-out. *Insurance Math. Econ.* **20**, 1–15.
12. Azcue, P. and Muler, N. (2005). Optimal reinsurance and dividend distribution policies in the Cramér–Lundberg model. *Math. Fin.* **15**, 261–308.
13. Bachelier, L. (1900). Théorie de la Spéculation. *Ann. Sci. École Norm. Sup.* **17**, 21–86.
14. Bacinello, A.R and Ortu, F. (1993). Pricing equity-linked life insurance with endogenous minimum guarantees. *Insurance Math. Econ.* **12**, 245–257.
15. Balkema, A.A. and de Haan, L. (1972). On R. von Mises' condition for the domain of attraction of $\exp(-e^{-x})$. *Ann. Math. Stat.* **43**, 1352–1354.

16. Bäuerle, N. and Rieder, U. (2005). Portfolio optimization with unobservable Markov-modulated drift process. *J. Appl. Prob.* **42**, 362–378.
17. Bellman, R. (1953). *An Introduction to the Theory of Dynamic Programming.* Rand, Santa Monica.
18. Bellman, R. (1957). *Dynamic Programming.* Princeton University Press, Princeton, NJ.
19. Bellman R. and Dreyfus, S. E. (1962). *Applied Dynamic Programming.* Princeton University Press, Princeton, NJ.
20. Bertsekas, D.P. and Shreve, S.E. (1978). *Stochastic Optimal Control. The Discrete Time Case.* Academic Press, New York-London.
21. Billingsley, P. (1979). *Probability and Measure.* Wiley, New York.
22. Bingham, N.H., Goldie, C.M., and Teugels, J.L. (1989). *Regular Variation.* Cambridge University Press, Cambridge.
23. Black, F. and Scholes, M. (1973). The pricing of options and corporate liabilities. *J. Political Econ.* **81**, 637–659.
24. Boudarel, R., Delmas, J., and Guichet, P. (1971). *Dynamic Programming and its Application to Optimal Control.* Academic Press, New York.
25. Boulier, J.-F., Trussant, E., and Florens, D. (1995). A dynamic model for pension funds management. *Proceedings of the 5th AFIR Colloquium* **1**, 361–384.
26. Brémaud, P. (1981). *Point Processes and Queues.* Springer-Verlag, New York.
27. Brocket, P.L. and Xia, X. (1995). Operations research in insurance: A review. *Trans. Act. Soc.* **XLVII**, 7–80.
28. Browne, S. (1995). Optimal investment policies for a firm with a random risk process: exponential utility and minimizing the probability of ruin. *Math. Oper. Res.* **20**, 937–958.
29. Bühlmann, H. (1970). *Mathematical Methods in Risk Theory.* Springer-Verlag, New York.
30. Cai, J., Gerber, H.U., and Yang, H. (2006). Optimal dividends in an Ornstein-Uhlenbeck type model with credit and debit interest. *North Amer. Actuarial J.* **10**, 94–108.
31. Cairns, A.J.G. (2000). Some notes on the dynamics and optimal control of stochastic pension fund models in continuous time. *ASTIN Bull.* **30**, 19–55.
32. Cairns, A.J.G. and Parker, G. (1997). Stochastic pension fund modelling. *Insurance Math. Econ.* **21**, 43–79.
33. Chen, D.-F. and Xiang, G. (2003). Time-risk discount valuation of life contracts. *Acta Math. Appl. Sin.* **19**, 647–662.
34. Chistyakov, V.P. (1964). A theorem on sums of independent, positive random variables and its applications to branching processes. *Theory Prob. Appl.* **9**, 640–648.
35. Chover, J., Ney P.E., and Wainger, S. (1973). Functions of probability measures. *Journal d'Analyse Mathématique* **26**, 255–302.
36. Cramér, H. (1930). *On the Mathematical Theory of Risk.* Skandia Jubilee Volume, Stockholm.
37. Cramér, H. (1955). *Collective Risk Theory.* Skandia Jubilee Volume, Stockholm.
38. Dassios, A. and Embrechts, P. (1989). Martingales and insurance risk. *Stochastic Models* **5**, 181–217.
39. Davis, M.H.A. (1984). Piecewise-deterministic Markov processes: A general class of non-diffusion stochastic models. *J. Roy. Stat. Soc. Ser. B* **46**, 353–388.

40. Davis, M.H.A. (1993). *Markov Models and Optimization.* Chapman & Hall, London.
41. Dayananda, P.W.A. (1970). Optimal reinsurance. *J. Appl. Prob.* **7**, 134–156.
42. Dayananda, P.W.A. (1972). Optimal reinsurance with several portfolios. *Scand. Actuarial J.*, 14–23..
43. Delbaen, F. (1990). Equity linked policies. *Bulletin Association des Actuaries Belges*, 33–52.
44. Dellacherie, C. and Meyer, P.-A. (1980). *Probabilités et Potentiel.* Hermann, Paris.
45. Delbaen, F. and Haezendonck, J. (1989). A martingale approach to premium calculation principles in an arbitrage free market. *Insurance Math. Econ.* **8**, 269–277.
46. Dickson, D.C.M. (2005). *Insurance Risk and Ruin.* Cambridge University Press, Cambridge.
47. Dickson, D.C.M and Waters, H.R. (2004). Some optimal dividend problems. *ASTIN Bull.* **34**, 49–74.
48. Doob, J.L. (1953). *Stochastic Processes.* Wiley, New York.
49. Doob, J.L. (1994). *Measure Theory.* Springer-Verlag, New York.
50. Dothan, M.U. (1990). *Prices in Financial Markets.* Oxford University Press, New York.
51. Duffie, D. (1996). *Dynamic Asset Pricing Theory.* Princeton University Press, Princeton, NJ.
52. Dufresne, F. and Gerber, H.U. (1991). Risk theory for the compound Poisson process that is perturbed by diffusion. *Insurance Math. Econ.* **10**, 51–59.
53. Dynkin, E.B. (1961). *Theory of Markov Processes.* Prentice-Hall, Inc., Englewood Cliffs, NJ.
54. Dynkin, E.B. (1965). *Markov Processes.* Academic Press, New York.
55. Embrechts, P., Klüppelberg, C. and Mikosch, T. (1997). *Modelling Extremal Events.* Springer-Verlag, Berlin.
56. Embrechts, P. and Veraverbeke, N. (1982). Estimates for the probability of ruin with special emphasis on the possibility of large claims. *Insurance Math. Econ.* **1**, 55–72.
57. Ethier, S.N. and Kurtz, T.G. (1986). *Markov Processes.* Wiley, New York.
58. Feller, W. (1971). *An Introduction to Probability Theory and Its Applications.* Volume II, Wiley, New York.
59. de Finetti, B. (1957). Su un' impostazione alternativa della teoria collettiva del rischio. *Transactions of the XVth International Congress of Actuaries* **2**, 433–443.
60. Fleming, W.H. and Rishel, R.W. (1975). *Deterministic and Stochastic Optimal Control.* Springer-Verlag, New York.
61. Fleming, W.H. and Soner, H.M. (1993). *Controlled Markov Processes and Viscosity Solutions.* Springer-Verlag, New York.
62. Fleming, W.H. and Zariphopoulou, T. (1991). On optimal investment/consumption model with borrowing constraints. *Math. Oper. Res.* **16**, 802–822.
63. Frisque, A. (1974). Dynamic model of insurance company's management. *ASTIN Bull.* **8**, 57–65.
64. Frolova, A., Kabanov, Y., and Pergamenshchikov, S. (2002). In the insurance business risky investments are dangerous. *Finance Stoch.* **6**, 227–235.
65. Furrer, H.J. (1998). Risk processes perturbed by α-stable Lévy motion. *Scand. Actuarial J.*, 59–74.

66. Furrer, H.J. and Schmidli, H. (1994). Exponential inequalities for ruin probabilities of risk processes perturbed by diffusion. *Insurance Math. Econ.* **15**, 23–36.
67. Gabih, A, Grecksch, W., and Wunderlich, R. (2005). Dynamic portfolio optimization with bounded shortfall risks. *Stoch. Anal. Appl.* **23**, 579–597.
68. Gabih, A, Richter, M., and Wunderlich, R. (2004). Dynamic optimal portfolios benchmarking the stock market. In: vom Scheidt, J. (ed.) *Tagungsband zum Workshop "Stochastische Analysis,"* 45–83. Technische Universität Chemnitz.
69. Gabih, A and Wunderlich, R. (2004). Optimal portfolios with bounded shortfall risks. In: vom Scheidt, J. (ed.) *Tagungsband zum Workshop "Stochastische Analysis,"* 21–41. Technische Universität Chemnitz.
70. Gaier, J. and Grandits, P. (2002). Ruin probabilities in the presence of regularly varying tails and optimal investment. *Insurance Math. Econ.* **30**, 211–217.
71. Gaier, J., Grandits, P., and Schachermayer, W. (2003). Asymptotic ruin probabilities and optimal investment. *Ann. Appl. Prob.* **13**, 1054–1076.
72. Gerber, H.U. (1969). Entscheidungskriterien für den zusammengesetzten Poisson-Prozess. *Schweiz. Verein. Versicherungsmath. Mitt.* **69**, 185–228.
73. Gerber, H.U. (1970). An extension of the renewal equation and its application in the collective theory of risk. *Skand. Aktuar Tidskr.* **53**, 205–210.
74. Gerber, H.U. (1973). Martingales in risk theory. *Schweiz. Verein. Versicherungsmath. Mitt.* **73**, 205–216.
75. Gerber, H.U. (1974). The dilemma between dividends and safety and a generalization of the Lundberg-Cramér formulas. *Scand. Actuarial J.*, 46–57.
76. Gerber, H.U. (1979). *An Introduction to Mathematical Risk Theory.* Huebner Foundation Monographs, Philadelphia.
77. Gerber, H.U. (1981). On the probability of ruin in the presence of a linear dividend barrier. *Scand. Actuarial J.*, 105–115.
78. Gerber, H.U. (1997). *Life Insurance Mathematics.* Springer-Verlag, Berlin.
79. Gerber, H.U. and Shiu, E.S.W. (2000). Investing for retirement: Optimal capital growth and dynamic asset allocation. *North Amer. Actuarial J.* **4**, 42–62.
80. Gerber, H.U., Shiu, E.S.W., and Smith, N. (2006). Maximizing dividends without bankruptcy. *ASTIN Bull.* **36**, 5–23.
81. Girlich, H.-J., Köchel. P., and Küenle, H.-U. (1990). *Steuerung dynamischer Systeme.* Fachbuchverlag, Leipzig.
82. Glynn, P. and Iglehart, D.I. (1989). Importance sampling for stochastic simulations. *Management Sciences* **35**, 1367–1392.
83. Goldie, C.M and Resnick, S. (1988). Distributions that are both subexponential and in the domain of attraction of an extreme-value distribution. *Adv. in Appl. Prob.* **20**, 706–718.
84. Goovaerts, M. J., Kaas, R., van Heerwaarden, A.E., and Bauwelinckx, T. (1990). *Effective Actuarial Methods.* North-Holland, Amsterdam.
85. Grandell, J. (1977). A class of approximations of ruin probabilities. *Scand. Actuarial J.*, 37–52.
86. Grandell, J. (1991). *Aspects of Risk Theory.* Springer-Verlag, New York.
87. Grandits, P. (2004). An analogue of the Cramér–Lundberg approximation in the optimal investment case. *Appl. Math. Optim.* **50**, 1–20.
88. Grandits, P. (2005). Minimal ruin probabilities and investment under interest force for a class of subexponential distributions. *Scand. Actuarial J.*, 401–416.

89. Hadwiger, H. (1940). Über die Wahrscheinlichkeit des Ruins bei einer grossen Zahl von Geschäften. *Archiv für mathematische Wirtschafts- und Sozialforschung* **6**, 131–135.
90. Hald, M. and Schmidli, H. (2004). On the maximisation of the adjustment coefficient under proportional reinsurance. *ASTIN Bull.* **34**, 75–83.
91. Harrison, J.M. and Kreps, D.M. (1979). Martingales and arbitrage in multiperiod security markets. *J. Econ. Theory* **20**, 381–408.
92. Harrison, J.M. and Pliska, S.R. (1981). Martingales and stochastic integrals in the theory of continuous trading. *Stochastic Process. Appl.* **11**, 215–260.
93. Harrison, J.M. and Pliska, S.R. (1983). A stochastic calculus model of continuous trading: Complete markets. *Stochastic Process. Appl.* **15**, 313–316.
94. Harrison, J.M. and Taylor, A.J. (1978). Optimal control of a Brownian storage system. *Stochastic Process. Appl.* **6**, 179–194.
95. Hinderer, K. (1970). *Foundations of Non-Stationary Dynamic Programming with Discrete Time Parameter.* Springer-Verlag, Berlin.
96. Hipp, C. (2004). Stochastic control with application in insurance. In: Stochastic Methods in Finance, *Lecture Notes in Math.* **1856**. Springer-Verlag, Berlin, 127–164.
97. Hipp, C. and Plum, M. (2000). Optimal investment for insurers. *Insurance Math. Econ.* **27**, 215–228.
98. Hipp, C. and Plum, M. (2000). Optimal investment for investors with state dependent income, and for insurers. *Preprint, University of Karlsruhe.*
99. Hipp, C. and Schmidli, H. (2004). Asymptotics of ruin probabilities for controlled risk processes in the small claims case. *Scand. Actuarial J.*, 321–335.
100. Hipp, C. and Taksar, M. (2000). Stochastic control for optimal new business. *Insurance Math. Econ.* **26**, 185–192.
101. Hipp, C. and Vogt, M. (2003). Optimal dynamical XL reinsurance. *ASTIN Bull.* **33**, 193–207.
102. Hoem, J.M. (1969). Markov chain models in life insurance. *Blätter der DGFV* **9**, 85–121.
103. Højgaard, B. (2001). Optimal dynamic premium control in non-life insurance. Maximising dividend pay-outs. *Preprint, University of Aalborg.*
104. Højgaard, B. and Taksar, M. (1997). Optimal proportional reinsurance policies for diffusion models. *Scand. Actuarial J.*, 166–180.
105. Højgaard, B. and Taksar, M. (1998). Optimal proportional reinsurance policies for diffusion models with transaction costs. *Insurance Math. Econ.* **22**, 41–51.
106. Howard, R.A. (1960). *Dynamic Programming and Markov Processes.* Wiley, New York.
107. Hubalek, F. and Schachermayer, W. (2004). Optimization expected utility of dividend payments for a Brownian risk process and a peculiar nonlinear ODE. *Insurance Math. Econ.* **34**, 193–225.
108. Iglehart, D.L. (1969). Diffusion approximations in collective risk theory. *J. Appl. Prob.* **6**, 285–292.
109. Irbäck, J. (2003). Asymptotic theory for a risk process with a high dividend barrier. *Scand. Actuarial J.*, 97–118.
110. Jacka, S. (1984). Optimal consumption of an Investment. *Stochastics* **13**, 45–60.
111. Jacobs, O.L.R. (1967). *An Introduction to Dynamic Programming: The Theory of Multistage Decision Processes.* Chapman & Hall, London.

112. Jacobsen, M. (2006). *Point Process Theory and Applications. Marked Point and Piecewise Deterministic Processes.* Birkhäuser, Boston.
113. Kallianpur, G. and Karandikar, R.L. (2000). *Introduction to Option Pricing Theory.* Birkhuser, Boston.
114. Karatzas, I. (1997). *Lectures on the Mathematics of Finance.* American Mathematical Society, Providence, RI.
115. Karatzas, I., Lehoczky, J., Sethi, S., and Shreve, S.E. (1986). Explicit solution of a general consumption/investment problem. *Math. Oper. Res.* **11**, 261–294.
116. Karatzas, I., Lehoczky, J., and Shreve, S. (1987). Optimal portfolio and consumption decisions for a "small investor" on a finite horizon. *SIAM J. Control Optim.* **25**, 1557–1586.
117. Karatzas, I. and Shreve, S.E. (1991). *Brownian Motion and Stochastic Calculus,* Second Edition. Springer-Verlag, New York.
118. Karatzas, I. and Shreve, S.E. (1997). *Methods of Mathematical Finance.* Springer-Verlag, Heidelberg.
119. Klüppelberg, C. (1988). Subexponential distributions and integrated tails. *J. Appl. Prob.* **25**, 132–141.
120. Koller, M. (2000). *Stochastische Modelle in der Lebensversicherung.* Springer-Verlag, Berlin.
121. Korn, R. and Korn, E. (2001). *Option Pricing and Portfolio Optimization.* American Mathematical Society, Providence, RI.
122. Lamberton, D. and Lapeyre, B. (1996). *Introduction to Stochastic Calculus Applied to Finance.* Chapman & Hall, London.
123. Lehoczky, J., Sethi, S., and Shreve, S. (1983). Optimal consumption and investment policies allowing consumption constraints and bankruptcy. *Math. Oper. Res.* **8**, 613–636.
124. Lehoczky, J., Sethi, S., and Shreve, S. (1985). A martingale formulation for optimal consumption/investment decision making. In: Feichtinger, G. *Optimal Control and Economic Analysis* **2**, 135–153. North-Holland, Amsterdam.
125. Lehtonen, T. and Nyrhinen, H. (1992). On asymptotically efficient simulation of ruin probabilities in a Markovian environment. *Scand. Actuarial J.*, 60–75.
126. Lindvall, T. (1973). Weak convergence of probability measures and random functions in the function space $D[0,\infty)$. *J. Appl. Prob.* **10**, 109–121.
127. Lundberg, F. (1903). ***I.** Approximerad Framställning av Sannolikhetsfunktionen. **II.** Återförsäkring av Kollektivrisker.* Almqvist & Wiksell, Uppsala.
128. Lundberg, F. (1926). *Försäkringsteknisk Riskutjämning.* F. Englunds boktryckeri A.B., Stockholm.
129. Markussen, C. and Taksar, M.I. (2003). Optimal dynamic reinsurance policies for large insurance portfolios. *Finance Stoch.* **7**, 97–121.
130. Martin-Löf, A. (1994). Lectures on the use of control theory in insurance. *Scand. Actuarial J.*, 1–25.
131. Merton, R. (1969). Lifetime portfolio selection under uncertainty: The continuous time case. *Rev. Econ. Stat.* **51**, 247–257.
132. Merton, R. (1971). Optimum consumption and portfolio rules in a continuous time model. *J. Econ. Theory* **3**, 373–413.
133. Møller, C.M. and Norberg, R. (1993). Thiele's differential equation by stochastic interest of diffusion type. *Working paper **117***, Laboratory of Actuarial Mathematics, Copenhagen.
134. Møller, T. (1998). Risk minimizing hedging strategies for unit linked life insurance contracts. *ASTIN Bull.* **28**, 17–47.

135. Møller, T. and Steffensen, M. (2007). *Market-Valuation Methods in Life and Pension Insurance.* Cambridge University Press, Cambridge.
136. Nielsen, J.A. and Sandmann, K. (1995). Equity-linked life insurance: A model with stochastic interest rates. *Insurance Math. Econ.* **16**, 225–253.
137. Nielsen, P.H. (2005). Optimal bonus strategies in life insurance: The Markov chain interest rate case. *Scand. Actuarial J.*, 81–102.
138. Nielsen, P.H. (2006). Utility maximization and risk minimization in life and pension insurance. *Finance Stoch.* **10**, 75–97.
139. Nielsen, P.H. (2006). Financial Optimization Problems in Life and Pension Insurance. Ph.D. thesis, Laboratory of Actuarial Mathematics, University of Copenhagen.
140. Niemiro, W. and Pokarowski, P. (1995). Tail events of some nonhomogeneous Markov chains. *Ann. Appl. Prob.* **5**, 261–293.
141. Norberg, R. (1999). A theory of bonus in life insurance. *Finance Stoch.* **3**, 373–390.
142. Norberg, R. (2001). On bonus and bonus prognoses in life insurance. *Scand. Actuarial J.*, 126–147.
143. Øksendal, B. (2003). *Stochastic Differential Equations.* Springer-Verlag, Berlin.
144. Øksendal, B. and Sulem, A. (2005). *Applied Stochastic Control of Jump Diffusions.* Springer-Verlag, Berlin.
145. Paulsen, J. (2007). Optimal dividend payments until absorbtion of diffusion processes when payments are subject to both fixed and proportional costs. *Preprint, University of Bergen.*
146. Paulsen, J. (2007). Optimal dividend payments and reinvestments of diffusion processes with both fixed and proportional costs. *Preprint, University of Bergen.*
147. Portenko, N.I. (1990). *Generalized diffusion processes.* Translated from the Russian by H.H. McFaden. *Translations of Mathematical Monographs* **83**. American Mathematical Society, Providence, RI.
148. Protter, P.E. (2004). *Stochastic Integration and Differential Equations.* Springer-Verlag, Berlin.
149. Revuz, D. and Yor, M. (1991). *Continuous Martingales and Brownian Motion.* Springer-Verlag, Berlin.
150. Richard, S. (1975). Optimal consumption portfolio and life insurance rules for an uncertain lived individual in a continuous time model. *J. Fin. Econ.* **2**, 187–203.
151. Rogers, L.C.G and Williams, D. (2000). *Diffusions, Markov Processes, and Martingales.* Cambridge University Press, Cambridge.
152. Rolski, T., Schmidli, H., Schmidt, V., and Teugels, J.L. (1999). *Stochastic Processes for Insurance and Finance.* Wiley, Chichester.
153. Samuelson, P.A. (1965). Rational theory of warrant pricing. *Industr. Manag. Rev.* **6**, 13–31.
154. Schäl, M. (2004). On discrete-time dynamic programming in insurance: Exponential utility and minimizing the ruin probability. *Scand. Actuarial J.*, 189–210.
155. Schmidli, H. (1994). Diffusion approximations for a risk process with the possibility of borrowing and investment. *Stochastic Models* **10**, 365–388.
156. Schmidli, H. (1994). Corrected diffusion approximations for a risk process with the possibility of borrowing and investment. *Schweiz. Verein. Versicherungsmath. Mitt.* **94**, 71–81.

157. Schmidli, H. (1995). Cramér–Lundberg approximations for ruin probabilities of risk processes perturbed by diffusion. *Insurance Math. Econ.* **16**, 135–149.
158. Schmidli, H. (1999). Perturbed risk processes: a review. *Theory of Stochastic Processes* **5**, 145–165.
159. Schmidli, H. (2001). Distribution of the first ladder height of a stationary risk process perturbed by α-stable Lévy motion. *Insurance Math. Econ.* **28**, 13–20.
160. Schmidli, H. (2001). Optimal proportional reinsurance policies in a dynamic setting. *Scand. Actuarial J.*, 55–68.
161. Schmidli, H. (2002). On minimising the ruin probability by investment and reinsurance. *Ann. Appl. Prob.* **12**, 890–907.
162. Schmidli, H. (2002). Asymptotics of ruin probabilities for risk processes under optimal reinsurance policies: the small claim case. *Working paper **180***, Laboratory of Actuarial Mathematics, University of Copenhagen.
163. Schmidli, H. (2004). Asymptotics of ruin probabilities for risk processes under optimal reinsurance policies: the large claim case. *Queueing Syst. Theory Appl.* **46**, 149–157.
164. Schmidli, H. (2004). On Cramér–Lundberg approximations for ruin probabilities under optimal excess of loss reinsurance. *Working paper **193***, Laboratory of Actuarial Mathematics, University of Copenhagen.
165. Schmidli, H. (2005). On optimal investment and subexponential claims. *Insurance Math. Econ.* **36**, 25–35.
166. Sethi, S. and Taksar, M. (1988). A note on Merton's optimum consumption and portfolio rules in a continuous-time model. *J. Econ. Theory* **46**, 395–401.
167. Shreve, S.E. (2004). *Stochastic Calculus for Finance.* Springer-Verlag, New York.
168. Shreve, S.E., Lehoczky, J.P., and Gaver, D.P. (1984). Optimal consumption for general diffusions with absorbing and reflecting barriers. *SIAM J. Control Optim.* **22**, 55–75.
169. Siegl, T. and Tichy, R.F. (1996). Lösungsverfahren eines Risikomodells bei exponentiell fallender Schadensverteilung. *Schweiz. Verein. Versicherungsmath. Mitt.*, 95–118.
170. Siegl, T. and Tichy, R.F. (1999). A process with stochastic claim frequency and a linear dividend barrier. *Insurance Math. Econ.* **24**, 51–65.
171. Siegmund, D. (1976). Importance sampling in the Monte Carlo study of sequential tests. *Ann. Stat.* **4**, 673–684.
172. Siegmund, D. (1979). Corrected diffusion approximation in certain random walk problems. *Adv. in Appl. Prob.* **11**, 701–719.
173. Steffensen, M. (2000). Contingent claims analysis in life and pension insurance. *Proceedings AFIR 2000*, 587–603.
174. Steffensen, M. (2002). Intervention options in life insurance. *Insurance Math. Econ.* **31**, 71–85.
175. Steffensen, M. (2004). On Merton's problem for life insurers. *ASTIN Bull.* **34**, 5–25.
176. Steffensen, M. (2006). Quadratic optimization of life insurance payment streams. *ASTIN Bull.* **34**, 245–267.
177. Striebel, C. (1984). Martingale conditions for the optimal control of continuous time stochastic systems. *Stochastic Process. Appl.* **18**, 329–347.
178. Sundt, B. (1991). *An Introduction to Non-Life Insurance Mathematics.* Verlag Versicherungswirtschaft e.V., Karlsruhe.

179. Taylor, G.C. (1976). Use of differential and integral inequalities to bound ruin and queueing probabilities. *Scand. Actuarial J.*, 197–208.
180. Thonhauser, S. and Albrecher, H. (2007). Dividend maximization under consideration of the time value of ruin. *Insurance Math. Econ.* **41**, 163–184.
181. Tobler, H.F. (1989). Über Barrierenstrategien und das Selbstbehaltproblem bei Risikoprocessen mit Dividenden. Ph.D. thesis nr. 8792, ETH Zürich.
182. Touzi, N. (2004). *Stochastic Control Problems, Viscosity Solutions and Application to Finance.* Scuola Norm. Sup., Pisa.
183. Veraverbeke, N. (1993). Asymptotic estimates for the probability of ruin in a Poisson model with diffusion. *Insurance Math. Econ.* **13**, 57–62.
184. Vogt, M. (2004). *Optimale dynamische Rückversicherung.* Verlag Versicherungswirtschaft GmbH, Karlsruhe.
185. Waters, H.R. (1983). Some mathematical aspects of reinsurance. *Insurance Math. Econ.* **2**, 17–26.
186. Wiener, N. (1923). Differential space. *J. Mathematical Phys.* **2**, 131–174.
187. Yong, J. and Zhou, X.Y. (1999). *Stochastic Controls.* Springer-Verlag, New York.

List of Principal Notation

$\mathbb{N}$	The natural numbers, $\mathbb{N} = \{0, 1, 2, \ldots\}$
$\mathbb{N}^*$	The strictly positive natural numbers, $\mathbb{N}^* = \{1, 2, 3, \ldots\}$
$\mathbb{P}$	The basic probability measure
$\mathbb{P}^*$	Changed probability measure
$\mathbb{R}$	The real numbers
$\mathbb{R}_+$	The positive real numbers, $\mathbb{R}_+ = [0, \infty)$
$\mathbb{Z}$	The integer numbers, $\mathbb{Z} = \{\ldots, -2, -1, 0, 1, 2, \ldots\}$
$\mathfrak{A}$	Generator of a Markov process
$\mathcal{D}(\mathfrak{A})$	Domain of the generator
$\mathcal{F}$	σ-algebra of the probability space
$\{\mathcal{F}_t\}$	Filtration
$\{\mathcal{F}_t^X\}$	Natural filtration of the process X
$\mathcal{S}$	Subexponential distributions
$\mathcal{S}^*$	Subset of $\mathcal{S}$; see page 223
$\mathfrak{U}$	Set of admissible strategies
$\mathcal{U}$	Decision space
$B(E)$	Space of bounded measurable functions on E
$D = \{D_t\}$	Accumulated dividends process
$M_Y(r)$	Moment-generating function of the random variable Y
$\mathrm{MDA}(F)$	Maximal domain of attraction of the distribution F
$N = \{N_t\}$	Claim number process
T_i	ith occurrence time
$V(x)$, $V_T(x)$	Value function
$V^U(x)$, $V_T^U(x)$	Value of the strategy U
$W = \{W_t\}$	Standard Brownian motion
$X = \{X_t\}$	Stochastic process
c	Premium rate
$x^+ = x \vee 0$	Positive part of x
$x^- = (-x) \vee 0$	Negative part of x

$\lfloor x \rfloor$	Integer part of x, $\lfloor x \rfloor = \sup\{n \in \mathbb{Z} : n \le x\}$
$\Delta X_t = X_t - X_{t-}$	Jump height
Ω	Event space on which probabilities are defined
$\delta(x)$	survival probability
λ	Claim arrival intensity
μ, μ_n	nth moment of the claim sizes
$\psi(x)$	Ruin probability

Index